PATERNALISM, CONFLICT, AND COPRODUCTION

Learning from Citizen Action and Citizen Participation in Western Europe

ENVIRONMENT, DEVELOPMENT, AND PUBLIC POLICY
A series of volumes under the general editorship of
Lawrence Susskind, *Massachusetts Institute of Technology, Cambridge, Massachusetts*

ENVIRONMENTAL POLICY AND PLANNING
Series Editor:
Lawrence Susskind, *Massachusetts Institute of Technology, Cambridge, Massachusetts*

THE LAND USE POLICY DEBATE IN THE UNITED STATES
edited by Judith I. de Neufville

CAN REGULATION WORK?
Paul A. Sabatier and Daniel A. Mazmanian

PATERNALISM, CONFLICT, AND COPRODUCTION
Learning from Citizen Action and Citizen Participation in Western Europe
Lawrence Susskind and Michael Elliott

Other subseries:

CITIES AND DEVELOPMENT
Series Editor:
Lloyd Rodwin, *Massachusetts Institute of Technology, Cambridge, Massachusetts*

PUBLIC POLICY AND SOCIAL SERVICES
Series Editor:
Gary Marx, *Massachusetts Institute of Technology, Cambridge, Massachusetts*

PATERNALISM, CONFLICT, AND COPRODUCTION

Learning from Citizen Action and Citizen Participation in Western Europe

Lawrence Susskind and Michael Elliott
Massachusetts Institute of Technology
Cambridge, Massachusetts

and Associates

PLENUM PRESS • NEW YORK AND LONDON

Library of Congress Cataloging in Publication Data

Susskind, Lawrence.
Paternalism, conflict, and coproduction.

(Environment, development, and public policy. Environmental policy and planning)
Bibliography: p.
Includes index.
1. Municipal government—Europe—Case studies. 2. Political participation—Europe—Case studies. 3. Transportation and state—Europe—Citizen participation—Case studies. 4. Urban renewal—Europe—Citizen participation—Case studies. I. Elliott, Michael. II. Title. III. Series.
JS3000.7.A15S94 1983 320.8/094 83-2282
ISBN 0-306-40963-1

A Division of Plenum Publishing Corporation
233 Spring Street, New York, N.Y. 10013

Printed in the United States of America

To our parents

CONTRIBUTORS

DONALD APPLEYARD, Late Professor at the Department of Urban and Regional Planning, University of California, Berkeley

MARC DRAISEN, Program Manager at the Mayor's Office of Policy Management, City of Boston

MICHAEL ELLIOTT, Fellow at the Joint Center for Urban Studies of Massachusetts Institute of Technology and Harvard University, Cambridge

DAVID GODSCHALK, Professor and Chairperson of the Department of City and Regional Planning, University of North Carolina, Chapel Hill

CHESTER HARTMAN, Visiting Fellow at the Institute for Policy Studies, Washington, D.C.

JANICE PERLMAN, Associate Professor at the Department of Urban and Regional Planning, University of California, Berkeley; on leave as the Executive Director of the Committee for a New New York

HANS SPIEGEL, Professor and Chairperson of the Graduate Program in Urban Affairs, Hunter College, New York, New York

LAWRENCE SUSSKIND, Associate Professor at the Department of Urban Studies and Planning, Massachusetts Institute of Technology; and Director of the Public Disputes Program at the Harvard Center for Negotiation, Cambridge

JOHN ZEISEL, Principal of Building Diagnostics, Cambridge.

PREFACE

A research team from the United States has completed an examination of citizen participation experiments in seven European countries. The team included Donald Appleyard, Marc Draisen, David Godschalk, Chester Hartman, Janice Perlman, Hans Spiegel, John Zeisel, and ourselves. This book is a product of our joint efforts.

Our studies are aimed at summarizing and sharing what can be learned from recent European efforts to enhance the effectiveness of local government through increased public involvement in the organization and management of public services and urban redevelopment. Almost a year was spent assembling the team, developing a shared framework for analysis and identifying appropriate case-study cities. European and American public officials and citizen activists helped us assess the potential impact of such a study on current practice. A second year was spent visiting the European cities and preparing the case-study drafts. Finally, team members gathered in Washington, D.C., with fifty American and European public officials, citizen activists, and scholars. A two-day symposium provided an exciting opportunity to present preliminary research findings and encourage an exchange of ideas between researchers, activists, and policymakers.

The final versions of the case studies that appear in this book, along with several commentaries by symposium participants, are written especially for city officials and citizen activists. We have tried to translate the results of our scholarly inquiry into pragmatic suggestions for officials and activists. We would also like to believe that our work will influence those scholars who seek to redefine the conventional wisdom about citizen participation.

By pinpointing ways in which client involvement and public participation have helped local governments to solve pressing problems and provide public services more effectively, we hope to intrigue elected officials in the United States who have been standoffish about participation. We have provided illustrations of successful participation efforts, although we have also pointed out the costs involved. The scholar–practitioners involved in the preparation of this book are convinced that participation can lead directly to improvements in urban services and the quality of redevelopment, even when all the costs involved are weighted against the apparent benefits. Our aim is to provide stories that are rich enough in detail to allow readers to weigh the full range of costs and benefits themselves.

Citizen participation exists when residents or consumers of public services supplement the normal machinery of representative democracy by their involvement in local planning or decision making. Participation implies a common ground on which public officials and citizens meet. Participation can take the form of blue-ribbon advisory committees or collaborative processes in which some actual sharing of power occurs. Local residents frequently do not wait to be invited to participate by public officials. When residents or consumers organize themselves to oppose the programs or priorities established by local officials or administrators, *citizen action* emerges. Citizen participation and citizen action, while springing from different political roots, are nonetheless linked. Not only do they sometimes exist simultaneously but one may stimulate and strengthen the other. In fact, there is some doubt as to whether participation in the absence of citizen action can lead to significant increases in resident satisfaction or to a real sharing of ideas, skills, or power.

This book, then, is organized around three patterns of citizen participation and action: *paternalism* (in which municipal decision making is highly centralized and advice giving by citizens is either discouraged or closely managed by government officials), *conflict* (in which centralized decision making is dominant but resident and consumer groups struggle openly to wrest control over certain decisions), and *coproduction* (in which decisions are made through face-to-face negotiation between decision makers and those residents claiming a major stake in particular decisions). Chapter 1 describes these patterns in detail, using examples from the stories told throughout the remainder of the book; analyzes how and why these patterns of participation arise and change; and summarizes the most important ideas for action that emerge from our reflections on the European experience.

The diverse and multifaceted participatory efforts described in the subsequent chapters do not fit neatly under our three headings. They do, however, exhibit dominant features that allow us to make approximate classifications. Chapters 2 and 3 discuss examples of paternalistic patterns of participation in Copenhagen, Denmark; Delft, the Netherlands; and the borough of Camden in London. Chapters 3 through 6 describe conflict as a long-range strategy (in the Docklands and Covent Garden areas within London and also in the city of Coventry, some 80 miles to the northwest) and as a short-range tactic (in Paris, Brussels, and Switzerland). Chapters 7 through 9 emphasize coproduction in Madrid, the Hague, Rotterdam, and Helmond (the Netherlands). The coproduction cases address the links between coproduction and both paternalism and conflict. The epilogue offers a very brief reflection on the problems of transatlantic comparative research—especially research that seeks to influence public policy.

This undertaking would have been impossible without the generous support of the German Marshall Fund of the United States. They have promoted dialogue between Europeans and Americans in many fields, and we are grateful for their interest in citizen participation. We also appreciate the help we received from literally hundreds of Europeans and Americans who contributed to our inquiry. They are acknowledged at the conclusion of each chapter. Finally, special thanks to Rebecca Black (symposium coordinator), Renate Engler (typist), and Peter Clemons (illustrator and designer), whose devotion to excellence has been a considerable blessing.

LAWRENCE SUSSKIND

MICHAEL ELLIOTT

CONTENTS

1

DENMARK
ENGLAND
THE NETHERLANDS
BELGIUM
SWITZERLAND
FRANCE
SPAIN

1

PATERNALISM, CONFLICT, AND COPRODUCTION

Learning from Citizen Action and Citizen Participation in Western Europe

LAWRENCE SUSSKIND AND MICHAEL ELLIOTT

Learning from Western Europe

For advocates of participation, citizen involvement in government decision making is synonymous with (1) *democratization* of choices involving resource allocation, (2) *decentralization* of service systems management, (3) *deprofessionalization* of bureaucratic judgments that affect the lives of residents, and (4) *demystification* of design and investment decisions.[1] These code words are, at the same time, anathema to a great many elected officials. Indeed, many public officials feel that only professionalization of service administration, centralization of bureaucratic structures (to ensure coordination and a clear "chain of command"), and implementation of the most sophisticated computer hardware for cost accounting and performance monitoring can enhance the ability of government to respond effectively to the needs of residents. This divergence of views accounts for much of the difficulty that has plagued citizen participation efforts in the United States in recent years.[2]

[1] Some examples of the writings of such advocates include Ivan Illich's *Deschooling Society* (New York: Harper & Row, 1970); Saul Alinsky, *Reveille for Radicals* (Chicago: University of Chicago Press, 1946); and Milton Kotler, *Neighborhood Government* (Indianapolis: Bobbs-Merrill, 1969).

[2] These difficulties are described more fully in the Advisory Commission on Intergovernment Relations' *Citizen Participation in the American Federal System* (Washington, D.C.: U.S. Government Printing Office, 1979); and Daniel Moynihan, *Maximum Feasible Misunderstanding* (New York: Free Press, 1969). For a general summary, see Stuart Langton, *Citizen Participation in America* (Lexington, Mass.: Heath, 1979).

Public participation programs have been undermined by both administrators and residents. In some instances, local public officials have limited federally mandated citizen access to government decision making by retaining control over the participatory process and undermining any attempt to redistribute power and responsibility.[3] Officials with patronizing attitudes toward citizen involvement have lost credibility with many residents, who prefer to remain uninvolved rather than permit themselves to be co-opted.[4] Other officials who have offered advice-giving opportunities in good faith have met with some success, but citizen action groups seeking a real sharing of power have, at times, spurned even these invitations to participate. Some consumer and citizen groups prefer to remain aloof from government-sponsored participatory programs, fearing that any form of collaboration might drain organizing energies and undercut long-term commitments to radical reform.[5] Finally, both residents and officials have, on occasion, backed off from commitments to participate for want of programmatic mechanisms that work.[6] Citizen participation remains something that many residents and consumer groups demand and most public officials prefer to avoid.

These tensions will not diminish. Resident and consumer groups will continue to push for a greater role in decisions that affect them—sometimes accepting formal invitations to participate, sometimes rejecting them in favor of "independent" activities. Public officials will continue to fend off pressure for a real sharing of power and responsibility, arguing that "extra-representational" participation by unelected groups is nondemocratic and nonaccountable. For even a temporary reconciliation to occur, public officials must be convinced that residents and consumers can participate in a responsible and accountable fashion. They must also be convinced that participation will enhance the efficiency and effectiveness of government operations and not merely create costly delays. Citizen action groups, on their part, must be convinced that involvement in "formal" participatory programs will not undermine their longer-term political objectives. Both groups must feel that there are actually models for participation that meet their needs and that effective arrangements for joint problem solving can be created.

[3] See footnote 2. These experiments are also summarized in a number of books, including Richard Cole, *Citizen Participation and the Urban Policy Process* (Lexington, Mass.: Heath, 1974); Peter Marris and Martin Rein, *The Dilemmas of Social Reform* (London: Routledge, 1967); and Lawrence Johnson and Associates, Inc., *Citizen Participation in Community Development* (Washington, D.C.: U.S. Government Printing Office, 1978).

[4] See Stuart Langton, ed., *Citizen Participation Perspectives—Proceedings of the National Conference on Citizen Participation* (Medford, Mass.: Tufts University Lincoln Filene Center for Citizenship and Public Affairs, 1979), especially "Participation from the Citizen Perspective" by David Cohen *et al.*, pp. 63–71.

[5] This is true in Boston, Atlanta, New York, Chicago, Washington, D.C., and other cities where ACORN, SECO, Mass. Fair Share, and other groups have pursued a broader agenda.

[6] Advisory Commission on Intergovernmental Relations, *Citizen Participation*.

With financial support from the German Marshall Fund of the United States and with considerable help from a number of European counterparts, a team of American researchers recently undertook an analysis of citizen participation experiences in more than a dozen western European cities. We deliberately selected cases that would shed light on the prospects for a reconciliation between activists and officials in the United States. Our intent was to search the European experience for patterns. Our cases—the Docklands, Camden and Covent Garden areas of London; Çonventry; Copenhagen; Brussels; Madrid; Delft; Rotterdam; The Hague; Paris; Helmond (the Netherlands); and Switzerland—provide valuable images of the patterns of participation prevalent in western Europe.

These thirteen cases suggest that public participation and citizen activism, while not costless, may be well worth the price. The ongoing struggle in American cities to escape paternalistic and conflict-laden patterns of interaction is mirrored in our European case studies, but in Europe some communities have achieved, if only temporarily, an effective reconciliation. Residents and consumers have shared responsibility with public officials for designing, managing, and evaluating urban services and neighborhood redevelopment efforts. They have "coproduced" changes in their neighborhoods while retaining their independence and authority. These attempts at coproduction, while difficult, have enhanced the ability of local governments to solve pressing problems and, in some cases, led to increased public satisfaction with public services.[7]

We will summarize what we saw in Europe, emphasizing the ways in which coproduction differs from the more typical patterns of paternalism and conflict. We will describe the mechanisms that have been employed to lower the barriers between public officials and citizen activists. In these pages, we will provide a brief sketch of how public participation and citizen action in Europe differ from those in the United States and how we in the United States might learn from the European experience. The full set of case studies is presented in the chapters that follow.

Three Patterns of Public Participation in Europe

We can look back on the evolution of public participation in the United States and note a rapid evolution over the past twenty-five years. Similarly, the attitudes of European public officials and citizen activists have not remained constant. The structures and mechanisms for involving consumers and residents in neighborhood and municipal decision making have continued to evolve.

Our current snapshot of these fluctuations reveals three distinct patterns of citizen participation in western Europe. We have named these *paternalism,*

[7] The concept of coproduction has been appropriated by a number of authors. See, for example, Stephen Percy, "Conceptualizing and Measuring Citizen Co-Production of Community Safety," *Policy Studies Journal*, Special Issue: Symposium on Policy and Law Enforcement Policy (1978).

conflict, and *coproduction*. These terms classify relationships between public officials and citizen groups. They characterize the attitudes that citizens and governments have toward each other, they hint at the quality and style of interaction, and they imply different sets of expectations regarding the outcomes of participation.

Paternalism is that pattern of participation in which municipal decision making is highly centralized and advice giving by citizens is either discouraged or closely managed by local government officials.

Conflict is that pattern of participation in which centralized decision making is dominant but in which resident and consumer groups struggle openly to wrest control over certain resource-allocation or policy decisions from elected or appointed officials. Conflict leads either to the fragmentation of political power (and from there to sustained confrontation) or to acceptance by public officials of the need to adjust relationships between citizens and government.

Coproduction is a third (and infrequently found) pattern of participation in which decisions are made through face-to-face negotiation between decision makers and residents claiming a major stake in particular decisions. While some cities have moved from paternalism through a pattern of conflict to coproduction (as if along a continuum), the quite different experiences of other cities suggest that these patterns are not phases that inevitably follow each other; rather, they are the by-products of antecedent conditions at the national and local levels, past experiences with public participation, and the confluence of personalities and random events. Each of the patterns we have identified can be found in several different western European countries. Although these patterns are in flux, it is possible to examine them.

Paternalism

In Denmark, Holland, and England, we have observed municipal decision making that remains highly centralized and in which advice from citizens is either restricted or closely prescribed.[8] In these cases, the rules indicating when and how citizens can participate are highly refined. Public officials agree that some direct involvement of residents or consumers is necessary to legitimize decisions that must be made, but these same officials are quick to point out that only they (on the basis of their election or appointment) are actually empowered to decide. Officials who subscribe to this highly republican view of democracy argue that legislative and executive control over the allocation of resources is crucial to the just and equitable distribution of goods and services. Direct in-

[8] Janice Perlman and Hans Spiegel, "Copenhagen's Black Quadrant: The Struggle Against Gentrification"; Donald Appleyard, "Citizen Participation in Delft, Camden and Britain"; and David Godschalk and John Zeisel, "The Anarchist and the Burghers: Co-Producing Urban Renewal in the Netherlands." German Marsall Fund project papers.

volvement of residents or consumers in public policymaking is viewed by these officials (and, in some cases, by the citizens they represent) as costly and possibly counterproductive, since the participation of self-interested and often uninformed individuals may sidetrack elected and appointed officials responsible for balancing the concerns of all interest groups.

In this context, citizen advice is viewed as a supplement to, but not as a replacement for, representative democracy. Comments from residents or consumers (after careful screening) may enable municipal governments to target public services more efficiently and in a manner more responsive to the needs of particular groups. From the standpoint of elected and appointed officials, better information exchange can also forestall citizen opposition to local government action, since residents sometimes protest out of ignorance or unfounded fear. Public policies enacted in response to the presumed desires of residents, without any effort to test the accuracy of such presumptions, often lead to delay or opposition. Many public officials now believe that if citizens had a better sense of what government had in mind, opposition might not develop. While information exchange sometimes masquerades as public education—sending information out with no expectation of a reply—many public officials now agree that a two-way exchange of information, which clarifies public desires as well as government's intent, is very valuable.

We do not assume that the term *paternalism* implies a particular outcome. A paternalistic pattern of participation can lead to whatever outcome the elected or appointed officials have in mind. From what we can see, however, paternalistic patterns of participation rarely lead to substantial alterations in the distribution of power or to the redistribution of goods and services. *Paternalism* describes the dominance of government officials in controlling when and how residents and consumers voice their concerns. Finally, paternalistic patterns of participation are not necessarily repressive; indeed, they may lead intentionally to the politicization of segments of the population or provide time for resident and consumer groups to organize and develop. Examples of paternalistic patterns of participation can be found in Copenhagen, Delft, Camden, and Helmond.

Copenhagen has elaborate procedures to achieve "ordered" citizen participation.[9] Government mechanisms permit citizens to complain, to help plan, and to give and receive advice. The Danes have a municipal planning law that requires city governments to initiate a process of providing information and encouraging discussions about proposed plans. Despite the elaborate channels for citizen complaints and advice giving, however, citizen input in Copenhagen rarely leads to substantial modification of city plans. Appointed agency staff resist advice that runs counter to their proposals. Many elected officials in Copenhagen remain largely closed to any real sharing of decision making power or responsibility,

[9] Perlman and Spiegel, "Copenhagen's Black Quadrant."

and concerted action by local residents poses only a meager challenge to the power of public officials.

In Delft and Camden, community residents have organized in response to invitations from local officials. Citizen participation in these cities is seen by public officials as one means of improving the design of local traffic schemes. Residential precincts have been created in which pedestrians have the right-of-way and in which residents can request that streets be landscaped and closed to through traffic. A group of young planners and a creative police chief developed this concept—called the *woonerf*—which was supported and publicized by the Royal Dutch Touring Club, a national citizen interest group. With over 800 implemented *woonerven* in 220 Dutch cities and residents demanding more, the concept of involving residents in street planning appears to have taken hold. However, while the participatory process serves an important function by legitimizing the selection of streets for inclusion in the program and by permitting fine tuning of each landscape design, the similarity of the various *woonerven* suggests that local residents play a rather limited role in making final decisions.[10]

In the early 1970s, Camden (a borough of London) tried to develop a comprehensive traffic plan to redirect traffic out of neighborhoods and onto arterial streets. While an attitudinal survey showed that a majority of residents favored the plan, public meetings were filled with people who opposed it. The plan was scrapped in favor of neighborhood improvement schemes. These schemes appear to be responsive to resident demand, not through the application of a standard solution but through participatory processes that invite residents in each neighborhood to indicate their preferences. Technical expertise and control of the design process, while still in the hands of public officials, were moved a step closer to residents.[11]

In the working-class areas of Helmond, an industrial town of 60,000 in the south of Holland, local government officials have sought to politicize a previously unorganized population, to alter the attitudes of fatalism and apathy deeply rooted in a history of industrial feudalism.[12] By demonstrating to residents that they can not only understand but also influence plans for their neighborhoods, public officials have used a paternalistic pattern of participation in an extremely progressive fashion.

Conflict

Decision making in many European cities remains highly centralized; in some cities, however, open struggle by citizen action groups or political parties has led public officials to change policies, redesign programs, or accede to other resident or consumer demands. Citizen action in these instances has involved

[10] Appleyard, "Citizen Participation."
[11] Ibid.
[12] Godschalk and Zeisel, "The Anarchist and the Burghers."

petitions, political organization aimed at influencing the outcome of elections, court action, demonstrations (some legal, some not), and, on occasion, violence. In cities where such struggles are under way, resident and consumer groups have banded together to contest government actions they think will affect them adversely. Demonstrations, political action, and especially legal maneuvering are commonplace in cities in which conflict characterizes the pattern of participation.

Political action groups sometimes choose to emphasize or sustain a pattern of conflict even after they appear to have won their demands for more extensive participation. Confrontation is, in and of itself, enormously important to the success of certain grass-roots political movements. A willingness to compromise and accept local government's invitation to participate in a government-sponsored participatory process may help to legitimize that process without ensuring the outcome that the action groups seek. When conflict itself becomes more valuable as a long-term organizing tool than even a short-term victory (i.e., a change in government policy), citizen action groups may choose to remain aloof from government-sponsored citizen involvement activities. In the Docklands (London), Coventry, Covent Garden (London), and Renbaankwartier in The Hague, continued patterns of conflict have resulted as much from decisions by resident groups to sustain them as from the unwillingness of elected or appointed officials to accede to citizen or consumer demands.

Frequently, an increase in the flow of information between officials and residents or consumers only serves to sharpen existing conflicts between groups with competing interests. Participatory techniques can sometimes be used to manage or vitiate such conflicts, but direct participation by self-proclaimed stakeholders may be unacceptable to officials who feel that it is their responsibility, and theirs alone, to weigh and balance the demands of diverse interests. Thus, patterns of participation characterized by continuing conflict may also be caused by the unwillingness of public officials to acknowledge the claims of groups asserting that they ought to have a major role, indeed a disproportionate say, in decisions that directly affect them.

When no government-sponsored forum exists or when there are no channels through which residents or consumers feel they can effectively express their opposition to government policies or programs, action groups may choose to utilize nontraditional means of making their views known. In Paris, Brussels, and Switzerland, citizen action groups organized around shared interests (but not place-specific concerns) have sought to involve themselves directly in shaping government policies. The conflicts that arose in these situations stemmed less from the groups' unwillingness to accept formal government invitations to participate, than from the lack of a forum in which the groups could contest particular policies without relying solely on general elections.

Citizen action groups will adopt a long-range strategy of opposition if they feel they have more to gain from conflict than from short-term victories. Even when opposition is limited to short-term objectives, public officials will accept continuing conflict if they feel that the cost of compromise is higher. Examples of both are presented below.

Conflict as a Long-Range Strategy. Just east of the Tower of London lies 5,500 acres of land slated for redevelopment. Stretching eight miles along the Thames, the area is the largest tract of urban development land in Europe today. At one time the "Hub of the Empire," these docks and industrial lands are now largely underutilized. While there is little disagreement about the desirability of redeveloping this area, the 56,000 residents of the so-called Docklands wish to preserve their homes in the process.[13] The Joint Docklands Action Group (JDAG) was formed to protest the first redevelopment plan proposed by the city. Over time, JDAG has developed its monitoring and planning capabilities.

While a city-sponsored citizen participation process has been developed, JDAG continues its attempt to act as an independent, informed voice of the community. Conflict dominates the pattern of relationships between citizens and local government in the Docklands area. Residents have formed an organization that seeks to wrest some measure of control away from elected city officials. While confrontation carries costs for both sides, it appears to be a functional, productive, and perhaps necessary step in leveraging an expanded role for residents in decision making.

Coventry is a center of manufacturing in England. Since the mid-1960s, the city's manufacturing employment has declined and housing for moderate- and low-income families has become increasingly scarce. In 1969 a national Community Development Project (similar to the U.S. Model Cities Program) was initiated and a Coventry neighborhood was selected as a pilot site. Out of this project grew an independent and autonomous organization whose mission was to call attention to and ultimately address the issues of poverty and deprivation in Coventry. This organization is the Coventry Workshop.[14]

The Workshop bills itself as "a local research and advisory unit for trade unionists and community organizations." Its concerns include unemployment, industrial democracy, occupational and environmental health, municipal services, and housing. The Workshop, while emphasizing technical assistance to other groups, is also willing to program activities of its own when necessary. The Workshop's fundamental concerns with employment and the quality of life in Coventry place it in a position to advocate the concerns of residents, especially tenants. The Workshop insists on avoiding government-sponsored participation processes. This open opposition to government policy enables the Workshop to sustain alliances with groups outside of government (such as trade unions) and

[13] Hans Spiegel and Janice Perlman, "Docklands and Coventry: Two Citizen Action Groups in Britain's Economically Declining Areas." German Marshall Fund project paper.

[14] Ibid.

to retain the freedom it needs to act on behalf of its constituency. Were the Workshop to accept local government's invitation to participate in a formal, city-run participatory process, it would compromise its independence and be forced to reallocate the time and effort of volunteers, thereby subverting its own action agenda. The Workshop's insistence on independence has gained it the respect of unionists and tenants.

Covent Garden is one of the best-known neighborhoods in London and the home of one of the most successful community groups in England.[15] The Covent Garden Association not only protested and halted a huge redevelopment project but successfully organized a number of creative social and environmental activities as well. The most visible of these was the cultivation of open spaces as community gardens. The Association has opted to remain apart from the advisory and review committees established by local government in the hope of retaining its identity as an independent pressure group.

The Renbaankwartier neighborhood in The Hague offers still another example of continuing conflict.[16] The City Council forced the aldermen and City Development Agency to enter into a participatory process with a citizen's group they neither respected nor wished to negotiate with. The process was brief (only six public meetings over a span of one month); little was accomplished. No work groups were established through which citizens and city officials could hammer out agreements on complex issues. A draft plan was produced, but residents and the city still disagreed on key points. The residents are appealing the plan while trying to stall implementation at the provincial level.

The participatory experiment almost broke down on at least two occasions. The level of trust between the city and the residents actually declined as a result of the process. Nonetheless, the residents used the brief encounter to win several significant victories by achieving changes in the content of the plan for their neighborhood. Public officials and resident groups were in an active state of conflict both before and after the formal participatory effort. The process was not a genuine participatory experiment but rather a continuation of the ongoing confrontation.

Conflict as a Short-Term Tactic. Paris, with one of the premier public transportation systems in the world, is also a city whose residential attractiveness is perhaps most threatened by the automobile. Citizen and consumer action in the transportation field has at times during the past decade focused on the conflict between the automobile and other means of transportation as well as on the conditions of the public transit system itself.

[15] Donald Appleyard, "Citizen Action in Covent Garden and Brussels." German Marshall Fund project paper.

[16] Marc Draisen, "Fostering Effective Citizen Participation: Lessons from Four Urban Renewal Neighborhoods in The Hague." German Marshall Fund project paper.

The politicization of transportation planning in the Paris region was part of the larger sweep of popular activism in the early 1970s.[17] During this period, a transit riders' union used a combination of confrontation and bargaining strategies to win several victories. More recent attempts to extend the scope of protest from questions of the price and quality of public transportation to questions of the automobile versus public transit have been less effective.

Inter-Environment of Brussels is a federation of some 55 citizens' groups. One of the most innovative and carefully constructed citizens' group coalitions in continental Europe, Inter-Environment is especially interesting because it is metropolitanwide and carries out a wide range of projects,[18] which range from protest to the generation of counterplans. This federation has the capacity to conduct technical studies on its own. It has become, at least to a limited extent, a citizens' planning agency that can assist citizens' groups in their efforts to challenge the technical credibility of government plans.

In Switzerland, where certain approaches to direct democracy in the electoral arena are well developed, the initiative process has been used by citizens' groups to challenge prevailing government transportation policies.[19] Some outstanding recent examples are the Albatross Initiative, which sought to set stricter emission standards for automobiles; the "Democracy in the Construction of the National Highways" initiative, which sought to reconsider Switzerland's twenty-year-old Autobahn plan and subject still unbuilt segments to voter approval; and the Burgdorfer Initiative, which sought to ban private automobile, boat, and plane traffic one Sunday per month. While all three initiatives failed, these instances of citizen and consumer involvement in public policymaking are nonetheless instructive.

In Paris, Brussels, and Switzerland, citizen and consumer groups sought to challenge prevailing government policies. Action groups, organized at the city and regional levels around issues of transportation and environmental protection, sidestepped government-run participation processes in favor of other means of direct involvement. Unlike the groups in London, Coventry, and The Hague, these citizen action and political mobilization efforts were not place-specific. These groups were not offered a chance to join in a formal participatory process since such processes are rarely available for issue-oriented (rather than place-specific) debates. The electoral process is typically thought of as the appropriate means for citizens and consumer groups to participate in citywide, regional, or national policymaking. There are, however, challenges to this conventional view. Indeed, citizen action groups are exploring various ways of interjecting their

[17] Chester Hartman, "Transportation Users' Movements in Paris in the 1970's." German Marshall Fund project paper.

[18] Appleyard, "Citizen Action."

[19] Chester Hartman, "The Voter Initiative as a Form of Citizen Participation in Swiss Transportation Policy." German Marshall Fund project paper.

views into areawide policy debates through referendums, public education, and technical assistance programs and through the formation of consumer unions that demand a seat at the bargaining table when contracts affecting them are negotiated.

Coproduction

Public decisions about some local matters have been made through face-to-face negotiations between public officials and residents or consumers. In such cases, public officials and citizen groups accept not only the legitimacy of each other's involvement in the process of deciding but also the possibility that residents or consumers might share responsibility (along with government) for the production of services or the management of the development process. Coproduction is most likely to occur when decision making has been formally decentralized—although this is not essential. Decentralization and coproduction of what were previously centrally managed public services have occurred when cities felt it was advantageous to "off-load" certain costs and responsibilities, thereby relieving overloaded municipal budgets. As concern about the levels of taxation and the intrusion of government into people's everyday lives spreads, decentralization of city services and the management of development decisions may win further support within the ranks of local officials.

Coproduction is not without its problems. Tensions between citizen or consumer groups and public officials always remain. Citizens try to avoid the cooptation often associated with paternalistic patterns of participation, while public officials try to balance their willingness to negotiate with their responsibilities to citywide interests. The constant threat of a return to previous patterns of conflict helps to keep both groups working in a coproductive fashion.

There are times when coproduction is seen as undesirable not only by public officials but by citizen groups as well. While the former may feel that it is inappropriate to share power, the latter may prefer to not accept responsibility or may feel that an agreement to negotiate will undermine their political credibility. A citizens' group might prefer to influence outcomes without seeking the right of self-management. The tensions and opportunities associated with coproduction are exemplified by the patterns of participation in the Dutch cities of Rotterdam and The Hague and in Madrid.

Coproduction has been used as a strategy for developing urban development plans in Holland.[20] Government agents and neighborhood residents work together to define problems, devise plans, and carry out renewal actions. Long-term resident satisfaction is frequently linked to high levels of cooperation between residents and government in the planning phases of redevelopment. Coproduction epitomizes the power sharing that residents in some cities are seeking.

[20] Godschalk and Zeisel, "The Anarchist and the Burghers."

In the Netherlands, citizens participate in urban renewal in diverse ways. The mechanisms used to achieve citizen participation turn disagreements into occasions for productive planning. The process is aimed at making planning real and immediate for local participants. The Dutch have emphasized (1) contextual diversity, which assumes that planning processes and organizations must vary to fit different local conditions; (2) self-organization, which encourages groups to organize themselves to achieve what they want rather than waiting to be organized by government; (3) extended doorstep planning, which is designed to mobilize participants around very small scale physical improvement; (4) co-production contracts, which are formal agreements between local government and citizen action groups spelling out the responsibilities of all the parties in the development process; and (5) give-and-take planning, in which resident groups develop plans by exchanging and commenting on documents.[21]

Madrid has adopted similar participatory strategies premised on bringing the consumers of services and local residents into coproductive roles.[22] The Program for Immediate Action (PAI) starts with the self-identification of neighborhood needs. This program was designed to utilize the capacities of resident associations both as a source of information and as a means for setting neighborhood priorities. In Madrid, the goal of such efforts is to achieve greater efficiency in the allocation of scarce resources as well as in the delivery of public services and the programming of development and to increase public satisfaction with the ways in which the city is run.

Cities that have adopted patterns of participation in which residents or consumers share responsibility for producing services or jointly managing the development process have done so for various reasons. Coproduction may be an expedient response to extended conflict or it may represent the achievement of a high level of trust on the part of both residents and public officials. Whatever the reasons, coproduction is a distinct pattern of participation characterized by the direct involvement of unelected interest group representatives in the operation of government—in some cases on an ad hoc basis and in others as the result of decentralization of municipal decision making.

The three patterns of participation that we have identified are defined by the relationship between elected officials and citizens. A paternalistic pattern of participation suggests that elected officials are unwilling to endorse direct citizen involvement in the design or management of services and development. Citizens and consumers in paternalistic situations appear to accept or acquiesce to the judgment of public officials. This may be because they have not had a chance to develop the capacity to participate. Participation characterized by conflict

[21] Ibid.

[22] Janice Perlman, "Citizen Action and Participatory Planning in Madrid." German Marshall Fund project paper.

suggests that elected officials are unwilling to permit the level of direct involvement that citizen are seeking. There are exceptions. Officials might be willing to endorse direct participation by citizen action groups, but the groups might be unwilling to accept responsibility for what they view as strategic reasons. Coproduction suggests that officials and citizens have reached accord on the appropriate role of residents and consumers in the design and management of service delivery systems and development. These patterns are not stable. They emerge and change in fairly complicated fashion.

Explaining How and Why Patterns of Participation Arise and Change

The patterns of participation in a municipality are shaped by both past and present conditions. The historical development of the civic, economic, and political culture; the institutionalization of planning and participation; and the tension between direct action and responsible but passive citizenship help explain why one city has adopted one pattern and a second city has opted for something quite different. Other factors—such as externally generated political movements, changes in communication technology or the organization of settlements, the attitudes of the electorate toward particular public officials, and unexpected events—also contribute significantly to variations in patterns of participation. Policy-oriented analysis seeking to define participatory strategies or techniques that are transferable must begin with a careful appraisal of the role these conditions play.

Cultural Antecedents

In seeking to understand the pattern of participation that has emerged in a particular city, we are drawn initially to an exploration of the attributes of the civic culture. The legitimacy of government decision making, and thus resident and consumer claims on resource-allocation decisions and management of government enterprise, must be viewed in the context of the political and economic organization of a country. The submissiveness that permits paternalistic patterns to survive or the willingness to sustain a pattern of continued conflict is also, in large measure, the result of culturally induced expectations.

The evolution of European political and economic structures has created a mix of government systems and spawned a range of participatory patterns. Copenhagen's elaborately formal process for citizen review, which remains closed to any substantial sharing of power and responsibility, is best understood as the product of an affluent planned economy. The welfare state of Denmark is served by an array of voluntary, quasi-voluntary, and public agencies. This bureaucratic structure is at least a century old. The elaborate machinery built to cope with complaints and head off potential protests resists change and blocks ad hoc attempts to alter the pattern of relationships between residents and their government.

The current vitality of citizen action groups in Madrid is partly a product of the vast economic inequities and the structure of self-help organizations that developed in the squatter settlements under the Franco regime. The coproduction arrangements in the Netherlands can be accounted for in part by the existence of government-supported housing associations that act as developers. A renewal process controlled by a quasi-public housing developer provides different opportunities for participation than one controlled by private contractors.

The norms of the civic culture and deeply rooted expectations and rules about acceptable behavior, proper deference to authority, the appropriateness of violence, and the balance between individual liberty and community stability all shape the attitudes of citizens and public officials toward the opportunities for participation that will be allowed in any city. The mass demonstrations and confrontation politics of Paris, the emergence of neighborhood–trade union coalitions and confrontation through research in the Docklands and Coventry, and the open negotiations of the Netherlands are conditioned by the prevailing assumptions in each country and each city about the legitimacy of government control and direct action. Similarly, the fatalism and apathy found in Helmond can be traced to a civic culture shaped predominantly by feudal industrialism.

The range of constitutional guarantees is also a powerful explainer of the pattern of participation that emerges in a city. Societies vary in the degree to which claims for direct access to resource-allocation decisions are given credence. Rights to land tenure provide one basis for such claims. To the extent that those citizens who do not hold property are accorded tenure rights, patterns of participation will change. The granting of full rights of tenureship to occupants of squatter settlements in Madrid, for example, should be compared with the appointment of government supervisors to manage a playground forcibly established by Copenhagen residents on a vacant city lot. While the lot remained in the use desired by the residents, control temporarily reverted to the city.

Attitudes about the legitimate use of confrontation tactics to achieve political ends vary throughout Europe. We compare a Paris demonstration on behalf of mass transportation reforms which drew 100,000 people onto the streets with the display of black flags in the Netherlands during a tour by the Dutch Minister of Housing. Both acts were unprecedented to the extent that thousands of residents chose to participate "directly" in efforts to shape decisions regarding urban development and service delivery. Demonstrations put residents somewhat more at risk than the hanging of flags. Both acts, however, must be understood within the context of the rights that citizens feel they have, the framework of constitutionally guaranteed rights to protest, and the norms of the civic culture.

While even superficial explorations of the interplay between cultural attributes and the emergence of participatory patterns are revealing, they cannot lead to firm theoretical understandings. We have not been able, in the course of our research, to delve deeply enough into the forces that shape the civic culture and the traditions that animate these forces. We are only in a position to suggest that patterns of participation arise and shift for many reasons. Antecedent conditions are clearly crucial. In attempting to transfer a pattern of participation from one culture to another, any effort that fails to take adequate account of antecedent conditions is bound to fail.

The Institutionalization of Participation in the Planning Process

While deeply rooted cultural attributes are important, the outcome of past experiments with citizen involvement in municipal decision making also shapes patterns of participation. The patterns that now exist are supported by institutional arrangements that have accumulated over time. Through trial and error, accretion and reform, institutional arrangements emerge. These arrangements evolve in a number of different sectors of community life: in the planning process, in intergovernmental relations, in the activities of political parties, and in the organization of action groups. The patterns of participation that we see at any point in time, as well as the changes that occur, are as much a product of the dominant institutional arrangements as they are a reflection of the cultural attributes of the population.

Systems for mounting and financing participation and for conducting municipal planning activities—whatever the pattern—reflect what these institutional arrangements can and cannot handle. In the Netherlands, interest groups, including groups opposed to prevailing government policy, are supported financially by the government. Citizen action groups have an impressive survival rate because they can count on government funding. These groups have, in some cases, been able to command coproduction contracts with local governments because they are credible as long-standing representatives of particular neighborhoods. Institutionalized support for citizen action groups appears to be a crucial factor in the emergence of a pattern of coproduction.

Both Delft and Camden have developed participatory processes that respond rapidly to resident complaints, monitor community feelings, and allow for the outright rejection of development schemes if majority support is not forthcoming. The tradition of town planning in England is built around certain feedback mechanisms making it hard for residents to complain that their concerns are not being heard. Under these circumstances, it is difficult to press for a shift to coproduction. Government planning may be ritualized; citizen participation in the planning process may appear to be extensive, but little in the way of real influence may in fact be guaranteed. Indeed, planning procedures—which include extensive formalized participation—may be nothing more than a means of strengthening the hand of those who traditionally make policy: political authorities (who overrepresent upper-class interests) and bureaucrats and officials (representing mostly middle-class views). We found this to be the case in Copenhagen and Paris, where well-established planning bureaucracies developed complex participatory systems that effectively blocked involvement in important and timely decisions.

Intergovernmental relations can constrain the pattern of participation that emerges at the municipal level. Support for local development programs frequently originates at the national level. Thus, the national government is in a position to undermine or encourage local attempts to increase participation. With grant-in-aid funds held hostage, local governments usually comply. The national government can also bypass more conservative local governments and funnel support directly into the hands of citizen action groups. National governments can assert procedural guidelines but do little, if anything, else to ensure that guidelines are enforced. In such situations, action groups must take the initiative to avail themselves of the leverage that national policy provides.

The organization of political parties can affect patterns of participation at the local level. In the Danish multiparty structure, the local party functionary becomes a de facto community organizer trying to persuade residents that their concerns are synonymous with the party's platform and proposed solutions. A vertical party structure, in which local parties operate under the umbrella of a national party, provides an opportunity for citizens to become involved along party lines but makes horizontal coalitions in which a variety of local action groups join together more difficult because of interparty competition. Also, rapid changes in national party platforms may make it difficult for neighborhood or consumer groups to sustain their involvement in certain local planning efforts. In England, community groups have attempted to build coalitions with trade unions rather than relying solely on alliances with national political parties. Finally, the presence of political parties organized at the local level often makes it difficult to build and sustain a voluntary consumer or neighborhood action group. Volunteers may be preoccupied with local party activities. Residents may have already committed their time and energies to a political party. In Madrid, we found party affiliation both an opportunity for and a threat to the viability of the neighborhood associations.

Citizen groups can also build on their own successes, institutionalizing citizen action from outside the government. In Madrid, independent citizen groups have served as a training ground for community leaders who, in turn, have helped to broaden public awareness of citizen responsibility. Over the past decade, activists in Brussels and Covent Garden have learned how to select projects for their symbolic and media-attracting value. The Coventry Workshop has meticulously examined its successes and failures in a conscious attempt to learn from past mistakes. The credibility of the Workshop is growing. In England and the Netherlands, some government-sponsored participation efforts show clear signs of maturation, spinning off a new generation of local leaders and implanting in the popular mind a clearer sense of how and why participation is important.

Mass demonstrations in Paris lead to immediate results, but the failure to build a durable constituency, to institutionalize opportunities for consumer and resident involvement, and to build the capacity of action organizations to get better at what they do undercut long-term efforts to affect policy.

While antecedent conditions affect both the patterns of participation and the outcomes, they are not insurmountable. While each nation and each city has developed unique traditions that hinge on these conditions, evolution can and does occur. Thus, efforts to transfer what we have learned about particular patterns of participation can be successful if initiated with careful attention to the need to compensate for differences in antecedent conditions.

The Tension between Direct Political Action and Passive Citizenship

Whether or not a city establishes formal channels for public participation in local decision making, resident and consumer groups may decide to press for direct involvement. This will depend, as we have seen, on the traditions of political action in the cities involved. Concerted political action implies conflict. Conflict puts individuals and groups at risk—not necessarily because government will employ repressive measures but because government is, at the very least, likely to take steps to undermine citizen protest through administrative retaliation. For this reason, urban residents who take action on their own are frequently not invited to participate by government.

We have discovered, however, that both citizen action and citizen participation can occur in the same city at the same time, even around the same issue. In these places, there is a tension between citizen action (direct political mobilization) and citizen participation (formalized procedures managed by local government). They may either reinforce or undercut each other. One view is that citizen action can heighten public concern and stimulate more extensive involvement in government-sponsored activities. A contrary view is that citizen action can sour people on formal participation, raising expectations that cannot possibly be met; thereby dooming government-sponsored participation to failure. The reverse is also a possibility—effective government-sponsored participation may thin the ranks of residents or consumers who are willing to become involved in direct action.

Our analysis of the European experience has produced competing illustrations of the relationship between citizen action and citizen participation. The presence of a strong, independent citizen action group may be a precondition to a successful citizen participation effort (if we define success as an increase in resident satisfaction or greater efficiency in the delivery of urban services). Citizen action can also create a demand for the institutionalization of participatory rituals which, over time, drive out citizen action, leaving the process sterile.

The participatory process associated with the design of the *woonerf* in Delft did not result from citizen action. The program is nationwide. Residents must approve of new street designs and changes in rights-of-way before they can be implemented. Because *woonerven* are of immediate concern to residents and modest in scale, many residents have become involved in their design. Yet, in the absence of pressure from one or more citizen action groups, the concept of the *woonerf* has become static and the process for planning particular *woonerven* rigid. The opportunities for resident participation during the design phase have diminished and the designs of many *woonerven* have become more predictable and less imaginative. It would appear that in the absence of an independent citizen action group, programs may tend to ossify. Citizen participation without continuing citizen action may be ineffective.

In Copenhagen, where elaborate channels for participation in municipal planning have developed over the past fifty years, there is some doubt as to government's willingness to make adjustments in response to what residents have to say. Recent redevelopment efforts have tested the efficiency of the participatory process in Copenhagen and led some groups to believe that nothing short of independent citizen action will cause government to listen. Citizen action groups appear to be having some difficulty, however, building coalitions; the tradition of independent citizen action has faded.

Citizen participation can drive out citizen action in several ways. Participation consumes time and money. A citizen action group that decides to participate in a formal participatory process may find itself short of the resources it needs to act independently in the future. Participation requires compromise; citizen action groups can sometimes splinter over the extent to which compromise is appropriate. Certain action groups have had to moderate their claims or their style of attack once they accepted a formal offer to participate. This may cause media coverage to diminish, thereby sapping some of the group's ability to apply pressure on the city. If an action group loses its autonomy, it may lose its ability to threaten resumption of conflict. Action groups may find that their membership has disbanded and that they cannot regain their former members if their decision to participate causes the residents or consumers involved to lose ground in their struggle with the city.

In Covent Garden, citizen participation and citizen action exist in parallel. The Covent Garden Community Association (CGCA) is an activist group serving the weakest segment of the community. Its independence has allowed it the freedom to be action-oriented. The Association has promoted the cultivation of open spaces as community gardens, established a social center, and rehabilitated several flats. For a time the Association participated in a government-sponsored Forum (advisory board). While the Forum was created in response to the pressure brought on local government by the Association, the CGCA ultimately resigned from membership in the Forum. It was losing its recognition, identity, credibility, and local constituency. In addition, the Association felt considerably weakened by its involvement in a formal public inquiry. Almost the entire budget of the Association went into the presentation of its case before the inquiry. While the effort succeeded in stopping a redevelopment plan that the Association opposed, the Association was financially drained. Participation in government-run processes may yield short-term victories and long-term losses.

Outside Influence and the Tendency for Citizens to Expand Their Claims on Government

Our research has not focused explicitly on the dynamics of social movements, but we do feel it would be a mistake not to comment on the ways in which international social pressures can accelerate the tendency for citizens to expand their claims on government. Residents in one city, apparently content (or unwilling to press for additional resources), hear the claims that residents in other cities are making and immediately step up their own claims on local government.

As residents and consumers living with paternalistic patterns of participation learn that conflict has been used as an effective tool to expand tbe rights and rewards citizens have achieved in other places, the shift from paternalistic patterns of participation to sustained conflict has accelerated. It is our assumption that the stories of successful coproduction in some cities and countries will, in similar fashion, accelerate a shift from conflict to coproduction in places where requests from citizens for a greater role in the design and delivery of urban services and the management of the development process have previously been rebuffed. In part, success stories will help citizen leaders mobilize more effective confrontations and cause some elected and appointed officials to yield to the expanded claims of residents. In part, success stories will provide officials with a way of justifying a move from conflict to coproduction without appearing merely to collapse under pressure.

Political philosophers have speculated for many hundreds of years about the innate propensities of human beings to strive for expanded personal liberty. The same kinds of speculations concerning the claims that citizens make on their government also abound. We are not in a position to contribute substantially to these debates, but we do note that in the cities studied, residents and consumers appear to be expanding their claims on government—claiming "rights" that go beyond the literal constitutional guarantees to which they are entitled. The claims of residents in the Netherlands and Copenhagen to a "right to a decent home at an affordable price" and the claims of Parisians to a "right to be paid for time in transit to work" are two such examples.

Country by country, we note additional groups demanding the right to control decisions involving resource allocation and development that affect them directly. These groups are not satisfied with the right to vote in democratic elections. They want a supplementary means of participating directly in government decisions. For some, this takes the form of a call for administrative decentralization of municipal services (or neighborhood government). For others, coproduction, in which the groups most concerned about the delivery of certain services or the redevelopment of particular sections of the city contract with the city government to handle service delivery or development, comes closest to what they have in mind.

The expansion of claims on government—which includes the "right" to control activities not only inside the household and the workplace but also in the neighborhood—has caused substantial consternation. No one seems to have figured out how to empower resident and consumer groups without creating severe obstacles to expanded development (which will cause unwanted changes but would be both useful and productive). How can local governments empower neighborhood groups to protect what they have now while at the same time ensuring that needed new development will be accommodated? How can local government correct vast inequalities in the financing and management of municipal services if the financial base for such services is restricted only to decentralized sets of service recipients? How can local government empower neighborhoods to maintain the quality of life they desire while at the same time ensuring that sites will be available for facilities that everyone in the city agrees are needed but no neighborhood wants in its backyard? How, in a period of fiscal stringency, can local government justify the cost increases that seem to accompany both decentralization and extensive participation?

The increasing claims on government in one country seem to feed on the experiences elsewhere. Yet, only those experiences that *justify* the expansion of claims tend to find their way through the informal international network. The claims that citizens are making on their government—especially with regard to the right to participate in decisions that directly affect them—are expanding constantly; there is no apparent end in sight.

Summary

Patterns of participation arise and change in response to many factors. Antecedent conditions—including the cultural attributes of the population, constitutional guarantees, past experience with and the accumulated institutional arrangements for planning, and the traditions of political action—shape the patterns of participation that arise and constrain the changes that occur. Influences that encourage citizens to extend the claims they make on government also account for changes in the patterns of participation.

Cities may move back and forth from one pattern of participation to another. Paternalistic patterns of participation may sow the seeds of conflict. Conflict can give rise to coproduction. For instance, neighborhood groups formed in Rotterdam to fight demolition. They drew their expertise from resident students and young professionals taking advantage of low rents in these areas. Informal efforts led eventually to the enactment of an ordinance for an Organization of Urban Revitalization to promote government–citizen coproduction in urban neighborhoods. In a second city, urban revitalization in the Painters' District (Schilderswijk) of The Hague evolved from government control to citizen-generated conflict to government–citizen collaboration. The approach to neighborhood renewal was completely "retooled," shifting from centralized government production to decentralized coproduction by municipal authorities, nonprofit housing associations, and a militant residents' organization.

During the 26 years of the Franco dictatorship in Spain, neighborhood associations could function only within tightly prescribed limits. These associations provided opportunities for collective decision making and limited action taking; they also encouraged the development of indigenous leadership. After Franco's death, neighborhood associations became one of the building blocks of the new democracy. In fact, the new metropolitan planning process in Madrid hinges on the assumption that these associations will have the capacity to represent their members adequately.

Not all attempts to move from conflict to coproduction succeed. In Paris, the 1970–71 demonstrations against the traditionally autonomous transit authority were directed by the major trade union confederations and a coalition of political parties on the left. The aim was to impress the government with the opposition and power of working-class transit users. This strategy led to several significant successes. At the same time, however, apart from the mass demonstrations, little was accomplished in the way of organizational development. Subsequent efforts in the middle and late 1970s to organize citizen action groups met with only limited success. The movement lacked the ability to turn out masses of people for the public demonstrations that characterized the earlier period, and the coordinated leadership of the major trade union confederations was no longer present. New groups continued to splinter from old ones. Attempts to inject the transit riders' union into more recent decision making have failed.

We have also seen examples of new levels of conflict and even new forms of paternalism emerging from efforts at coproduction. In The Hague, for example, plans to demolish housing in the Schipperskwartier in order to increase access to the central business district were proposed by city government. Under the leadership of a cadre of social workers from a local community house, residents protested this action. The city eventually agreed to revise the plans with input from neighborhoods. A process was developed to involve social workers, residents, local business people, city workers, and the housing corporation. The social workers, however, dominated the process. City workers became nervous that the new plan (especially given its emphasis on demolition and new construction as opposed to rehabilitation) would not conform to the desires of the residents. It appears that the goals of the social workers displaced those of the residents. The coproduction process replaced the paternalism of the city with the paternalism of the social workers.

Additional cases in a number of countries must be examined carefully before we can say more about the forces that cause patterns of participation to change.

Initiating the Coproduction Process

Attempts by citizen groups (largely with the financial support of the federal government and foundations) to strengthen neighborhood and consumer control over local government decisions were most successful in the United States during the 1960s. Community organizations sought and in some cases won the administrative decentralization of city government. However, the relationship between public officials and citizen activists has never been anything other than adversarial. Citizen advisory groups were invariably thought of as a political threat to public officials. General Revenue Sharing, Community Development Block Grants, and other elements of the "new federalism" that began in 1970 were, in large measure, a backlash against the citizen participation attempts of the 1960s. Local officials wanted to be free from federal pressure to ensure participation and to experiment with decentralization. The new federalism of the 1970s signaled a retreat from the experiments with direct democracy and a renewal of paternalistic efforts to manage public involvement through tightly prescribed procedures controlled by city hall. Some consumer and resident action groups have fought this shift, but they have not had much success. They have tried, through political pressure and direct confrontation, to retain or expand opportunities for direct involvement. The citizen participation experience in the United States continues to oscillate between paternalism and conflict.

In the current period of cutback planning, new opportunities for direct citizen and consumer involvement in local decision making may arise. Public officials seem more inclined to share responsibility (and blame) for deciding how cuts will be made than they are for deciding how to allocate "new" revenue.[23] Neither liberals nor conservatives in the United States are especially happy about our current approach to building consensus on spending priorities in this period of fiscal austerity. An opportunity may exist to experiment with coproduction.

Some techniques for promoting and sustaining citizen participation might be transplanted from Europe to the United States. In attempting to cull such strategies and approaches from our case studies, however, we must stress the difficulties that haunt all efforts to transfer ideas and experience from one country to another. Some approaches may need to be recast before they will make sense in an American context. This is not a reason to ignore them. Indeed, as America searches for ways of achieving a reconciliation between public officials and action groups, radically different concepts ought not be rejected out of hand. What we may need is not just minor adjustments in past strategy but an entirely new approach. We offer these observations on the European experience with a host of caveats but with some hope that cross-cultural comparisons will allow us to see our own situation in a new light.

[23] Lawrence Susskind, "Public Participation and Consumer Sovereignty in an Era of Cutback Planning," in Edward Hanton *et al.*, eds. *New Directions for the Mature Metropolis* (Cambridge, Mass.: Schenkman, 1980).

Five strategies emerge from our reflections on the European experience. None will necessarily lead to reconciliation or coproduction, but each could help to break down some of the forces that maintain paternalism and conflict.

1. *Local political parties can be used to provide more effective opportunities for residents and consumers to participate in the design and delivery of urban services and the management of redevelopment.*

Political parties organized around local issues are more prevalent in Europe than in the United States. Coalitions between political parties and citizen groups offer both promise and problems. The Federation of Residents' Associations in Madrid, for example, includes some 90 community groups, of which 15 are core. In April 1979, the Federation conducted a mobilization for the municipal elections. These elections, the first held in Madrid since 1931, brought to power people either highly sympathetic to the neighborhood movement or part of it. The Federation was at least partially responsible for the success of the Communist and Socialist parties.[24]

The relationship between local resident associations and local branches of the Communist and Socialist parties is at times difficult. While the parties would prefer that resident groups become part of the official structure, the resident groups believe that their survival depends on their independence. The neighborhood groups see democracy as having emerged from their struggles rather than from the efforts of the political left. The left is seen as a useful ally, but not one to which autonomy should be sacrificed.

The loss of autonomy to political parties may undermine the success of some local citizen action groups. The initial creation of resident groups in the Black Quadrant of Copenhagen emerged not only as a spontaneous response to local needs but also as a result of the deliberate intervention of the leftist political parties seeking to build contacts with the "working class."[25] While these groups might not have formed without the intervention of local political parties, their long-term survival was undermined when the parties shifted their focus to struggles in the workplace rather than the neighborhood. In addition, interparty feuding has made coalitions between local action groups difficult to achieve.

Citizen action groups in the United States might similarly seek to exploit the support of local parties, to the extent that political parties are organized and interested in municipal issues. In the short-run, alliances with political parties can be helpful to residents' organizations by supplying resources, continuity, and a well-developed viewpoint. With care, citizen groups can retain their autonomy while expanding their influence through coalitions.

2. *Consumer unions organized locally around the delivery of urban services may be able to win a place at the bargaining table when elected officials*

[24] Ibid.

[25] Perlman and Spiegel, "Copenhagen's Black Quadrant."

and service providers discuss policy questions that directly affect public service consumers.

Consumer involvement in decisions about the delivery of urban services is not prevalent in either the United States or Europe. In Paris, attempts to forge a union of transit riders in 1971 were somewhat successful. The union succeeded in affecting some decisions because it was able to orchestrate mass demonstrations. It has been less successful in holding a coalition together.

When public transit fares increased dramatically in the early 1970s, eight political parties and trade union confederations interested in the quality and price of public transit formed an umbrella group: the Cartel.[26] Their goal was to influence public policy by exercising political pressure for programmatic changes in the delivery of services. Changes in the quality and the cost of public transit were effected, but little was accomplished in the way of constituency building between demonstrations. Thus, a permanent political force for implementing transit users' demands was not created. Attempts since the mid-1970s to organize a more permanent constituency have met with limited success.

Despite many shortcomings and a still uncertain future, the French users' movement offers a model that citizen activists in the United States might emulate. Local consumer unions could be organized around any of several urban services. Service-specific unions could band together to provide a unified consumer voice in negotiations with public officials.

3. *Referendums and initiatives can be used at the local level to provide additional public involvement in citywide policy decisions.*

The Swiss use referendums or initiatives as a means of pushing certain issues higher up the agenda of public concerns. In recent years, Switzerland has seen a marked increase in the use of voter initiatives introduced by groups who were out of power in Swiss society. By and large, these groups have been too large and too heterogeneous to develop formal organizations that can survive in the highly organized country of Switzerland. Since it is relatively easy to get an initiative on the ballot and since the Swiss system provides a three- to four-year period for government reaction, an issue can easily be pushed into the limelight for a protracted period.

The Burgdorf Initiative was written and the campaign organized by an ad hoc group totally independent of any political party.[27] The initiative, which called for one auto-free Sunday per month, encouraged widespread debate on the advantages and disadvantages of the automobile. Newspaper coverage was extensive. A proposal to mandate at least eight auto-free Sundays was submitted

[26] Hartman, "Transportation Users' Movements."
[27] Hartman, "The Voter Initiative."

to and rejected by the parliament. By rejecting this compromise, the parliament allowed the matter to go on the ballot without a counterproposal. In addition to widespread debate in the newspapers, the central Burgdorf group established regional and local organizations to raise money, generate publicity and make contacts with other groups. About ten thousand people were active in some phase of the campaign, although not one was paid.

Even when initiatives fail to win voter approval, they serve to focus political and public attention. The Burgdorf Initiative led to serious consideration of an annual one-day ban on private automobile use. The establishment of the Verkehrs-Club der Schweiz, a transportation club aimed at helping pedestrians, cyclists, and train riders as well as car drivers, is another spinoff of the initiative. The initiative provides a relatively low-cost means of involving citizens in local policymaking.

4. *Public officials may find it advantageous to encourage resident and consumer groups to do the research needed for the preparation of counterplans.*

Systematic information gathering and problem documentation are tasks that citizen action groups in Europe engage in more often than do their counterparts in the United States. Action research and the preparation of counterplans by citizen groups are encouraged by some public officials in Europe because they help to focus negotiations.

The economic decline of the Docklands area of London was so severe that neither residents nor politicians would have dared to press for the maintenance of the status quo.[28] Rather, the key question was: Whose interests should be served by redevelopment? The Joint Docklands Action Group organized smaller groups within the Docklands. JDAG is a hybrid of the direct-action coalition (Alinsky type) and the research and training organization (Center for Community Change type).[29] JDAG prepared alternative plans, offered in 9- to 60-page mimeographed, attractively bound pamphlets. These were carefully researched and closely reasoned tracts criticizing existing plans and proposing alternative solutions. The alternatives were meant to promote redevelopment that would favor the interests of current residents.

JDAG's pamphlet *The Engineering Industry in Docklands* carefully analyzed the engineering jobs tending to leave the area and offered explanations that took account of international market shifts as well as private investment goals and public policies. JDAG and the people who helped to put together a

[28] Spiegel and Perlman, "Docklands and Coventry."

[29] Saul Alinsky, *Rules for Radicals* (New York: Random House, 1971); the Center for Community Change is a nonprofit organization in Washington, D.C., that provides technical assistance to urban and rural community groups throughout the United States.

proposed strategy for revitalizing a whole industrial sector in Docklands had done their homework. This point was not lost on the policymakers toward whom their statements were directed.

Coventry Workshop is organized even more directly for the purpose of conducting and disseminating research.[30] The Workshop collects information and analyzes trends in an effort to help workers and community groups better understand possible changes in their workplaces and neighborhoods and how economic and political decisions might affect their jobs and community.

The Workshop was an offshoot of the government-sponsored Community Development Project. Several of the original staff of the Community Development Project in Coventry realized that locally sponsored planning efforts were addressed to only a limited set of issues. They were convinced that only an independent and autonomous organization could and would address questions of poverty and deprivation in Coventry. By carefully collecting and analyzing data about many facets of industrial and community life, the Workshop seeks to inform its constituency about the effects of economic and political decisions. Dozens of its publications—including *Chrysler's Crises: The Workers' Answer, Unemployment and the Multi-Nationals in Coventry,* and *Making Urban Renewal Decisions*—have dissected such issues. Technical advice and assistance to community and labor groups are key to the operation of the Workshop.

Inter-Environment of Brussels emphasizes counterplanning and information sharing. The organization has developed a network of friendly "informers" in the bureaucracy.[31] Information about new projects is gathered and disseminated through a citywide newsletter. A well-documented and professionally produced white paper was used in the fight against a freeway. Intensive, week-long seminars are held once a year to help educate community leaders. In Neder-over-Heembeek, a suburb to the north of Brussels, Inter-Environment developed a community development plan in opposition to the city's plan to construct a freeway through the town. The design ideas for the quarter, depicted in a 4-page leaflet, received widespread support. The city abandoned the freeway and plans for high-rise buildings.

Advocacy planning and community design centers flourished in the United States in the 1960s. Public officials saw these as a threat. The European experience suggests that public officials ought to encourage (and, indeed, provide financial support for) citizen-based research and counterplanning efforts because they help to focus and narrow the points that need to be negotiated.

5. *Citizen action groups and public officials can seek stable financial arrangements (including "fee-for-service" contracts) that will enable citizen organizations to become more effective.*

[30] Spiegel and Perlman, "Docklands and Coventry."
[31] Appleyard, "Citizen Action."

The Joint Docklands Action Group is financed by a £24,000 yearly appropriation from the national government. Funds are allocated to selected community resource centers. JDAG resource centers are supported because the action group is seen by the government's development coordinator as playing a vital part in helping people to articulate their views, even if these views are in opposition to local policies and plans. JDAG helps to focus debate on well-researched issues rather than on emotional appeals. The coordinator exercises no control over the JDAG centers except to require a regular audit of their financial records. JDAG is also seen by local officials as a vehicle for cumulating contradictory views and providing a single and credible voice representing community concerns.

In the Netherlands, local and national governments encourage groups to organize themselves and support them when they do. In Rotterdam—in neighborhoods such as Crooswijk, Oude Westen, and Feijenoord—the government has funded action groups organized to fight rent increases and improve local buildings. The city funds local action groups in eleven neighborhoods, enabling them to hire their own planning experts. The experts, often design students and architects living nearby, work as employees of the action groups to prepare revitalization proposals. In the smaller and less well-to-do town of Helmond, where hardly any action groups exist, the national government funded an independent citizen participation organization, Workgroup 2000, to help promote participation. In The Hague community of Schilderswijk, community organizers hired by the city work with citizen groups to help them express their attitudes and secure changes in the official plans.

The Coventry Workshop has been reluctant to accept government funds for fear that this might compromise its advocacy role. As a result, the Workshop has been operating on a meager allowance and still depends on foundation grants. Workshop leaders have been trying to develop an alternative source of funding; they have asked trade unions to donate a penny a day for each member in exchange for services provided. This is still an unrealized goal. Some citizen action groups may refuse government grants. Many, however, will use these funds (as long as there are not compromising conditions attached) to stabilize their organizations. Greater stability and the accumulation of expertise can enable these groups to represent citizen concerns more effectively. This, in turn, is something that public officials should applaud.

Conclusions

The five strategies presented above would support a shift to coproduction. Each day, however, public officials and citizen activists make decisions that can facilitate or hinder increased cooperation. Eventually, coproduction will emerge when citizen groups and public officials both seek ways to convert confrontation into opportunities for joint gain. As the Madrid and Netherlands cases show, such partnerships may lead to both increased participation *and* increased efficiency in the use of scarce resources.

Local authorities should not equate increased participation with decreased efficiency, as they so often do in the United States. The Madrid case shows that the two can be mutually reinforcing and promote the best use of scarce resources.

To make effective use of the skills and knowledge of local residents, Madrid is divided into twenty-five planning districts. Each district has a professional staff paid by the planning authority and chosen by the district to work with local groups to prepare detailed inventories of short- and long-range problems. A local Commission of Participation is self-organized in each district and, with the help of the paid staff, identifies community needs and suggests how they can be met. Programs are coordinated throughout the metropolitan area by a commission composed of representatives of the participating organizations. The goal, according to the director of this program, is to "achieve greater efficiency in the use of our limited resources and to increase public satisfaction with the way resources are utilized and distributed." Madrid has worked to build a partnership between elected officials and residents.

Citizens' groups, private developers, city planners, and quasi-public building corporations are all involved in urban renewal in Holland. These groups begin work together by agreeing on what their mutual responsibilities are. Agreements include explicit declarations of mutually agreed upon principles that form the basis for coproduction and of responsibilities each group expects the other to meet. Such "coproduction contracts" help to ensure that the diverse groups involved do, in fact, work toward agreed upon outcomes.

Coproduction contracts can be formal or informal. Workgroup 2000, an independent consulting group, enters into a working relationship with a city only if the city will agree to allow municipal planners to perform services for residents as well as the government, stipulating in a policy document the limits to the planning process, encouraging the development of alternative plans, and sharing information with all the parties involved. In the Schilderswijk neighborhood of The Hague, a nonbinding letter from the City Council stating its desire to help current residents to remain in the neighborhood following redevelopment was a sufficient accord to encourage residents to coproduce the redevelopment plan.

The process for devising a new "structure plan" for Schilderswijk in The Hague is a good example of the give and take crucial to effective partnership. First, city planners conducted an in-depth survey to determine neighborhood views. After this, planners met with citizens in a series of subject-specific work groups (e.g., housing, traffic, and schools) to hammer out agreements on the needs of the neighborhood. Then, city planners fleshed out the details of these agreements for ten districts in Schilderswijk. These more detailed plans were presented to ten district "consultation groups" for discussion, consensus building, and revision. The final plans went back to the consultation groups for one final review, after which the City Development Agency put the plan into shape for presentation to the City Council, which approved the plan without amendment. The process lasted for only one year and featured numerous deadlines for both planners and citizen participants, placing the burden for producing the plan on both sides.

The approaches described above could be used effectively in the United States. Effectiveness would depend, of course, on the willingness of residents and consumers to accept the government's invitation to participate as well as the willingness of public officials to endorse direct public involvement in the design and delivery of public services and the management of development.

2

2

COPENHAGEN'S BLACK QUADRANT

The Facade and Reality of Participation

JANICE PERLMAN AND HANS SPIEGEL

Introduction

Citizens in Copenhagen

Copenhagen is what it seems to be—a vibrant, livable "city that works"—unless you are a citizen trying to make your voice heard in city planning. Despite the elaborate mechanisms for citizen participation in the planning process, the impact of such involvement is minimal. In fact, according to Vilo Sigurdsen, Deputy Mayor in charge of planning, the elaborate mechanisms for participation have produced not one single instance in recent years when a plan has been substantially modified by citizen input.

The lack of consequentiality of citizen participation in Copenhagen stands in stark contrast to the numerous provisions that Denmark makes for its people's well-being. To many observers, Denmark is the prototype of the modern, affluent welfare state. Per capita income is the third highest in the world (preceded only by Switzerland and Kuwait) and welfare services are among the most generous anywhere. There is free medical care, education is available without charge through the university level and includes student stipends, unemployment payments range between 60 and 90% of salary, and old-age pensions start at age 63. As one Dane summarized it for us, "In this country it is against the law to be poor." This impressive package of services is financed by high taxes about which Danes grumble but which they generally seem to accept.

The images of Denmark as a highly civilized and socially oriented welfare state and of Copenhagen as a well-planned metropolitan area are further reinforced by the settlement innovations since World War II. Many American visitors are charmed by "wonderful, wonderful Copenhagen," with its pedestrian shopping streets, its historic center and harbor, its gleaming new suburbs built along rapid transportation "fingers," and, of course, its Tivoli Gardens. This physical attractiveness is due in part to Copenhagen's having been spared from bombing in World War II and also to its good fortune of having been too poor in the immediate postwar years to develop a massive steel-and-glass downtown like Stockholm's. By the time the resources would have permitted such architecture, disastrous experiences elsewhere were already well known and the city's leaders could choose the conservation/preservation path without having to learn from their own mistakes.

Aside from being famous for physical livability, Copenhagen is particularly well-known in the world of citizen involvement. First, this is due to innovative planning and management of daily-life activities in some of the new suburban settlements surrounding the city. Housing and community design experiments have been launched in a number of places; in these, future residents collaborated with architects and planners to develop physical and social life-styles involving increased communal spaces and functions as a tradeoff against individual space and privacy. Second, the city has recognized Christiania, a squatter-occupied, authority-defying, and self-governing commune on the site of an abandoned military barracks. Christiania is one of the largest urban communes in Europe and probably has the widest local support (12,000 people from all over Denmark defended it from a 1976 threat, and the planning director of the city government has written about it with respect and admiration).[1]

The Danes are a well-organized people. Danish bureaucracy seems to be fine-tuned to meet the needs of the population through numerous governmental agencies and programs. Supplementing the government's efforts are a series of private organizations that involve numerous people in adult education, sports and recreation, and cultural activities (70% of Danish youth between ages 15 to 20, are members of such groups).

The Danish welfare state aims to meet the housing needs of its population by providing a dwelling for all "married or unmarried couples" and all adult single people. An intricate system of guaranteed loans, private investments, outright grants, direct and indirect subsidies, and controls are provided by private and public organizations and quasi-public corporations.

A fundamental resource for producing rental housing are the nonprofit housing associations that have a 65 year history in Denmark and now own more than one-third of the country's rental housing stock. These are private cooperatives which own and operate housing but do so with considerable public assistance and oversight. For example, "social housing" is developed by the nonprofit association in partnership with the local and federal governments through the

[1] Kai Lemberg, "A Squatter Settlement in Copenhagen: Slum Ghetto or Social Experiment?" *International Review*, U.S. Department of Housing and Urban Development, Vol. I., 1, 1978.

mechanism of a jointly created "public condemnation company" and a building and renovation corporation.

Once the housing has been built, the tenants are eligible for graded rent subsidies based on financial need, the level of market rent, and the number of children. These subsidies are paid directly to the households and can be used in both nonprofit "social" and private properties. In situations of particularly great need, as for a number of elderly people, a special housing allowance can be made available that is higher than the maximum allowable subsidy.

On the surface, the conditions for broad-based citizen participation seemed almost optimum in Denmark. At first it was difficult to understand why some residents in Copenhagen were so bitter about the supposed intransigence of city

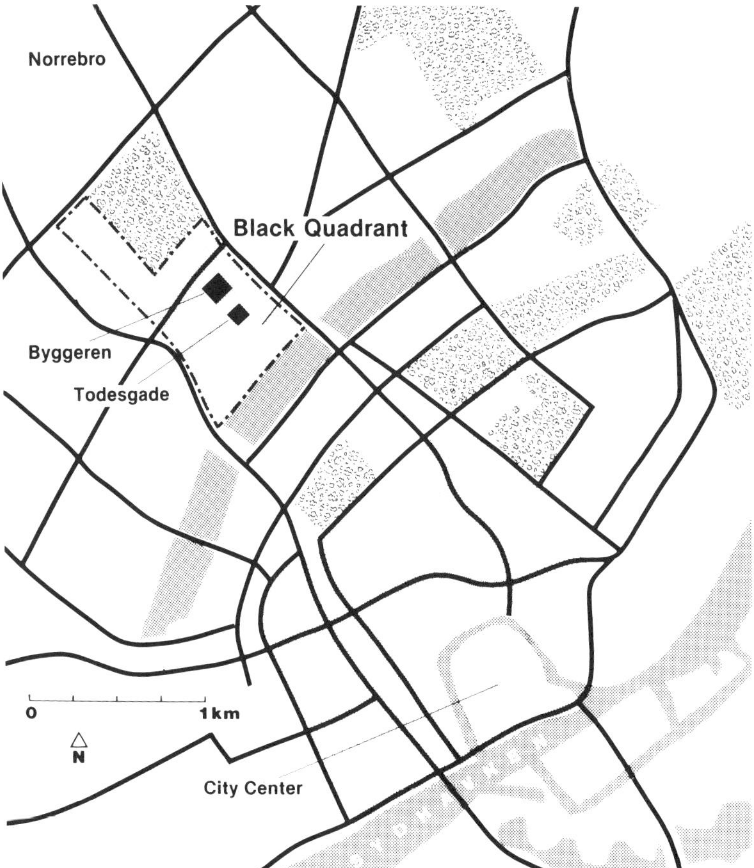

Figure 1. Copenhagen center, showing the location of the Black Quadrant relative to the old central city.

government and the majority party, the Social Democrats. Did the highly structured and ordered form of citizen participation actually thwart meaningful involvement around consequential issues? A federal official may have given us a clue when he pointed out, "In a formal way, we're democratic. But our bureaucratic organizations are slow to change. They were vivid a century ago, but now they are institutions. If you want to get them to do something different, they get offended."

We soon discovered that, indeed, some Danes want to do things differently, especially in urban centers. The issue of urban neighborhood renewal illustrates the struggle between the more traditional forms of welfare-state operations and the newer challenges, mostly from the left, pertaining to distribution of resources and decision making authority for residents. The case we are about to describe deals with a resident movement called Norrebro Beboeraktion (Norrebro People's Action) that sparked considerable citizen action in the Sorte Firkant (the Black Quadrant) neighborhood of Copenhagen. It describes the controversy between this resident action group and City Hall—a controversy that sometimes erupted into open hostility and massive police intervention.

The reality of citizen participation in the Black Quadrant centers on the question of "who gets what" as this neighborhood is being renewed. The resulting dialogue between the residents and the various levels of government (and the quasi-public renewal agencies) is not all benign Danish pastry and cream. It is a struggle of ideological, political, and sometimes physical dimensions. The battleground is principally in the Black Quadrant, but—because of partisan politics in Denmark's multiparty system—the struggle also takes place in Copenhagen's ornate brick City Hall and even in Parliament, which meets in the old royal compound.

The Local Setting

As visitors from abroad leave Copenhagen's old central city and walk for about a kilometer on Oorrebrogade over the bridge that leads to the Black Quadrant, they may well be surprised. Facing them is not the place they had imagined as "one of the worst areas of the city" but a picturesque, tree-bordered area on the opposite shore of Peblinge Lake. Entering the middle of the Black Quadrant, one is aware that renewal action has begun and is now taking place, but the area is too tidy and clean, the people too well dressed, and the shops too amply stocked to qualify the area as a slum in the American sense of the term.

The Black Quadrant consists of less than one square kilometer composed of 25 blocks, five of which have been demolished. Its rather somber name comes from the factories that were established in the area over a century ago and belched black soot all over the neighborhood. Most of the factories are gone today, but some of the old houses still stand. Some of them still do not contain interior toilets, forcing occupants to use outhouses in the courtyards. A number of old apartment houses, with their dingy, unlit hallways that have not seen a coat of fresh paint for decades, are occupied by foreign workers and their families.

The Black Quadrant has had more population density and poorer housing quality than the rest of Copenhagen. Ten years ago the area had 18,000 inhabitants, but it has only an estimated 10,000 today. A decade ago, 70% of the dwelling units had no central heat, while only 32% of units in Copenhagen as a whole were without this amenity. The latest statistics about the work force living in the area involve people in the five large blocks that were demolished. According to these prerelocation figures (1973–74), 13% were civil servants, 9% skilled workers, 36% unskilled workers, 3% self-employed, and 39% un-

Figure 2. The Black Quadrant. View from across the Peblingeso (canal). (Courtesy of the City of Copenhagen, Office of the Lord Mayor.)

employed. (It should be added that a sizable number of the unemployed were reportedly students who had moved into the area in rather large numbers starting in the 1960s because of the relatively low rents available there and its convenience to downtown and educational institutions.)

Today, even though urban renewal activities are far from complete, the socioeconomic character of the population has already shifted. Not only are there fewer people in the Black Quadrant but they tend on the average to be wealthier. A number of large apartment houses have been rehabilitated, a middle-income structure built, and a public housing project (including a senior citizen component) has been constructed. In addition, a few condominiums with relatively high rentals have sprung up in the area. Renewal activities are slated to continue in the Black Quadrant for several years, and the resulting market forces are likely to continue the upgrading trend. If unchecked, these trends will probably cut the number of unskilled workers in the Black Quadrant still further and considerably increase white-collar and middle-class inhabitants.

This shift in the population composition is at the heart of the citizen action. In a word, the burning issue is *revitalization*—the so-called upgrading of land, buildings, and people. The working-class people now living in the Black Quadrant fear that they may be excluded from their old neighborhood; they know that they will be offered decent accommodations elsewhere in the Copenhagen area but probably at considerably higher rents, and they want to stay in their old, familiar surroundings.

The Planning Context of Neighborhood Development

To fully appreciate the irony of citizen participation in Copenhagen one must first realize how elaborate the established procedures actually are. Resident participation in neighborhood development and urban renewal in Denmark involves a three-level planning process. The most global level is the Regional Plan, a second level is the Municipal Plan, followed by the third level, the Local Plan. A recent reform of planning and urban renewal law clarifies this three-level process and also provides for considerable citizen involvment. For example, the creation of the Municipal Plan now requires a first draft that is to be thoroughly discussed by citizens. Information must be widely distributed and discussion meetings arranged for three or more months. Then, according to the new regulations, a reformulated draft plan is to be publicized together with alternative views which, however, can only be proposed by members of the city council. Following this publication and distribution to all affected residents, an additional period, this time four months, is set aside for citizen objections and comments.

It is significant that in the case of the Black Quadrant, the Copenhagen municipality avoided some of the stringent requirements for citizen participation and other approvals of the new law (such as the sign-off from the minister of the environment) by creating an intermediary form which they called an Overall Plan, which falls somewhere between the Municipal Plan and the Local Plan.

They submitted it one day before the new regulations were to go into effect. This Overall Plan covered the whole of Inner Norrebro (about twice the land area of the Black Quadrant) and was based on earlier planning regulations, thus cleverly bypassing the new participatory mandates.

Not surprisingly, a number of residents and their organizations were rather upset by the creation of the plan, for which no provision had been made in the new planning legislation. The Overall Plan immediately became known as Egon's Plan, after Egon Weidekamp, the Lord Mayor. His strategy was to move quickly into urban renewal activities before political obstruction could materialize. Such opposition was looming on the political horizon and materialized some time later in 1977 by the appointment—through a coalition of leftist groups—of one of the six mayors (or deputy mayors in American parlance) who, as circumstances would have it, had local planning in his municipal portfolio. The appointment of Vilo Sigurdsen, from the Left Socialist party, encouraged champions of neighborhood control to think that "the little man" living in the poorer sections of Copenhagen had, at long last, a strong voice in city hall. However, it soon became apparent that the Social Democratic–Conservative coalition had the votes to override Sigurdsen.

The defenders of Egon's plan insist that there was nothing sinister in their procedures. They wanted to avoid the delays that would come with extended citizen debates and time-consuming central government approvals. They wanted to "get on with the job" and not call upon their own heavy political artillery to push through every detail. Furthermore, according to the lord mayor's staffers, the Overall Plan provided for citizen information and discussion and an orderly, collaborative execution of the planning process. They felt strongly that the Black Quadrant needed to be "de-densified" and upgraded for the good of the city and region, and they did not welcome attempts to torpedo these efforts.

The Popular Response

After toiling diligently to learn the myriad rules and regulations governing citizen participation in the planning process—the requirement for hearings, the number of days in advance that information must be posted, the submission of alternative plans, and so on—it was a rude shock to discover that across the entire city of Copenhagen there had not been a single substantive change in any proposed plan over the previous fifty years! This is especially notable in light of the 1977 law which mandated public hearings and formalized one of the most elaborate citizen participation mechanisms imaginable.[2] Partly in response to the

[2] As Peter Bjerrum pointed out, however, the lack of citizen impact does not imply the hegemony of planners but rather the responsiveness of city hall to downtown business interests as opposed to neighborhoods. He pointed out that "city hall has never refused any wish from any kind of private or public corporation or institution, even if in *direct conflict* with the plans, so long as it was in the interest of developing a modern, central business district."

growing frustration with this state of affairs, a number of independent community groups have formed in an attempt to exert more direct pressure on the planning process. They are called Beboeraktion (People's Action), and there are about 20 to 25 of them in Copenhagen today. They are all part of a citywide coalition called FKB (Copenhagen Federation of People's Action), which meets every third week to discuss common problems, tactics, and strategies. These issues most frequently revolve around maintaining the character of moderate-income residential neighborhoods in the face of urban renewal and the pressure to convert them into expensive office space and luxury condominiums. They often involve struggles over family-oriented services such as a day care, nursery schools, kindergartens, schools, and playgrounds; over lessening the danger and noise of through traffic; and over the upgrading and modernization of the existing housing stock. The Federation not only discusses these issues but also dissiminates information, takes public political positions, and engages in joint actions.[3]

There are also two nationwide coalitions which include members of people's action groups, the SME, and the LLO. The SME is the coalition against condominiums. It unites the neighborhood groups, progressive political parties, and labor unions in a common effort to prevent the conversion of rental units into condominiums. The LLO is a nationwide tenants' organization, closely allied with the Social Democratic party and mostly concerned with the legal protection of tenants' rights. But, for example, it joined the fight against condo conversion and in various local areas has played a more independent and progressive role.

In addition there are some 12 local councils within Copenhagen which are self-initiated and have no legal status but serve as an unofficial government.[4] They are composed of merchants' associations, industrial organizations, recreation clubs, parent–teacher associations, and the people's action groups representing the citizens in the area. They deal with basic local problems such as slum clearance, traffic patterns, or local plans and also run neighborhood social events, parties, and entertainment to raise funds for social service delivery. Although city hall regards these councils with some animosity and often sees them as a threat, it *does* pay for their basic secretarial expenses and in some cases provides meeting space in the area. The councils often maintain a storefront operation which is open a few hours a week and has a telephone answering service. Attempts by the Copenhagen city government to absorb these councils under the political party rubric have been opposed by both right and left—the right on the grounds that the city council already does the job and no recognition of local councils is called for and the left on the grounds that the councils' most important asset is their autonomy and independence.

[3] Unfortunately, the coalition is weak and many of the delegates are overburdened with too many meetings of their own neighborhood groups, action committees, political parties, and labor unions as well.

[4] Interview with Kai Lemberg, director of city planning, July 1979.

Citizen Action in Norrebro

Within Norrebro, aside from the local council, there is a longer-standing tenants' union called Griffenfeld, a people's action group called Red Rose, and the best-known direct-action group in Copenhagen—Norrebro Beboeraktion (NB). We will summarize the first two briefly and describe the third in greater detail.

Two Antecedent Groups

The Griffenfeld Tenants' Union was founded in 1941 and was part of the Copenhagen Tenants' Organization (CTO) until four years ago. At that time the chairman of the board of the Renovation Company (which does urban renewal) was named head of the Copenhagen Tenants' Union, creating such an obvious conflict of interest that Griffenfeld withdrew from CTO in protest. They have a staff of nine who maintain their storefront office, have about 300 dues-paying members, hold a monthly general assembly, send out a newsletter twice a year, and work to create and strengthen their local block clubs. Like most other tenants' organizations, they work mainly *within* the legal process, informing people of their legal rights and helping to enforce them. Only as a last resort do they consider disruptive actions such as rent strikes.

Together with Norrebro Beboeraktion, the Griffenfeld Tenants' Union recently won a victory which set a critical precedent in the area. It concerned an elderly woman pensioner who had been displaced and who had occupied a vacant flat in the neighborhood cooperative social housing with the help of the tenants' group. She stayed there for three years while the tenants fought for reform of the pension. Even for the small two-room flat the woman was living in, the rent was 1,900 kroner as compared with 300 kroner, which is 15% of her pension. Under the law, she could not be expelled and is still living there at 300 kroner per month.

The second citizen action group, Red Rose, was founded about 10 years ago and defines its membership as everyone living in the area. It has 30 to 40 activists in the core group and is best known for its manual entitled *The Red Rose Book: Greetings to Neighbors, Friends, and Enemies from Inhabitants of the Area*. It is a handbook on community organizing—both the philosophy and the techniques—and goes into great detail in documenting what the group has done and what can be learned from it. In fact, the group put so much energy into the writing, publication, and dissemination of the book that the organization itself almost withered away.

The major internal achievements of Red Rose have been developing a collective leadership (without any one person's domination), adding new members and socializing them into the organization, avoiding partisan politics, and recognizing they can be provocative and stubborn in their demands without necessarily taking to the streets. They complain that it seems to take a crisis to mobilize people but are finding ways to work around that,

Their major external achievements are (1) detailed traffic, parking, foliage, and amenity plans for each subarea in the neighborhood (including, for example, tables and benches in Balders Plads as well as a small theater backdrop and a special box for dog waste); (2) forcing compliance with the rules regarding parks, open spaces, and amenities in demolished areas (which are usually dropped once the housing has been built); (3) forcing a timetable for implementation of local plans and clear designation of responsibility for each aspect; (4) getting community proposals to city hall before developers made bids; (5) setting up their own kindergartens, nursery schools, and child-care center when the municipality failed to provide them (including finding a site, getting children signed up, recruiting teachers, and publishing a pamphlet telling others how to do this); and (6) sensitizing local residents to their own needs, rights, and ways to become informed of these and fight for them.

Norrebro Beboeraktion (NB)

NB was founded on January 13, 1973. Its origins can be traced back to the 1960s, when renewal plans for Norrebro began to go into effect and a squatter movement formed in response to the pressing problems of people without a place to live on the one hand and increasing numbers of empty flats on the other. The squatters were mostly students of middle-class origins. At the beginning they remained fairly isolated from the other residents, but gradually they began to reach out and make connections. They started a People's Cafe on the first floor of one of their occupied buildings, a People's House in an abandoned factory, and a People's Park in the front of the latter. They decided to try to mobilize the community to defend its interests, starting with the fight to turn a slum-clearance lot into an "adventure playground" for neighborhood children rather than allowing it to serve as a parking lot, its current use, or to become the site of a block of flats being planned for the area. The first meeting held in the People's House was attended by about 15 people and represented the first "people's action" (Beboeraktion) in Copenhagen. One year later, the core group had grown to 50; at its height, it included 120. As in all these groups, however, most of the members were students or young people and only a small fraction were old-timers in the area.

The first action of the group was to occupy the parking lot designated for new flats. They knocked down the fence, let the cars remain on one half the site, and encouraged children to use the other half as a play area. They held a huge street party to commemorate the event. After a few days, the children were so rough on the cars that drivers simply stopped parking there and the entire space became a playground. Dozens of volunteers came to the site to help erect tepees, slides, jungle-gyms, and a temporary crafts space for the children—all in crude natural materials and of spontaneous design. Parents volunteered to supervise the children while pressure was put on the municipal government to legitimize the venture by supporting a five-person professional staff to run the playground and adding a mobile unit with toilets, office space, and an indoor recreation area. This campaign was carried out through letters, local media, and a massive publicity effort.

Eventually city hall hired a staff to run the playground, but it avoided employing the local residents who had been active in creating and running it. This, of course, raised the fundamental question of whose playground it was and who had control over it. The NB members told the children not to pay any attention to the government-appointed supervisors and to harrass them. The children responded with considerable enthusiasm, with the result that one staff member had a nervous breakdown, another quit, and the residents' association was able to negotiate for its own staff accountable to the neighborhood. This was a real victory for the new organization and set a precedent throughout the city: a local group succeeded in converting land use in its neighborhood to its own purposes and in getting the government to pay for the maintenance of the facility while retaining control for themselves.

Since then NB has taken on a number of issues and campaigns. One of the first things they did (May–June 1973) was to publish a pamphlet entitled *Speculation and Slum Clearance Go Hand-in-Hand* and distribute it free to everyone

Figure 3. The Byggeren playground. Built in 1973 by local residents and the People's Action Committee of Norrebro on a lot previously cleared for apartment construction. Copyright 1980 by Heine Pederson. Reprinted by permission.

Figure 4. The city of Copenhagen recognized Byggeren as a temporary city institution in May 1974 and provided a grant of a half-million kroner to pay for playground attendants. Copyright 1980 by Leif Gram. Reprinted by permission.

in the neighborhood. It describes the history of Norrebro and shows how its present conditions were the consequences of the 1972 district master plan promoting renewal. It urges residents to unite and propose an alternative plan. NB demanded a rethinking of the urban renewal process, focusing on (1) modernization rather than demolition, (2) spot demolition (rather than massive demolition) only where necessary, and (3) rehousing of displaced residents within the neighborhood. They set up a slum-clearance patrol which went through the neighborhood building by building, reporting code violations and registering lack of maintenance by specific landlords. At the same time they began forming block clubs to work out detailed local plans.

In January 1974 the first alternative was formed to fight the proposed removal of 500 of the 850 families on the block. Residents met in the People's House and developed an alternative plan which they presented to the slum-clearance company with 605 signatures. They were successful, and within a few months eight block clubs with similar strategies had formed. Simultaneously, NB as an organization was fighting for the right to use a vacant storefront owned by the slum-clearance company as an office and local information center.[5] They occupied the back bottom flat of a condemned building for $1\frac{1}{2}$ years and tacitly won the right to use such space free of charge.

Other struggles focused on personalizing an attack on the two major slumlords and property speculators in the neighborhood—Nielsen and Donatsky—and launching a poster campaign against them. Nielsen is unaffectionately known in the area as "jaws."

In July 1976, NB split into two factions—NBB (Norrebro Beboeraktion Blagardsgade) and NBT (Norrebro Beboeraktion Tomrergade)—each named for a different street in the neighborhood where their respective offices are located. There are many explanations for the group's failure to hold together, most focusing on the large size it had reached and fundamental internal differences in ideology and orientation. NBB favored the creation of a "mass base," with popular participation and widespread mobilization as a high priority, and therefore saw education and consciousness raising as essential. NBT, on the other hand, thought the most efficient organizational form was lean and quick-acting and that it could best accomplish its goals through the "exemplary actions" of a well-trained, closely-knit cadre of leaders from whom others could learn. Furthermore, whereas NBB favored rehabilitation and upgrading, NBT supported demolition and new construction. Behind these differences are the more general disagreements within the left—the new left (or revolutionary left), which is more closely represented by NBB, versus the old left and Communist party, which is closer to NBT.[6]

Since the split, both groups have been severely crippled in their effectiveness, and there has been a good deal of confusion, both within the neighborhood and in the larger citywide and national coalitions and meetings, as to who really represents the citizens of the area. (For example, NBB and not NBT is the Copenhagen Federation of People's Action, and NBT is more active than NBB in SME, the anti-condominium coalition.) Yet each has continued to struggle with neighborhood issues and local political problems in creative and often impressive ways.

NBT—Norrebro Beboeraktion Tomrergade

NBT members stressed that Norrebro's Black Quadrant had always fostered a tradition of militancy. They pointed to the 1930s, when the area was the focal point for the anti-Nazi movement. They saw themselves in that tradition and

[5] Ultimately NB occupied the slum-clearance company's office (March 1976), calling it the "nationalization of the slum-clearance company."

[6] The larger labor constituency of the Communist party favors demolition and new construction, for example, in the belief that it creates more jobs than does rehabilitation. This is not necessarily a correct assumption.

wanted more direct action rather than consciousness-raising drives. They believed that people learn best through observing the results of successful actions.

They have about 15 to 20 people full time (living on welfare and unemployment benefits), 20 others part time, and a large number of "one-eighth militants" who come out for actions and social occasions. They observe that despite widespread support in the neighborhood, people say they "don't have time to get involved." They hold a general assembly twice a year, an executive committee meeting once a week, and working group meetings once a week. The working groups are organized around specific issues such as traffic, condominium

Figure 5. Following the February 1980 city decision to begin construction of the apartment building originally planned for the Byggeren lot, a physical blockade was called by the People's Action Committee. The group successfully prevented demolition of the playground during March and April. Copyright 1980 by Alfa Foto. Reprinted by permission.

conversion, urban renewal, law, and so on. The executive committee is composed of one representative from each such working group plus officers elected at large by the general assembly.

While the issues of NBT are generally similar to those of NBB, some of their activities have a different flavor. For example, they engage in street theater, presenting skits in parks and plazas that depict—in high melodrama, black humor, and satirical songs—money-grabbing landlords, menacing tenants, aloof city hall bureaucrats, and innocent citizens in distress. One of their theatrical groups has made a nationwide bus tour, putting on various skits and shows. They have excellent media relations and are well covered in the press.

They also run the famous "fighting high school"—an organizer-training program which has attracted students from all over the country. They have a three-person core staff to run this school and offer 14 day-and-night courses ranging from a weekend to a few weeks. This has been a very effective way of developing leadership; there are over 100 graduates around the country already.

Furthermore, NBT is involved in "public interest research"—locating scandals in the government, housing societies, renewal companies, and so on, which are publicized through the press; NBT also undertakes independent investigations and calculations on the value of houses to be demolished or the cost of government projects. This gives the group added credibility in challenging official claims and proposing alternatives.

Finally, they are very active in the creation of alliances and public pressure in regard to the urban renewal law. They claim to be taking a leading role in demanding public input in the law and putting together coalitions which could be active in doing that. The new law was proposed by the minister of housing in October 1979 and, in preparation, they have been meeting since Christmas 1978 on a weekly basis with the cooperative housing societies and other relevant parties. They held a large conference in May (attended by the minister) which put forth a series of proposals about (1) relocation and rehousing, (2) residents' influence, (3) economy, (4) speculation, and (5) financing. Under each category, they spelled out what the law stipulates at present, what is needed, what should be taken into consideration, and what possible solutions would be.

As of November 1980, this proposed bill has still not been submitted to the parliament and is unlikely to be officially introduced as legislation due to massive opposition from the Conservative and Liberal parties.

NBB—Norrebro Beboeraktion Blagardsgade

The overarching goal of NBB as expressed by Michael Steffensen (one of the founders of the original NB and present leaders of NBB) is to "fight for better housing and a better neighborhood for people living in Norrebro."

NBB has no single leadership. It has a shared decision making process among a core group of eight people. They are elected every second year by a general assembly. They have no formal membership, but their constituency in the neighborhood is composed of about one-third students and unemployed young people, one-third elderly long-term residents, and one-third low-paid unskilled workers. There are decreasing numbers of children in the neighborhood, as couples tend to move elsewhere, if possible, to raise their families. The leadership is mostly students and young couples without children, generally in the range of 25 to 35 years of age and almost equally divided as to sex. Their average time in the community is about 5 to 10 years. It was a common complaint that there are very few activists below the age of 25. It seems that young people, not involved even tangentially in the political ferment of the 1960s, are much less politicized than the older generations.

There are 7,000 people on the NBB mailing list, 50 to 100 dues-paying contributors, and a handful of dedicated militants who staff the office. This is deceptive, however, in that at peak times—such as the demonstration in front of the city hall—they are able to bring out 2,000 to 3,000 people.

Aside from dues, newsletter ads, and subscriptions, NBB gets some revenues from slide shows and speeches given at schools and from the sale of posters, buttons, and so on. During the playground struggle they received support and donations from the teachers' union, the child-care workers' union, the Federation of Building workers, and from a multitude of small, informal sources

Figure 6. On April 22, under protection of 800 policemen, workers removed playing materials and structures from half the playground. After the police left, a few hundred local residents moved the materials and structures back to their original position. Copyright 1980 by Lars Bahl. Reprinted by permission.

including a knitting club of five elderly women who sent in 50 kroner. They also receive modest funds from private foundations.

The major issues NBB had dealt with since 1976 are (1) firetraps, (2) traffic patterns, (3) the Todesgade day-care center, (4) housing legislation, and (5) Lord Mayor Egon Weidekamp's overall plan. A brief discussion of the events and outcome of each is in order.

1. The *firetrap* issue was fueled by the Stengade fire, in which a disastrous loss of life and property was directly traceable to poor maintenance by the landlord (previously pointed out by neighborhood residents in their slum-clearance report). The massive protest which resulted led to a much more stringent law regarding fire hazards, including the mandatory installation of metal fire doors on each wing and each floor of multifamily dwellings.[7] In September 1976, NBB printed and distributed a leaflet entitled *Protection of Firetraps Is Rubbish,* pointing out that the new law had not been respected by local landlords and bringing pressure to bear on this.

2. *Through traffic* on certain streets in Norrebro had created an intolerable level of noise, pollution, and danger to children and the elderly. The noise was up to 75 decibles and there were more traffic accidents per capita in Norrebro than in any other area of Copenhagen (or all Denmark, for that matter). During September 1976, NBB organized a series of demonstrations highlighting this issue. They put up barricades on the main thoroughfares to block traffic during rush hour, permitting only buses to pass through and only at reduced speeds. They had gone all over the city stealing traffic signs of every type—"street closed," "no through traffic," "dead end," "no right turn," "left turn prohibited"—and posted these everywhere, creating total chaos. They set out tables, fruit stands, and flower stalls on the streets; played music; and gave speeches pointing out the desirability of rerouting the traffic. Three hundred people were enjoying the street activities when the police rushed in, harrassing, beating, and making arrests on the ground that the participants had no permit to close the street. NBB had indeed applied for a permit but received no reply, which they took as a go-ahead. After some negotiations about this, the police left, only to return the following day in even greater numbers. After six days of confrontation, the city agreed to the people's demands for an experiment, closing both major arteries in question, except for buses. This was extremely successful, cost the municipality virtually nothing, and significantly enhanced the quality of life in the neighborhood.

3. In August 1977 an institution called Todesgade, including a child-care center, kindergarten, and youth club (with lots of outdoor space and very good access for Norrebro residents) was to be closed down as the first of three insti-

[7] It is interesting that this amendment to the housing legislation was pushed through parliament after the Stengade fire not through the force of citizen protests alone but in large measure through pressure by the Social Democratic party on grounds of its job-creation potential.

Figure 7. Police returned to Byggeren on April 29. This time, workers demolished the structures with bulldozers. Copyright 1980 by Henrik Saxgren. Reprinted by permission.

tutions to be phased out according to the overall plan. NBB organized a blockade to prevent its destruction and the demolition crew refused to cross the picket lines.[8] When police accompanied by attack dogs forced their way through the picket lines, the people became enraged. They ripped up bricks from the old streets and threw them at the police, turned over cars, barricaded the streets, and occupied the child-care center. Later in the month, on August 15 at about 7 o'clock in the evening, the police once again tried to force their way through the blockade, and this time they managed to remove a few sections of the Todesgade institution. When this was announced on the 7:30 P.M. television news, hundreds of people, not only from Norrebro but also the entire city, flocked into the neighborhood, barricaded the street, and engaged in a nine-hour fight with the police.[9] As a result, the Todesgade youth facilities were reopened under

[8] Workers cannot be forced to work under police protection on grounds that it is a threat to their safety and well-being. A recent amendment to the labor law makes it illegal to prevent anyone from crossing a picket line if they have a legitimate reason for so doing.

[9] This provoked a strong counterreaction in city hall, facilitating the passage of a law against picket lines and clarifying the fact that citizen residents engaged in acts of protest do not have similar protection under law as do labor union members during strikes. Those on picket lines can be jailed, whereas striking union members cannot.

community control for six months. When the first-day enrollments were opened, 80 children signed up, belying city hall's insistence that there was no longer a need for this type of service in the neighborhood.

A half year later, very early one morning, a massive police battalion stormed the picket line and closed the area completely, physically removing the Todesgade buildings and trucking the sections out of the area. It also became clear that labor union support would no longer be forthcoming, since the unions favored

Figure 8. Several hundred local residents tried to prevent demolition of Byggeren by embracing the structures, while thousands gathered to barricade the streets. Violence ensued. Copyright 1980 by Sonja Iskov. Reprinted by permission.

demolition and new construction rather than preservation or upgrading—on the grounds of supposedly increased potential to create jobs.

4. In terms of the overall impact of NBB, clearly one of its major achievements was modifying the *national housing legislation*. They waged a long campaign against both nonutilization of vacant flats and rapid turnover speculation, and—along with various allies, including strong labor support—succeeded in amending the housing law so as to (a) provide a six-week maximum period of vacancy for apartments (even if they are condemned for demolition), forcing the renting of empty flats on temporary contracts between renewal plans and actual construction,[10] and (b) ensure a five-year ownership period before resale of an apartment.

5. Egon Weidekamp became lord mayor in 1977 and proceeded to produce an *overall plan* for Norrebro that involved considerable demolition and reconstruction. In May 1977, NBB began a campaign against the plan, calling for "people's proposals" for alternatives. They were enraged that Weidekamp had spent 2 million kroner, contracting a private firm to draw up the plans, without *any* attempt to solicit input from area residents. NBB has been protesting the entire planning process as well as pushing for specific changes in open-space, housing heights and densities, traffic patterns, and so on. They have gradually sensitized the planning body to such issues as, (a) better informing the public of plans in advance, (b) using spot demolition instead of massive destruction, and (c) enabling residents to remain in their neighborhoods through special subsidies. On the other hand, the limitation has been that they would not stop demolition or the gradual gentrification of the neighborhood, and they have not been able to legitimize precedents for citizen involvement in the planning process.

In the interim between the research and publication of this chapter, a final struggle took place which pretty well decimated NBB. It was over the adventure playground Byggeren (literally the "building playground"), the most popular and colorful symbol of the people's action movement. The playground is just across the street, kitty-corner from the Todesgade institution. It had been occupied, fenced in, creatively built up (including a teepee, log house, wooden locomotive, etc.), and brightly painted at the very beginning of demolition in Norrebro. Throughout the years, it was widely used by people of all ages (even as a pleasant place to read while sitting in the sun). It had been constantly and imaginatively changed and had been defended numerous times from threat of demolition.

Vilo Sigurdsen, the progressive deputy mayor for planning, along with Norrebro parents and NBB, fought for the preservation of the playground both on the grounds of opposition to the proposed housing project for the site and of the need for open space. They demanded at least an equally large and accessible alternative location. On the other hand, the unions, NBT, and the Social Dem-

[10] The Municipal Council apparently enforces this law very selectively, only under sustained organized pressure from people's action groups.

ocrats wanted the playground relocated as quickly as possible so that new housing construction could begin and jobs be created.

In March 1980, as tension mounted on both sides, NBB organized a picket line to protect the playground. On April 22, some 800 policemen forced their way through the picket line, bringing in a mobile crane and a work squad which removed half the playground (enough to begin construction of the housing pro-

Figure 9. That night, a thousand children and parents rebuilt Byggeren with materials from a nearby building site. The protesters removed their blockades after police promised to refrain from further action. Copyright 1980 by Heine Pederson. Reprinted by permission.

Figure 10. But the police and construction workers returned in two days, and Byggeren was demolished one more time. Copyright 1980 by Sonja Iskov. Reprinted by permission.

ject). NBB organized a phone chain to draw out neighborhood support but was unable to break through the police line. They called the union to ask them to come and instruct their members not to work under police protection, but the response was "we don't have time just now." When the union leaders arrived, it was too late; half the playground had been demolished.

Later the same day, when the police had accomplished their mission and gone, NBB and the neighborhood residents rebuilt and reopened the playground, protecting it with a round-the-clock blockade.

A week later, the 800 policemen returned in full battle dress and this time destroyed the entire playground. The people erected heavy barricades in the streets, using everything from sandbags to overturned buses, and heavy fighting ensued. Many people were wounded and arrested, but they succeeded in driving the police out of the area by evening. To quote from Søren Johnson, a community activist,

> During the night the playground was rebuilt by a thousand hands, using tools and materials from the building sites in the area. The Black Quadrant was in revolt. Not just because of the playground, but because of the many years of suppression. We were fighting against Egon's plan, the police violence and the arrogance and hostility that has been characteristic of City Hall towards Norrebro and other slum areas. The citizen participation law hasn't been working at all in Copenhagen.[11]

On Saturday May 3, the police succeeded in destroying the playground for the third time, again with heavy fighting. Johnson continues:

> In these situations you find out what the constitution is worth—nothing! . . . We were beaten up by the police; they followed us up the stairways and inside the flats; they drove down the streets on their motorcycles randomly beating up people with their clubs. A legally assembled protest demonstration was violently smashed—people were held isolated for weeks in prison. If you were taking pictures, your film was confiscated and sometimes the camera was smashed. It was civil war in Norrebro.

According to a Gallup poll conducted in the area, 80% of Norrebro residents backed the playground. Nonetheless they lost. Construction workers proceeded to work under police protection, contending that the union "couldn't see" the police. With the defeat of the playground, NBB was defeated as well. Johnson writes:

> It was fantastic to experience the solidarity and enthusiasm among people fighting for the Byggeren (adventure playground), but depressing and very instructive to experience the violence and brutality from the police and the politicians at City Hall.

[11] Letter from Søren Johnson, May 26, 1980.

Figure 11. The neighborhood exploded with anger. Clashes between the police and a few thousand demonstrators ensued throughout the day. Copyright 1980 by Heine Pederson. Reprinted by permission.

Outcomes

Most core members of both Norrebro citizens' action groups felt that their respective associations had little long-term future and that they would probably dissolve as the old neighborhood residents were expelled and new, wealthier residents took their place. There was a sense of some good issues having been raised and some important inroads made, but both were aware that even as of summer 1979, they had limited contact with the grass roots in their own neighborhood. Both groups acknowledged that many long-time neighborhood residents had stopped coming to meetings, intimidated by the more articulate young people, turned off by their radical rhetoric, and skeptical that they could have an impact on the area. Both groups mentioned the difficulty of keeping residents active without a continual stream of crises to mobilize them and without an ideological stance that fosters activism for longer-range societal goals. Both groups, despite ideological differences dating from the split, ended up with a de facto small inner circle making all the decisions.

Furthermore, it was clear that the left itself was beginning to become disenchanted with community organizing and was directing its efforts to issues of wage and price policy, turning back to more active involvement with labor. The initial growth of Beboeraktion had come not only as a spontaneous response to local needs but also as a deliberate strategy of the left to build contacts with the "masses" and the "working class." Ten years before, they had decided to direct their work to the neighborhood level, not within existing organizations like the long-standing tenants' unions but through the building up of new independent groups—the Beboeraktion. At that time, many felt it was a waste of time to work with labor unions directly and that the most effective strategy would be to organize support groups for them in a wider base in the community. But the present severity of the economic crisis and the difficulties and relative impotence of community organizing has led to a renewed focus on struggles at the workplace rather than in the neighborhood.

As for what is happening in Norrebro, a clear paradox emerges. The goal of city hall had been to transform Norrebro into a desirable location for white-collar workers in downtown Copenhagen. However, the massive demolition and new construction in the area is likely to lessen the appeal and charm of the neighborhood to the professionals and administrators of the central business district. Thus, the apartments will have to be rented mostly by displaced residents from other neighborhoods, who will require enormous government subsidies to compensate for the difference between the enormously inflated prices and their ability to pay. The paradox is that if the renewal plan had followed the lines that NBB was pushing, it would have involved spot demolition, sensitive modernization of old buildings, lots of open space, and excellent community facilities. This would have enhanced the existing charm of the area and attracted precisely the white-collar professionals that the city was interested in.

Figure 12. Between ten and fifteen thousand people demonstrated in the City Hall Square against the elimination of Byggeren. The majority of the Council remained firm in their support of the renewal plan. On May 13, 1980, the People's Action Committee gave up the physical blockade of the new construction site. Copyright 1980 by Sonja Iskov. Reprinted by permission.

As Kai Lemberg, the progressive and very perceptive director of the Copenhagen planning department, pointed out in a recent speech, the key to understanding this sort of mishap is to look at "who is planning the planning." His answer was, first, the "market forces," which lead to plans desired by developers, builders, and the construction industry; second, the political authorities—the national parliament and local city councils—which egregiously overrepresent upper-class interests; and, third, bureaucrats and officials representing mostly middle-class views. All these forces are institutionalized through legislation, resulting in a series of laws overrepresenting men over women, owners over renters, shopkeepers over consumers, and the upper and middle classes over the working class. It is what he calls "participation from above." Grass-roots participants, he contends, have little chance of being heard, and the "victims of planning" are mostly silent altogether—the elderly, foreign workers, and the very poor. The Norrebro example provides an excellent case in point.[12]

Conclusions and Reflections

The Copenhagen case encapsulates many of the themes of citizen action which have emerged in other cases and contexts: (1) the vast difficulty of affecting policy from the bottom up; (2) the necessity for disruptive tactics when permissible channels for participation are ineffective; (3) the critical importance of a solidly organized base within the community; (4) the internal contradictions within citizen action groups (and consequent difficulties in creating effective alliances or coalitions), and (5) the potentials and limitations of party politics and electoral activity for citizen action groups.

Today in Copenhagen, no one, not even city hall, supports massive urban renewal in the Norrebro style of demolishing entire blocks. Policy now favors spot demolition and on-site upgrading of the housing stock. One could claim that this contradicts the first point above and that indeed citizen action, especially the incredible struggles in Norrebro, changed government policy. This would be only partially true, although there is no doubt that the growing public opposition to massive demolition and the publicity created by the Norrebro case made it much more difficult and costly for that policy to be pursued. However, the real reason for the policy shift was the cost factor of the destruction itself, the growing scarcity of resources, and the severe shortage of social housing. If the government found it expedient to pursue their earlier policy, they certainly

[12] As this case description was about to be sent to the editor, we received an unconfirmed report that the Danish parliament had passed a law that would permit residents of urban renewal areas to remain in their old neighborhoods. Unfortunately, our inquiries for further information could not be answered and, therefore, we do not know what mechanisms may be employed to implement this piece of legislation. If the law has indeed been passed, it might be interesting to conjecture whether Norrebro's citizen action had achieved, after all, some legislative impact.

made it clear that they would go to any extreme to do so, even making use of the tactics and repressive measures of police action.

This raises point 2, the necessity for disruptive tactics. Denmark has developed elaborate procedures for ordered citizen participation. By law, citizens are informed of plans, may voice criticisms of these, and may propose alternatives to them. But in practice, only the most minor modifications are acceptable through these procedures; in fact, the supposed citizen "check valves" are really escape valves for releasing community anger and frustration into meaningless hearings and paperwork. The only time the city government seems to respond to community input is when it comes in the form of civil disobedience, confrontation, or disruption of daily life. Our own American experience is similar. The mandated citizen participation of the War on Poverty was ineffective until it turned into confrontation, and then it was rapidly and forcefully repressed. As Sherry Arnstein concluded in her well-known article on citizen participation, "In most cases where power has come to be shared, it was *taken* by the citizens, not *given*." She went on to say:[13]

> Partnership can work most effectively when there is an organized power-base in the community to which the citizen leaders are accountable; when the citizen group has the financial resources to pay its leaders reasonable honoraria for their time-consuming efforts, and when the group has the resources to hire (and fire) its own technicians, lawyers and community organizers.[14]

Since all these elements are lacking in government mandated citizen involvement, it is not surprising that even as objective a work as the *Encyclopedia of Social Work* concludes that "in general, citizen participation has not been an effective means of achieving social reform."[15] It cites as evidence Roland Warren's study of 54 agencies involving citizen participation in 6 cities. Although citizens achieved 606 cases of "innovation," they had little impact on agency programs because 559 of these involved superficial modifications of organizational structure or procedures.[16]

This reinforces the importance of well-organized, independent, mass-based community associations. Point 3 to be drawn from the Norrebro case is that the groups, while understanding full well the need for disruptive tactics, did not really have the residents organized behind them. In the case of the adventure

[13] This section is based on Janice Perlman, "Grassroots Participation for Neighborhood to Nation," in Stuart Langton, ed., *Citizen Participation in America* (Lexington, Mass.: Heath, 1978) chap. 6.

[14] Sherry Arnstein, "A Ladder of Citizen Participation," *Journal of the American Institute of Planners* 35 (July 1969), 221.

[15] Peggy Wireman, "Citizen Participation," in *The Encyclopedia of Social Work* (Washington, D.C.: National Association of Social Workers, 1977), pp. 175–180.

[16] Roland Warren, *The Structure of Urban Reform* (Lexington, Mass.: Heath, 1974).

playground, for example, instead of filling the picket line from their own ranks, Norrebro activists recruited picketers from throughout Copenhagen through advertisements in the architecture schools. If they had worked through existing neighborhood groups or even local political parties, they would have had a much greater capacity to sustain their struggle. As it was, the neighborhood was divided, with NBT supporting the playground's demolition and many of the residents unaware of the issue altogether. In the United States, too, we have observed the fragility of building a base from the top down, whether it is done by students, a political party, a professional organizer, or a charismatic leader.

The affected grass-roots populations can originally be organized by other people than "their own" (such as Debs and the railroad workers, Alinsky and the Back-of-the-Yards residents, or Martin Luther King and a civil rights coalition), but only indigenous people themselves can *sustain* a viable grass-roots organization and a struggle over time. To create this takes years, not months, and must embody the process of "building people" and "building community" as well as "building power."

One of the things which makes citizen action organizing so difficult—as we have observed it throughout the 1970s in the cities of advanced industrial nations—is point 4, that it is deeply divided by *internal contradictions*. It was no accident that in the small Black Quadrant neighborhood of Norrebro these contradictions surfaced as well, with the residents split over wanting better housing but at affordable prices; wanting *more* social housing but not at the sites of parks or open spaces; or wanting more jobs but not at the cost of being displaced from their own communities. Since many people in Norrebro are construction workers, wearing their union hats gives them the vision that demolition will help create jobs; but since they are also residents of a low-income neighborhood, wearing community hats shows them that they will be displaced to more expensive housing elsewhere if they demolish their own community. Likewise with open space, with job competition, or with housing supply and demand. It is not simply a question of one group of residents with a clear interest opposed to another group with a different interest but of these competing interests, *within* each group and often within the *same individual*.

In the most successful cases of citizen action, such as the "self-reduction" movement in Italy, the groups managed to organize people simultaneously as producers and consumers. With the organizational help of the extraparliamentary left, they organized residents in nine Italian cities to withhold increases in their utility bills. When, finally, the company ordered power shutoffs, all the utility workers walked out and went home, since they too had been part of the self-reduction movement from the start.

Since citizen action groups alone will always be too weak to challenge unjust programs and policies effectively, it is imperative that—while retaining their own identity and vitality—the various groups find ways to enter into coalitions with each other and with labor in order to build enough clout to make a difference. In Copenhagen, some attempts to form people's action federations were made, but they were too weak and the links with labor remain as yet undeveloped. In the United States, we are now witnessing huge steps in this area, with the Citizen/Labor Energy Coalition, for example, and the newly emerging Citizen Action Coalition, an alliance of seven statewide citizen organizations that celebrated its first anniversary in Cleveland in November 1980. Scheduled for January 1981 in Chicago will be the first joint meeting of all the community organizing networks, a testimony to their increased sense of capacity and flexibility.

One constantly striking and major difference between European and American citizen action groups is their relationship to the political party structure and to electoral politics in general. Clearly the presence of left-wing political parties in European cities provides a context, a resource base; and a personnel pool for activism which is lacking in the United States. This can be a mixed blessing, however, when the parties do not respect the integrity and autonomy of community groups or wish to utilize them for other agendas. The Spanish groups' struggle for autonomy from both the Communist and Socialist parties in Madrid provides a case in point, as does the heightened conflict between NBB and NBT in Norrebro. Perhaps the split could have been avoided or softened in the absence of clearly defined party positions.

United States community groups of the direct-action type have traditionally remained aloof from electoral politics altogether. They have argued that even electing a favorable councilman or mayor makes little difference, since their powers are limited and they cannot readily be held accountable. The case of Vilo Sigurdsen, Copenhagen's fourth (deputy) mayor, would confirm this view, as did the election in the United States of a number of black and progressive mayors over the past few years without notable change in the distribution of goods or services within their respective cities. Election to office by itself is no assurance that political changes favorable to the deprived constituency will follow. Still, even in an unfavorable political climate, it is far better to have a person in office who is committed to grass-roots neighborhood development than one who is not. Given the recent electoral gains of the right wing in the United States, many of the citizens' action groups are therefore reconsidering their anti-electoral bias and intending to engage in voter education, hold accountability sessions, and even run their own candidates in the coming years.

To conclude, citizen action, like the Chinese character for *crisis,* is constantly pulled in opposite directions by its dual components—danger and opportunity. Its danger is that it will fail because it can only engage people at the most immediate and narrowly construed base of their self-interest: "What have you done for me lately?" If this is the case, it ultimately wastes people's energies, burns them out, creates little if any sustained change, and reinforces a passive view of people as objects being acted upon. Its *opportunity* is that through collective effort, starting from where people are, citizen action groups will develop their capacity to see themselves as efficacious actors shaping their own lives and formulating an ever clearer vision of what shape they would like those lives to take.

ACKNOWLEDGMENTS

Our principal research counterparts in Copenhagen were Peter Bjerrum and Søren Johnsen, both of the Royal Danish Academy of Fine Arts. They generously gave of their time and professional knowledge before, during, and after our fieldwork; our greatest debt of gratitude must go to them.

Numerous people from government, labor, community groups, political organizations, the unemployed, the field of education, and simply from the city itself helped us in our data gathering. Among them we are especially grateful for the help of Michael Steffansen, leader of NB; Tarju Gronberg of the Danish Building Research Institute; Vilo Sigurdsen, the fourth (deputy) mayor of Copenhagen, who is responsible for local planning; Jorgen Erik Fokdal, planning researcher; Egon Pederson of the Uno-x Oil Corporation; Carl Nissen of the Ministry of Education; Kai Lemberg, who is Copenhagen's director of general planning; Torben Egede of the Ministry of Housing; Jan Tfelt-Hansen of the Public Condemnation Company; Keld Absdol of the Cooperative Housing Society; and Per Krough and Preben Eilessen of the Copenhagen lord mayor's staff.

3

SCOTLAND
NORTH SEA
RELAND
WALES
THE NETHERLANDS
ENGLAND
London
Delft
Brussels
BELGIUM
LUX.
FRANCE

3

CASE STUDIES OF CITIZEN ACTION AND CITIZEN PARTICIPATION IN BRUSSELS, COVENT GARDEN, DELFT, AND CAMDEN

DONALD APPLEYARD

Introduction

I have approached these case studies primarily with the intent of finding innovative ideas that might be applicable to cities in the United States. My case studies have an environmental focus, although they have looked at social action as well. My principal theme has been: *Which aspects of the neighborhood environment, and which environmental actions, have encouraged citizen action and participation and have led to political success?* By *environmental actions* I mean actual changes to, or management of, the physical environment—buildings, housing, open spaces, streets—plans and designs for environmental changes, or other visible actions such as festivals and demonstrations, all of which are as social as they are environmental. Purely social actions are less visible and may cover efforts to implement rent control or stimulate political mobilization.

Evaluations of participation and degrees of political success are relatively crude. The apparent survival of a citizens' group or citizen acceptance of planned actions may not always produce the greatest benefit for the greatest number in the long run. The evaluations of active participants, however, are one valuable form of evaluation.

I have looked at case studies of *citizen action* in Brussels and Covent Garden and *citizen participation* (government-sponsored involvement) in Delft and Camden.

Citizen Action Groups in the Urban Environment: Confrontation or Cooperation?

These first two case studies—Brussels and Covent Garden—focus on the *styles and types of actions,* environmental and social, that have achieved some political success. Political success, obviously, depends on political purposes. In the case of these citizen action groups, their purposes were to bring pressure (through protest) on local government to change its policies and ultimately to create and develop their own counterplans, to create and maintain a core of political activists who would engage in social and environmental actions and commit themselves to the drudgery of keeping the group alive, and to create a broad constituency among both the local working-class residents and the moderate to progressive middle classes. Political success can be measured, in terms of these goals, by determining whether the local government changed its policies, whether the citizens' group survived and developed, and whether it achieved a broader constituency of working- and middle-class members.

The Character of Successful Actions

The environmental and social actions that appeared to achieve most political success were:

1. *Those that exposed the policies and actions of local government, private developers, and business as self-interested, callous, and fallible.*

One of the principal aims of citizen action groups is to question the credibility of those in authority, who have the weight of governmental power, professional strength, and technical sophistication to support their proposals. It is also, often, to expose the hidden linkages between local government and business interests. Without these exposés, supporters see government policies as immutable and beyond question.

2. *Actions of immediate and visible benefit to the local community, with which the local residents can identify.*

Citizen action is often aimed at defending the community from destruction or at improving environmental conditions close to home, on the doorstep, or in places with special meaning to the community. Successful actions do not push the working-class residents' sense of appropriateness and respectability too far.

3. *Actions consistent with the world view and style of young activists.*

Activist leaders, staff, and participant members are usually small in number, limited in financial resources and frequently time. They prefer to engage in vigorous expressive actions characterized by creativity, imagination, humor, and a minimum of red tape. Activists range from those ideologically committed to broader political movements to those more oriented toward neighborhood issues.

4. *Actions that were media-attractive, "reasonable," educational, and entertaining.*

To communicate with a larger, liberal, educated constituency, activists must work through news media, their own and broader newspapers, and television communications systems. Actions that receive the most support are "reasonable," frequently conservation-oriented, and aimed at saving a commonly perceived heritage of historical buildings, natural places, or the integrity of a local community.

Conflicts

Actions aimed at opposing local government and business interests and at constituency-building among local working-class residents as well as the larger middle class demand a sophisticated guidance system. Special steps must be taken to avoid alienating any part of the constituency. Too radical an action, such as squatting, can turn off local residents. Citizen groups are in constant danger of being torn apart by factions with competing views of appropriate action. Such conflicts are sometimes resolved by taking "symbolic actions" that may seem more radical or more respectable than they really are but which capture media attention at low cost. Figure 1 sketches these relationships.

ARAU and Inter-Environment, Brussels

Inter-Environment, Brussels, a federation of some 55 citizens' groups, is one of the most innovative and systematically thought out citizen action groups in continental Europe. Inter-Environment is especially interesting because it is metropolitanwide. It carries out a surprising range of interesting projects, including the preparation of counterplans, seminars, and an array of educational programs.

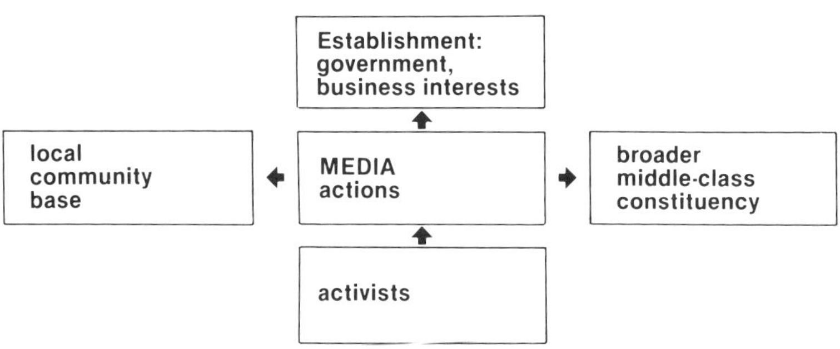

Figure 1. Activists use the media and "symbolic actions" to further their goals while building their constituency.

History

ARAU (Atelier pour Recherche et Action Urbaines) was formed in 1968, when a group of sociologists, planners, and architects held a press conference in a famous Brussels coffee- and beerhouse, *La Mort Subite* (Sudden Death), to propose a plan for the vast empty parking lot in the center of Brussels that was, ironically, called "the crossroads of Europe." During the economic boom of the 1960s, Brussels, the capital of Europe by virtue of the Common Market, which makes its headquarters there, had encouraged large-scale urban-renewal schemes and high-rise office developments. ARAU selected a series of government projects for which they drew up imaginative counterplans, including freeway routes, urban renewal programs, the construction of cultural facilities, office buildings, and the like, accompanied by devastating esthetic critiques of the designs of the establishment proposals and social critiques of the displacement of low-income housing.

ARAU's principal methods of action were analysis of official and developer projects, counterplans by ARAU, presentation of these projects to the press and public, and negotiation.

As their actions spread, their designs and modes of operation became more crystallized. Counterproposals were developed by different architects; the group often produced alternative schemes for the same project. None of the participating architects were to accept commissions while they voluntarily gave their time to ARAU. They all adhered to the principle of restoring the continuity of the original urban fabric, of using existing buildings where possible, and of encouraging mixed commercial–residential uses. The style of these schemes tended to be close to the indigenous or nineteenth-century architecture of Brussels. As such, it became quite popular with a wide audience, contrasting with the disruptive nature of establishment architecture. ARAU then combined with others to form the federation of groups called Inter-Environment to combat some of the major metropolitan-scale actions being taken in Brussels.

Inter-Environment Brussels

In 1972 a large number of resident action committees engaged in the struggle against pollution, nuisance, redevelopment, and in the defense of nature and historic conservation decided to join a federation under the auspices of a national organization carrying the name Inter-Environment–Union for Environmental Quality. By 1974, this national organization was transformed into a number of regional federations—a change prompted by the complexities of administration, the difficulties of dealing with two languages and cultures (French and Flemish), and the fact that the political institutions being pressured by Inter-Environment were also evolving into regional organizations. The existence of two cultures in Belgium creates severe stresses in that country as symptomized by the recent

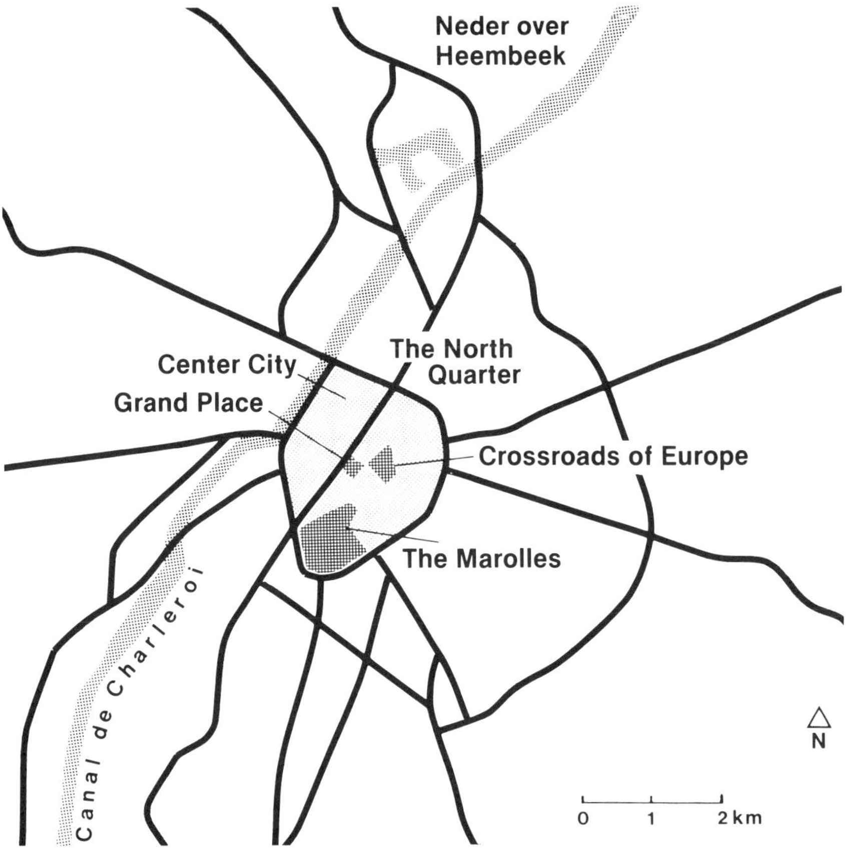

Figure 2. Metropolitan Brussels, showing the location of several of the communities in which Inter-Environment has worked.

riots. In general, the Flemish (Walloons) are in administration and business, while the French are workers and intellectuals. On the other hand, this fragmentation of the Belgian political structure at the city, metropolitan, regional, and national levels makes it possible to have allies at every level. The members of these Inter-Environment groups thus come from all sectors of the population—from political parties, trade unions, and sociocultural groups of various kinds. And they find friends those who think *à gauche* ("on the left") in all political parties.

Four autonomous regional federations were therefore formed: Inter-Environment Wallonie, Inter-Environment Brussels, Bond Beter Leffmilieu-Vlaanderen, and Raad Voorhet Leefmilieu to Brussel (BRAL).

In Brussels (1 million people), where there is a strong Flemish minority

(20%), there are two federations—one French (Inter-Environment Brussels), the other Flemish (BRAL)—who collaborate on certain issues like the protection of working-class quarters but differ on other issues. For instance, the Flemish groups do not want the metropolitan area to grow beyond 1 million, while the French favor growth to 1.5 million. On the other hand, the Flemish groups favor the peripheral freeway, while the French groups oppose it. This is a key issue for the French Inter-Environment because the last link of this freeway would displace at least 20,000 French people.

Inter-Environment, Brussels, consists of some 55 groups, of which about 45 are neighborhood groups. They include the Comités des Quartiers; the Comité d'Action des Transports Urbain (CATU), calling for public transport action; the Entente National pour la Protection de la Nature, a coalition of one hundred groups who are doing an inventory of open space in the city; Aves, a group trying to save birds, which in Belgium are killed in nets; Habitation Familiale; and Friends of the Earth.

Membership

To become members of Inter-Environment, organizations must have existed for one year and be against the peripheral freeway. This is the touchstone of membership and demonstrates that citizens can comprehend a larger-scale problem than their own neighborhood.

The Inter-Environment board of directors has a total of 16 members; four from the central working-class quarters of the city, four from the residential periphery, four from the nature and nuisance groups, and four from special groups such as CATU and ARAU. They are elected for two-year terms. Inter-Environment has three full-time staff members—an architect, an engineer/architect, and a typist with an office, paid for by the Ministry of Culture (at a cost of 1.5 million francs, or $120,000 per year). In addition, there are three conscientious objectors and six otherwise unemployed persons. Other special projects bring in extra funds. For instance, an urban open space study is being carried out for $70,000.

The guiding force for this group comes from ARAU. ARAU is a self-formed group of 21 active lawyers, sociologists, architects, politicians, and a trade unionist. They say that they are elected to ARAU by their wives and husbands—that is, they do not represent a constituency much larger than themselves. Their philosophy affects the long-range objectives of Inter-Environment. The core of their ideology concerns the form of the city they would like to have.

> The city is a center of power; the power the city creates must be returned to its inhabitants.
>
> Schoonbrodt, 1979

They want "a city with people in it, because the city is a means of development (individual and societal), whereas the suburb is an isolated society, each family in its clean little house and garden." (Schoonbrodt, 1979). They propose an "other" image of the city to that of CIAM,[1] and they are against secret decisions.

Two of the leading members, Rene Schoonbrodt, the sociologist, and Maurice Culot, the architect, write prolifically about their ideas. These articles continually and eloquently attack the precepts of CIAM, modern architecture, zoning, and bureaucracy.

Schoonbrodt is president of a housing commission for the south of Belgium, has been teaching in an architecture school (Le Cambre), and has written a book—entitled *The Sociology of Public Housing*—that comprises a detailed critique of the spatial discontinuity brought to the city by the modern housing movement. A sociologist with a primary concern for social space and territory, he spends a great deal of his time working voluntarily for Inter-Environment and ARAU. He does not espouse the more rigid leftist ideologies, favoring a mixed-economy state.

Culot, a teacher of architecture, theorist, and publisher, is developing a wholly new vision of architecture and planning expressed in numerous publications. He edits the *Archives d'architecture moderne,* a beautifully illustrated journal of early modern architecture as well as the projects of his students and fellow architects. He has recently published some books on nineteenth-century architects and, with Leon Krier, an important volume entitled *The Reconstruction of the European City*. With Schoonbrodt, he is founding a new school—School for the Reconstruction of the City—in Brussels.

Culot bitterly attacks modern architecture and planning for destroying the old city quarters and creating anonymous buildings and monocultural complexes. At the same time he rejects the excessive individualism of the American postmodernists. This is developed as a Marxian view that emphasizes a return to the elements of the eighteenth- and nineteenth-century city: the boulevard, the street, the square, and the quarter. More recently Culot has taken an "antiindustrial" stance, rejecting the use of high-energy-consuming metals like steel and aluminum while espousing the resurrection of stone as a basic building material and the revival of artisans who can use it. His students recently developed a design for a modern stone quarry.

While ARAU and Inter-Environment do not discuss the detailed style of projects, Culot's architectural philosophy fits in with the broader urban conservation ideas of Schoonbrodt and Inter-Environment, a group remarkable for its clear sense of mission.

The policy consequences of this urban-oriented philosophy are an emphasis

[1]Congrès Internationaux de l'Architecture Moderne. An international group organized in 1928 by Le Corbusier and others to promote modern architecture.

on housing and jobs for lower-income groups in the inner city, an emphasis on conserving the old city, an antagonism toward suburbia and the automobile, and espousal of transit. The joining up with many ecology groups has clearly added a new dimension to ARAU's original urban and social interest. It may have diffused their focus, though this is not apparent, since ARAU still seems to have firm control over meeting agendas. On the other hand, it has gained the group a much broader constituency.

Major Activities

Being Informed. One of the most innovative activities undertaken by Inter-Environment depends on the network of friendly informers that it has developed within and outside the establishment bureaucracy. Once a month one staff member spends three days telephoning all their connections within the ministries and municipalities to find out what policies and projects are being developed. In addition, they obtain information on a continuous basis when important projects are proposed. A bimonthly information list, *Les fiches d'information* (shortly to become monthly), is sent to 160 members of Inter-Environment (three copies to each member).

For instance, for May–June 1979, 17 proposals were reported. Each project is treated on a single sheet with standard format outlining the proposal, the zoning, the interested committees, dates of public inquiries, and suggested reactions. A report on the previous months' activities appears at the front.

In January 1980, Inter-Environment's sources of information were doubled. The new *Plan de secteur* for Brussels, a land-use plan, requires that all new projects be publicized. One is able to say on the notices at building sites "here is a preview of the building. Someone wants to build offices. If you disagree, go tell the Commission that examines the application." These notices are now published in *La ville et l'habitant,* the monthly newsletter of Inter-Environment. To have obtained this commitment to public notice of projects is the most significant victory Inter-Environment has had in its battles with the establishment.

Preparing Reactions, Publishing, and Holding Press Conferences. In selected projects, staff or volunteers develop counterplans to those originally proposed. These are professional pieces of work, not done by students, and usually have a dramatic quality.

Neder-over-Heembeek. In a developing suburb of 2,500 persons north of Brussels, the city proposed a 60-meter-wide freeway through its central area, with large-scale development around it. The residents protested (300 went to city hall), and the local committee (Comité d'Information et d'Action pour l'Amenagement de N.O. Heembeek) developed a plan asking instead for a swimming pool, a rapid tram service, housing, small shops, accessible green spaces, zoning for employment in the quarter, nonpolluting factories, ateliers, offices, and—in the environs of the existing Halles des Producteurs—a quarter of housing and enterprises linked to these halls, together with a district heating system.

The design ideas of the quarter, depicted in a four-page leaflet, show three- and four-story houses lining the streets and hemispherical squares, with a civic center and swimming pool as community landmarks.

This plan was announced at a press conference and received wide community support. The city has now abandoned the freeway and high-rise buildings but does not yet favor industry.

La Périphérique Sud—A White Paper. A major effort has gone into the production of a 32-page glossy white paper illustrated with photographs and colored diagrams. The report criticizes the proposed south peripheral freeway for being unfunctional, ambiguous in some aspects, destructive both of the natural landscape and of social communities, as well as increasing the level of air pollution. Aerial photographs show the proposed alignment of the road with threatened landmarks; ecological maps show the natural features destroyed. This is a well-produced and impressive report with probable appeal to a wide range of people. It looks sensible and sensitive.

Newsletter, La Ville et l'Habitant. The monthly newsletter of 4 to 8 pages deals with particular projects, presenting counterplans through perspective drawings. These projects show ways of maintaining existing streets, conserving historical buildings and bridges, using the army barracks for housing, and so on. In addition, there are interviews with local action committees, reports on Inter-Environment conferences, illustrations of protest posters, and an editorial. Coming events and new publications take up most of the last page. The format is restrained and the drawings calm, quite unlike the more blatant graphics of some community newletters. The newsletter, which will soon become semi-monthly, is distributed to some 2,000 individuals and organizations.

Education. Since its leaders also teach, and many of its members come from union organizations, ARAU has quite naturally taken an active educational role. Usually this has taken the form of intensive week-long evening seminars held once a year. The tenth urban school, announced for March 1979, covered such subjects as "Urban Space Theater of Struggles," "What Economic Future of Which Inhabitants?" "Streets and Trams for the Residents of Brussels," "What Housing . . . ?" and "A New Urban Charter." Films are shown and documents displayed. Inter-Environment held a conference on energy related to urban struggles in 1978 and an annual congress on the future of the Brussels region in early 1979, when many groups reported on their particular issues.

The Tour of Urban Battles. One of the most unique activities of ARAU is the weekly tour it organizes for visitors to Brussels, sponsored by the Ministry of Culture. The tour is entitled "Brussels seen by its inhabitants: discover the 'millenium city' with other eyes." The tour covers both the inner city and some of the suburbs, with 38 sites. Most of these are sites of urban struggles—for example, the Northern Quarter, "where 10,000 persons" lost their homes, which

Figure 3. In the center of Brussels, the "Crossroads of Europe" lies empty.

is now "50 hectares of waste land"—and other quarters menaced with destruction, buildings worth preserving, parks, avenues, and churches. Alternative plans for many of the sites are depicted beside the map of the tour.

Symbolic Actions, Murals, Demonstrations. ARAU has not generally engaged in street demonstrations, but the committees of various quarters have, and photographs of protestors holding large banners appear in some of the monthly newsletters. One unique action was a symbolic "funeral" held in some apartments left by one of the ministries to deteriorate. One Sunday a group with black plastic sheets entered the vacant buildings and covered the windows in funeral black. This, of course, was picked up by the news media.

Street murals were first painted several years ago at the ARAU headquarters in the Marolle district. In 1979, a committee in the Northern Quarter had produced a spectacular set of murals, the result of a competition. A small area of picturesque old buildings remain, fronted by a wide arterial highway and a vast wasteland awaiting development and backed by modern glass towers. Surrounded thus by the products of urban renewal, this small oasis already has a certain symbolic intensity. The murals, depicting the demise of high-rise buildings in one case and a utopian garden square with small houses in another, are highly effective, especially since they are visible from a major commuter highway. In addition, an energy enthusiast has erected a windmill and series of posters, arranged Burma-Shave style, to tell sequential stories for passing motorists and pedestrians.

Buildings and Other Actions. ARAU has been working for nearly 10 years in the Marolle District next to the massive Hall of Justice. This is a quarter of old houses, many dilapidated, others occupied by Middle Eastern immigrants.

The local neighborhood committee now has a budget from the government and the European Economic Community of $830,000. This has been used to develop a vocational school in order to prepare the inhabitants for jobs. A workshop based on breaking down old television sets and reusing parts for television repairs has also been started. About 150 people in the quarter are now employed in cleaning offices. The local committee now controls the local urban renewal and rehabilitiation program.

One new apartment building in the Marolle epitomizes the esthetic of ARAU. A building by Marc Wolfe, it is four stories high, has traditional vertical windows

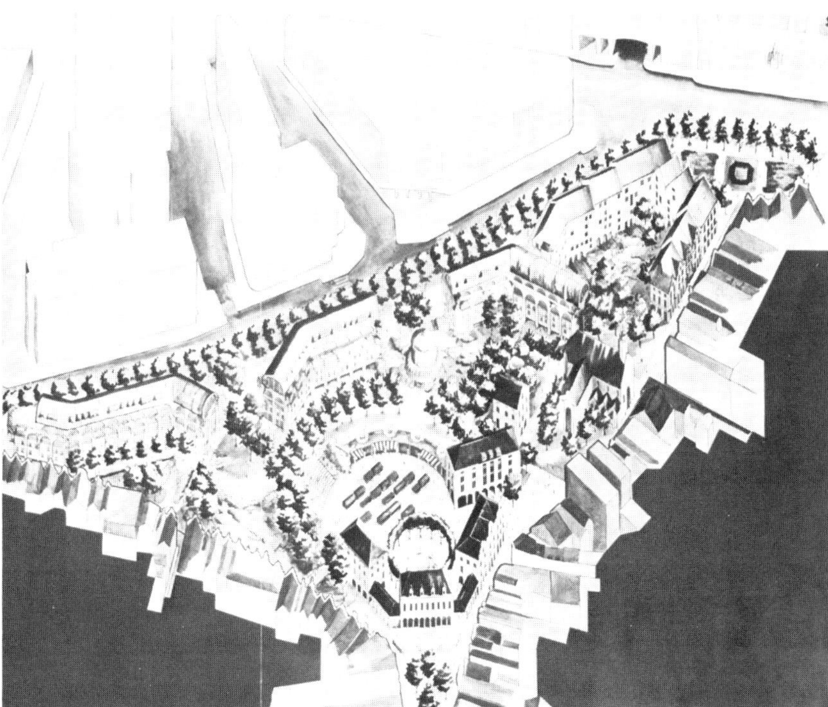

Figure 4. ARAU developed a plan for converting the vacant "Crossroads of Europe" into a revitalized center for Brussels. Traditional materials and styles were used in an effort to restore the urban fabric of the city.

with a black leaning bar across each, a sloping roof, and somberly-colored stucco in browns and ochres. According to the Inter-Environment bulletin (1975), it is

> a rigorous application of the principle of reconstruction of the city extolled by the committee of residents: the maintenance of housing, respect for the alignment and *gabarits* of the street, building continuity, constructed in concert with the inhabitants.

A serene but luminescent mural in pink and blue, showing a bird in a stony landscape, faces the intersection. Still, down the street is a small, modern, brightly colored apartment building in the style of Le Corbusier. The antithesis of the traditional building, it is also liked by the residents, probably for its bright colors and balconies.

Research and Planning. In 1980, Inter-Environment was funded to conduct a study of urban open space in Brussels. The first phase was an inventory of all open spaces, private and public, which are potentially available, together with the flora and fauna that they contain. Particular attention was to be paid to those spaces that might be lost to the south peripheral freeway.

In the second phase, recommendations were to be made for public acquisition and preservation, including proposals for a continuous network of touring trails.

Figure 5. Symbolic funeral, in which protesters covered windows with black plastic sheets. The apartments, allowed to deteriorate by one of the ministries, were declared dead.

The "Plan de Secteur," 1980. In January 1980, the Brussels city administration came out with the "Plan de Secteur," a general land-use plan for each zone of the city. The principal feature of this plan favored by Inter-Environment is the omission of the Périphérique Sud. As already noted, the plan also stipulates a much higher degree of public participation than previously. Any project that deviates from the plan must be subjected to public consultation. However, there is still no direct consultation with residents; it is all to take place through public representatives. Neither is there any assurance of saving buildings in the central pentagon of the city.

Inter-Environment's reaction to the plan was initially very positive. However, it was later discovered that a number of major highway projects had been approved a few days before the plan was made public. Accusations of deception filled the second issue of *La ville et l'habitant* in January 1980.

Advice on Rehabilitation. A new activity starting in 1980 will be an advice bureau for those residents who wish to rehabilitate, transform, or modernize their homes. Public agencies have the ability to give loans for such rehabilitation, but the administrative procedures are so complex that loans are difficult to obtain.

Conflicts

Whereas ARAU is a small, closely knit group, Inter-Environment covers a much wider range of opinion. Therefore, although all agree on major issues such as halting the peripheral freeway, conserving natural environments, and maintaining housing and industry in the inner city, there are differences of opinion. Some favor the proposed jumbo trams; others feel that a system of smaller vehicles traveling all night will better serve a fine-grained urban environment. There are differences on how much to support suburbia through park-and-ride facilities. And ARAU is under attack from architects for halting the construction of too many buildings.

The Lessons of ARAU and Inter-Environment

Inter-Environment is a group with a clear philosophy and a certain style which has been extremely effective. The style emphasizes clarity of thought, specificity, imaginative ideas, humor, and wit. There is little of the heavy-handed, humorless, and harsh rhetoric of much European left-wing community-action literature. On the other hand, there is a much broader urban and societal view than is to be found in most American community groups.

ARAU has been in existence for 14 years, since 1968. Their and Inter-Environment's success has been mostly in the preservation of a number of inner-city districts, in the slowing down of several new projects, and in the creation of public inquiries which must now be held for all projects. They are now consulted by the political parties when programs are drawn up, and the main topics of discussion now are urban housing and transit rather than suburban development and the automobile. They have been prime catalysts for change in the way people think about planning in Brussels. The expansion to a federation with a large number of environmental groups has strengthened Inter-Environment's power and utility. This federation is now able to counter major government plans, such as the construction of the peripheral freeway, and to conduct technical studies of its own. It has become a citizens' planning agency in a limited sense.

And what of the future? The Plan de Secteur relieves Inter-Environment of the need to have a spy network, because notices of projects will now be made public. This information will make the neighborhood committees better armed to act alone. The plans for the future include developing individual membership as well as group membership in the organization, intensifying the use of *La ville et l'habitant,* developing direct services to inhabitants, treating problems that are too large for the neighborhood groups alone, intervening in programs of urban renovation, working for better public transport, and continuing the action of the "Boutiques urbaines." Financial support will be difficult to obtain. The group has depended on the devotion of its members, and—despite efforts of the public agencies to suppress the movement—it has so far survived.

Covent Garden Community Association

Covent Garden is one of the best-known neighborhoods in England, and it was the home of one of the most innovative community action groups in the country. This group not only protested and halted a huge redevelopment program but also successfully organized a number of creative social and environmental activities.

History

Originally the convent garden of a monastery just north of the Thames and the Strand between Westminster and the City, the area was the subject of London's first major piece of town planning when laid out by Francis, fourth Earl of Bedford, "with a spaciousness and opulence previously unknown." The grand seventeenth-century piazza was surrounded by arcades of townhouses which, however, were never completed. At the end of the piazza one of London's first and finest Renaissance churches was constructed: St. Paul's, Covent Garden, designed by Inigo Jones. But this was followed by the market and then two theaters, as the affluent residents moved north. From this time on the reputation of the area declined, but the construction of London's main fruit and vegetable market and the many theaters that ringed it on all sides created one of the most colorful neighborhoods in London, though it did not become a center for restaurants, like the Les Halles area in Paris. Publishing houses and low-income tenements mixed in with the supporting activities of the market and the entertainment industry as the offices spread around its north, east, and south flanks. The area only suffered slightly from bombing during the war, neither was it affected by large-scale housing projects; therefore the network of intimate streets around the Market retained its continuity with the original seventeenth-century layout. The resident population were mostly low-income immigrants from the nineteenth-century Irish potato famine, with names like Keeley, Toomey, Driscoll, and Sullivan, most of whom worked in the markets.

By the late 1960s, huge continental trucks were jamming up the streets to serve the food market, and plans to move it out to Nine Elms were underway. At the same time, Odhams Press, a publishing house occupying two large blocks, was bought out and had to move away.

The Greater London Council, Westminster, and Camden

Jurisdiction over Covent Garden is complicated, because the southern part is in Westminster, with a traditionally Tory Council, while the northern part is in Camden, which has alternated between Labour and Conservative. Overseeing both, the regional authority, the Greater London Council (GLC), has changed political color from Labour in the 1960s to Tory in the 1970s. It can intervene in areas of regional significance. In the 1970s the GLC made a deal with Westminster which gave Westminster primary jurisdiction over Piccadilly to the west, while the GLC took over the comprehensive redevelopment of Covent Garden. Camden has always been unhappy about this and withdrew from the planning operation.

The GLC is a very unwieldy body with large numbers of councilors who have to leave much of the planning to their staff. The staff chief planner, Geoffrey Holland, has been remarkably persistent, surviving four administrations, and still believes that professionals should take care of planning problems. He has an office in the area.

The First Plan and Emergence of the Covent Garden Community Association (CGCA)

The Greater London Council immediately saw the area as a dramatic opportunity for comprehensive redevelopment, and produced an ambitious plan for a multilevel complex of high-rise office buildings, hotels, a conference center, and high-rise apartments with an underground east–west arterial route and several levels of parking along one of the street alignments. The plan was based, ironically, on a careful survey of the character and image of the area, acknowledged in what was designated as a pedestrianized east-west "line of character" through the old Market Site between the two complexes of theaters at each end. The rest of the scheme depended on the destruction of much of the remaining blocks.

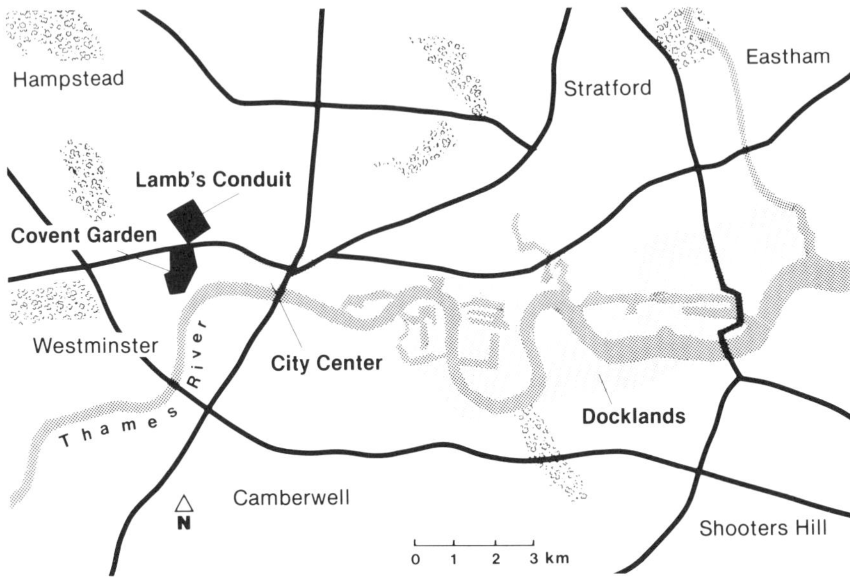

Figure 6. Central London, showing the location of Covent Garden and Lamb's Conduit (Camden) relative to the city center and the Docklands.

On April 1, 1971, the Covent Garden Community Association (CGCA) was formed to fight the proposals, led by architecture students from the nearby Architectural Association School, including Jim Monaham and local residents. With this and other opposition, a public inquiry had to be called by the Department of the Environment. An inspector was appointed and the inquiry held; his report, which endorsed the GLC plans, went to the then-new Conservative Minister for the Environment. No one expected any recommended changes, but to the surprise of all, Mr. Ripon, while giving the GLC a continued right to plan the area, "listed" 250 more buildings on top of the previous 50 to be preserved, effectively fragmenting their grandiose scheme. He substantially enlarged the conservation area, rejected the sunken roads and overhead walkways, and reduced the amount of office and hotel space. It appears that the CGCA had gained widespread support for preserving the character of Covent Garden, which is known and loved by all kinds of people of different political orientations. The only people in favor of the comprehensive redevelopment proposals were the developers and the GLC planners, some of whom defected or changed position when the battle erupted.

Historic conservation, however, is different from social conservation. Not long after, the GLC decided to organize a program of public participation around a Forum representing all parties in the community.

The CGCA was at first ecstatic at its victory, and began to participate in the Forum. The Forum, however, also included merchants, developers, and others and became bogged down in endless meetings; after one year and some debate, the CGCA walked out of the Forum, changing their constitution to exclude Forum members from being on their executive committee. The endless discussions did not suit CGCA's style of action, and they were being outvoted. There was also a split in the CGCA at this time, between those who wanted to continue confrontations, including occupation of the GLC Information Centre, versus those who favored environmental improvements even with GLC money. The CGCA and the GLC hammered out an agreement on the general goals: to double the resident population, restrict office growth, attract new small-scale uses, and create some open space. The GLC planners developed a new set of three scenarios: redevelopment, partial redevelopment, and rehabilitation. The CGCA soon began to realize it needed professional help to produce effective counterproposals based on realistic constraints and opportunities, "to know the area better than the planners." They were responding too late and in insufficient detail. They therefore raised money and, with a grant from the Rowntree Trust (1971–1973), hired two full-time architects to evaluate GLC proposals and work out realistic alternatives.

This soon began to pay off. For instance, the GLC draft plan divided Covent Garden up into different parcels. The CGCA, working with residents, workers, and owners in each parcel, became much more aware of existing social and planning restraints, while the GLC did not even know the number of employees in each building. Their parceling system also restricted them from making trade-offs. In the end, the GLC politicians accepted 8 of the 9 CGCA parcel proposals over their own staff. During this time CGCA branched out into a number of innovative activities.

CGCA

The CGCA has about 1,500 members who pay a $2 yearly membership fee. Anyone can become a full member if they live or have lived in the area; if they work, study, or conduct business there; or are sponsored in writing by 10 full members. Anyone can be an associate member provided they are approved by a simple majority at a general meeting. The CGCA executive committee of 10 to 12 members, elected by the membership, meets once a month. Its organizational motto is "small is beautiful." The meetings have "no weighty agendas," and decisions are made quickly.

Begun at a meeting of 500 people to protest the GLC plan in 1971, the CGCA has always been short of funds. A report at the end of 1972 showed their total assets at just over $2,000, much of which was spent on the public inquiry. Fund raising came from rummage sales, suppers, and a theater production. They received a grant of $5,000 a year for two years from the Rowntree Social Sciences

Trust. But this situation became worse in 1979, when they had to lay off the two paid staff members.

The enduring leaders of the CGCA have been John Toomey, a printing worker born in the area; Jim Monaham, an architect, and his wife Jane; Sam Driscoll; Brian Lake; and others.

The constituency of the CGCA has comprised both the young active professionals, artisans, and others who moved into the area over the last 10 or 15 years and the older working-class residents who have lived there all their lives. Beyond these, there has also been that larger constituency of Londoners and people from all over Britain who know Covent Garden through the theaters, the market, or its architecture. There is a significant absence of young working-class people in the area. The actions of CGCA have had to keep these three quite different constituencies in mind: the young residents, mostly from outside, "brave young Turks," active, display-oriented, flamboyant (especially due to the presence of theaters and designers), confrontational; the older residents, reticent, traditional, conservative in life-style, unused to speaking out, but with the most serious needs; and the outsiders, relatively ignorant, detached, but willing to participate in festivals or save a heritage.

Major Activities

In the first years, the major activities of CGCA involved making their cause visible through events, festivals, press conferences and releases, posters, leaflets, banners, exhibitions, and fund-raising activities; setting up tenants' associations to forestall or fight evictions (six were set up in the first year and a half); and their active participation in the public inquiry, which constituted the biggest expenditure of their resources. They hired a solicitor and a barrister and lobbied the Civic Trust and Town and Country Planning Association, both voluntary planning organizations, to object. A candlelight march was held on the eve of the inquiry, followed later by demonstrations outside. Ten CGCA witnesses took 10 days to state their case. Street theater in different pubs in Covent Garden explained their case.

During this year CGCA also demonstrated over a plan for Piccadilly and were the subject of programs on British and European television. Meetings were held with MPs, unions, and community groups around the country. The fame of the area attracted a number of students and academics, and even Ada Louise Huxtable of the *New York Times,* to write papers and theses which were generally supportive and sometimes useful.

After the CLC plan had been stopped, the CGCA undertook a number of specific projects while keeping up with its newsletters, criticism of the GLC, annual festivals, and other events. These activities were unique achievements for a community group with scarcely any resources. Deeds as well as words marked their style.

Establishing a Social Center. It took two years to find a place for the social center and build it with voluntary help. It has a licensed bar run as a club, a snack bar, and a meeting hall in the back, whcrc bingo, judo, and jazz as well as meetings take place. This bar brought "locals" as well as activists into the center.

Rehabilitating Five Flats. The Community Housing Association from Kentish Town rehabilitated some flats, which were allocated and managed by a Housing Cooperative set up by CGCA. This was a symbolically significant project, having rehabilitated two buildings at a total cost of $30,000 or only £1,200 (about $2,000) per person rather than a sum that the GLC had said would be "prohibitive." Five flats and three new shops were rehabilitated just up to standard: instead of slate, felt roofing was used, which would have to be renewed again after five years. It was described by CGCA as "the first major rehabilitation scheme in Covent Garden . . . since the war." The project took only six months, and although it was very small, it gave the CGCA a significant amount of credibility with the politicians at the GLC and elsewhere. These people clearly were committed to creative action, not simply to protest.

Rehabilitation was the most meaningful activity CGCA could encourage for the working-class residents. On the other hand, the small scale of the project necessarily limited it to a symbolic gesture, an example of what a larger program could do. The most was made of this through publicity, but it was not too visible to those visiting Covent Garden and took a great deal of effort.

The Housing Association then tried to get as many sites from the GLC as possible (they owned 15 acres); but since most of the residential space was set above shops, the (residential) Housing Association was not permitted to raise money. Therefore a Commercial Premises Association, represented by a developer, started buying houses for rehabilitation.

The Gardens. A number of open sites were created when the Odhams Press buildings were torn down, and—in the interim period between destruction of the old and construction of new development—CGCA has slipped in to cultivate these spaces as community gardens. These have been highly visible, relatively low-cost activities serving a wide range of people. They have become extremely successful examples of what CGCA is all about.

The first garden, the Japanese Water Garden, was laid out, ironically, by a Chinese architect who was on the Open Space Committee. Very much an informal "designed" garden, it was built with broken concrete paving slabs, plants, and a pond. It survived about two years before construction on a new GLC housing project began. The plants and turf were moved to other small gardens, and work began on the Italian Garden, resplendent with a few "classical" statues and columns, next to the Covent Garden Opera House. Its survival has been threatened by the desire of a nursery firm to build a garden center shop on the site.

CGCA finally negotiated with the Metropolitan Estates Property Company

Figure 7. The wall that surrounds the Covent Garden Community Garden is brightly painted and has peepholes through which onlookers can peer. Above is the billboard forecasting the use that eventually displaced the garden.

for the large empty basement site in the middle of Covent Garden, just north of the piazza. This took six months to negotiate, since the Property Company was advised against it by their consultants. They finally agreed to it because they had seen the previous garden and thought they might get good publicity, but they would not allow CGCA to apply to the GLC for planning permission. The GLC said there was no need for planning permission, but CGCA had to agree to dismantle the garden on six months' notice. The site was "real derelict," but there was a bulldozer still around after the demolition of the buildings, money was raised from some events, and a hard-core group of 20 transformed the derelict basement into a delightful oasis.

The garden was well below street level. Soil and turf were brought in to form rolling grass hillocks where children play. There was a stage at one end, a pergola with secluded sitting places, a jungle gym, a barbecue pit, and a vegetable garden with a greenhouse. A young part-time gardener looked after it. The whole was surrounded by a brightly painted wall of "flowers" and "plants," with numbers of viewing holes along the sidewalk that continually enticed passersby to stop and peep in. There was a gate with a sign announcing its origins, forbidding photographs, and stating the rules for use.

After the garden was finished, it took some time for local residents and office workers to use it, for it still seemed the exclusive territory of the CGCA construction group. But after a year, the garden was beginning to be used by older residents. At first, the informal ragged landscaping style seemed to attract more of those who let their children go naked than properly suited office workers. These could be seen looking (enviously?) in through the peepholes. Later many ventured inside and, in the final summer of 1979, the garden was overwhelmed with them.

The symbolism of this walled countercultural "Garden of Eden," with its peepholes, was not lost on the opera- and theatergoers who passed it by on the way to and from the adjacent tube station. But neither could they miss the billboard announcing the future construction of offices on the site. The paradise was temporary; CGCA had already been given notice, and, like nomads, they had to look for new space. By summer 1980, an office building was under construction; today, the only reminders of the garden's existence are tubs and window boxes scattered throughout the area containing some surviving plants and shrubs.

While the Community Garden was enclosed and had an entrance gate, the Italian Garden is on an open corner opposite a pub and is used by all kinds of people, including winos and office workers. Here vandalism is a problem, and it has become more of an anonymously owned public space. There are plans to build a protective fence to allow children to play there more peacefully.

The "Big Squat" in Trentishoe Mansions. The GLC evicted the tenants from this old public housing project and then systematically smashed the plumbing in the bathrooms to prevent squatters from entering. Nevertheless, the CGCA was party to a squat. This action offended many local working-class residents, even though they needed cheap housing. It was seen as self-oriented rather than community-oriented and too radical, even though the legality of squatting in some cases has been upheld. But the squatters have remained and, after four years, are now accepted by the established community.

Jubilee Hall. A much more successful activity was the conversion of the old Jubilee Hall, built in 1897 to commemorate Queen Victoria's jubilee and later commissioned as the Foreign Flower Market. The building, owned by the GLC, was empty, with no plan for the development of the site before 1982. The CGCA launched the Jubilee Hall Recreation Centre Limited, a private charitable company, supported by its members. They started off with less than $10,000 in the bank, found a contractor willing to start with this, and obtained a five-year lease from the GLC to use the hall temporarily, with a clause that allowed the GLC to give them six months' notice to vacate. Taking over in May 1977, they completed new flooring, changing rooms, and changes to meet fire and safety regulations for the opening in January 1978; the cost was $150,000—compared with new centers which are in the region of $1.5 million per facility.

This recreational center provides facilities for indoor football; net-, volley-, and basketball; badminton; roller skating; table tennis; trampolining; aikido; judo; tae kwon do; weight training; yoga; paddle tennis; acrobatics; indoor hockey; unicycling; and trapeze. In 1979, 1,600 people per week used the facility, which had 1,000 members. Between 8 and 10 P.M. the hall is reserved for residents. The nearest similar centers are three or more miles away. It therefore has a large catchment area and brings in a wide range of people: respectable gray-haired gentlemen, young residents, local nurses, office workers, and also some former manual workers who, with new technologies in their trades, now need exercise. Local schools make use of it.

This is a thriving, well-managed operation that serves a large constituency. How is it different from an official borough recreation center? It is less formal, there are no uniformed guards and fewer rules, it is more relaxed, and it hardly costs the borough anything. It is an exceptional example of what a community can do. The Duke of Edinburgh came to visit it.

But the GLC has plans to demolish the building by 1982 for a mixed-use complex of shops, offices, and flats with an underground garage for 250 cars. They argue that it is necessary to recoup the original purchase price of the building and that the present building spoils the "symmetry" of the piazza. The CGCA has a counterplan to save the building and redevelop the other half of the site. That battle is just beginning.

The Forum

After the minister's admonition to engage in community participation, the GLC set up the Covent Garden Forum. The 30 members were to be elected by residents and workers as a balanced group, representing 10 different sections of the community: residents (nine), business (nine employees, owners, or managers), services (nine employees or owners), and three property owners or leaseholders. Of the 40,000 potential voters, all 3,500 residents and 1,500 workers were registered as electors in 1978. The poll in earlier elections was around 27 to 34%, the norm for local elections. Up to three candidates out of 49 could be voted for. At the first election, half the members were from the CGCA. The Forum has worked closely with the GLC planners and has reviewed all new proposals for the area.

According to a CGCA member, the GLC created the Forum in its own image—with committees, endless meetings, and piles of paper. It also managed to get quite a high representation of businesses to counterbalance the residents. Some of the larger businesses had their employees vote for them, and some of the "resident" members also had businesses. It became, in CGCA's eyes, more like a chamber of commerce—an exaggeration, but with some truth in it.

The Future

Despite the initial success of CGCA in halting the massive redevelopment plan and their later general agreement with the GLC on the land-use mix—more

Figure 8. The Community Garden, sunk below street level, contains a greenhouse, vegetable garden, community stage, play equipment, and places to sit.

housing, fewer offices and hotels—by the summer of 1979 things looked rather gloomy, both for their specific projects and for the general trends in the area.

First, the Jubilee Hall and the Community Garden, the two most prized and successful projects, were faced with elimination.

Second, the new public housing under construction, the result of a local victory, was now under threat from the Thatcher government. It would be sold to the highest bidder, that is, become luxury housing. All other housing projects in London except for Covent Garden, Thamesmeade, and Docklands are to be stopped. This at least says something about the significance and perhaps the effectiveness of the Covent Garden community.

Third, the careful rehabilitation of the old Market was nearing completion, and a list of 112 boutiques of various kinds was published. The boutiques include shops for arts and crafts, books, fashions, food, toys, and a wide range of other specialties. Clearly, this place will become a fabulous success, not unlike Boston's Faneuil Hall Market. Already boutiques on Long Acre and the pubs and restaurants on the north side of the Market attract crowds swelled by opera- and theatergoers as they quaff beer and sip wine against the crude relics of squatters' murals, all that may soon be left of the Covent Garden protest movement. As one CGCA member said, it is easier to fight office buildings and hotels than 400 boutiques.

Fourth, important conservation sites are crumbling; speculation and property values continue to escalate.

Figure 9. Covent Garden Market, 1980, newly completed with its restored structure and boutiques.

Fifth, since CGCA lost their private funding, they are down to one paid staff member, but they are raising funds to employ another one.

Lessons of Covent Garden

1. Covent Garden epitomizes two styles of citizen participation: the CGCA, the focused activist advocate group that is totally committed, serving the weakest in the community—residents, long-established traders, and light industry; and the Forum, a representative low-key advisory and review committee. These organizations fit the two groups well, each with its own strengths and weaknesses.

2. The Covent Garden experience shows how a smart, creative community group can organize and stimulate a constituency and can hold together for 10 years, almost without funds in the face of huge opposition, through a combination of social, environmental, and symbolic actions. However, it depended a great deal on specific projects. By concentrating its energy on these projects, it gained recognition, credibility, and a local constituency.

3. Whether the CGCA was wise to leave the Forum has been a matter of debate. It seems that it was. It did lead to an isolation from the internal decision making process, but this reinforced its identity and independence. Camden continues to recognize the CGCA, not the Forum, as the local consultative group in planning matters. Whether to join or confront is a dilemma that faces all neighborhood action groups. The Forum has its own weaknesses. Dependent on the GLC, it has no final say and is ignored in matters where the GLC takes a different view. It is subject to the GLC's manipulation; when and if the GLC's local team pulls out, the Forum is likely to wind down also.

4. Covent Garden emphasizes the weaknesses, but also some strengths, of the public inquiry system. The public inquiry comes too late in the planning process. At the time, there was no prior opening for participation. The system assumes that an inspector, an establishment planner, appointed by the minister can be a neutral figure; it demands a high level of professional and legal expertise to submit the opposition case and is therefore costly. Almost the entire budget of the CGCA went on the public inquiry. On the other hand, it did stop the plan. Public appeals were listened to.

5. A letter from CGCA members who read this report pointed out that two "global" factors overwhelmingly influence whatever goes on in Covent Garden: the general economic context and the pattern and power of land ownership.

Citizen Participation in Neighborhood Transport Planning: From Construction to Participatory Management

The principal theme of the next two case studies is that medium-size cities or boroughs can develop successful participatory programs in the planning of local transportation systems. This depends, first, on a reorientation of traditional engineering attitudes. There needs to be a shift from new construction to a focus on managing the complex requirements of local citizens, motorists, transit users, and others.

Success, in the sense that a majority of citizens are satisfied with planned actions, depends on the *particular context,* the *style of the actions taken,* and the *character of the process*. Participation programs have been most successful on the small neighborhood scale.

Political Context

In both Delft and Camden, the local council had a high proportion of progressive council members, socialists or liberals, who represented the interests of local residents. The programs of participation have been on the whole much more successful than those, in Britain at least, that have been organized by regional authorities like the Greater London Council. And in Camden earlier efforts to plan for traffic at the boroughwide scale met with large-scale opposition. The borough therefore resorted to small-scale incremental planning, where there was a clear demand and where opposition was likely to be slight. This could be seen as avoiding the larger-scale, more difficult planning issues, but this is not quite the case. In 1978, Camden carried through a new district plan at the borough scale, engaging in a large-scale participatory program. However, the policy nature of this plan aroused less intensity of feeling than the earlier comprehensive traffic management plan.

These studies focus on the participatory management programs of the planners and traffic engineers who run them. The political context of their actions is formed by the following groups (see Figure 10).

Higher Level of Government. National ministries and their local representatives (and, in London, the GLC) who monitor local government programs from above and are the focus of many appeals and public inquiries.

Local Politicians. This group comprises council members from different political parties who make the decisions about which schemes are to receive priority and who, while keeping their constituencies, want something to show for their tenure of office.

Emergency and Other Services. These services (fire, police, ambulance, etc.) are important because they impose constraints on any change in local street patterns.

The Public. This body comprises both interest groups and individuals. The most salient groups concerned with local transportation issues are *neighborhood and street groups,* who are sometimes in conflict with each other; *merchants,* who are almost always in conflict with the residents; *transport groups or motorists,* who seldom congeal as a pressure group; and some *transit groups* such as the taxi drivers in Camden. In conflicts like those in Covent Garden, *the news media* become involved also, but in most neighborhood planning the media are, at best, local news sheets.

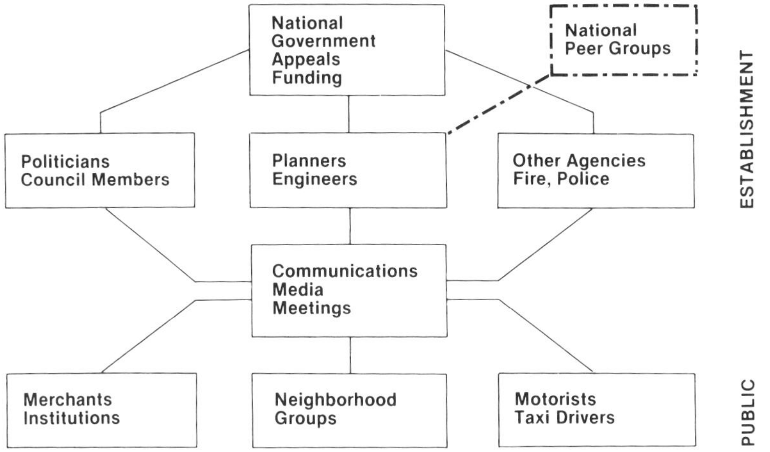

Figure 10. Groups politically active in participatory traffic management plans.

Local planners have to consider all these groups as they develop programs for local transportation. This context affects the character of their proposed actions and the programs they run. They also have their own professional peer group, both in planning and engineering, in which they have reputations to maintain as national leaders.

Character of Successful Actions

The political purposes of the local government planners in Delft and Camden appeared to be to develop environmental schemes and concepts that would be acceptable to substantial numbers of citizens, either as a result of their wish to see these concepts realized or to keep people satisfied; to respond to local community problems emerging from residents' complaints; to distribute the benefits and costs of schemes in an equitable manner between different interest groups and between streets and neighborhoods; to develop interesting, innovative solutions that would maintain leadership in the professional field and support the city's leadership among other cities; and to carry out schemes with minimal resources of cost and time. These cities were not always successful in achieving these goals.

The types of environmental action proposed by the planning agencies affected the success of their programs. Acceptable proposals had an immediacy and visibility to the local community, with obvious local benefit; they were based on simple, understandable concepts that could take different forms well fitted to local conditions; they had multiple benefits and no obvious losers, or only a small minority of losers; they were modest and flexible schemes for small areas and used simple devices that could be altered if they did not work; and they were media-attractive schemes that could be communicated graphically and through photography.

Characteristics of a Successful Participation Program

Both Delft and Camden have evolved their own processes over several years of experience. One fundamental characteristic of these programs, quite different from what one finds in most American cities, is the *close collaboration of planners and engineers.* In Camden the planners and engineers are in the same department, *The Department of Planning and Communications.* This reflected until recently the integration of transportation and planning at the national level (in the Department of the Environment, which has since been split apart again). The engineers in Camden, one of whom is the deputy planning director, appear to be completely unhooked from "automobiles" and dedicated to managing the borough's transport system in the interests of all, both travelers and bystanders.

The successful participation program has early open meetings to receive residents' complaints. Complaints are received by the planning agencies or by a traffic committee, and at an early stage meetings are held to discuss problems. Questionnaire surveys are carried out to gain a sense of overall community feeling and to hear from "the silent majority." Proposals are made public and, in the case of Camden, sent to every address in the community for comment. Final proposals are presented at open meetings. Citizens have a right of appeal to political representatives or, in Camden, to the national government. A majority vote or evidence of majority support is necessary before implementation of the scheme. Modification or elimination of schemes is possible if they are found unsatisfactory. Finally, in the Netherlands, neighborhood organizations are subsidized by grants from the national government.

The *Woonerf* in Delft

The *woonerf,* a concept that creates residential precincts in which pedestrians have the rights of way and where the streets are changed into landscaped residential environments, has been enormously popular in Holland—with over 800 implemented *woonerven* in 220 cities—and residents are demanding more. The concept is now spreading to Germany, where examples are being implemented in 30 cities. It has also been reported on with approval in England, where the "environmental areas" of the Buchanan report have been somewhat less popular.

This is an example of how an active interest group, Stop de Kindermoord (Stop the Killing of Children), together with a group of young planners and a creative police chief, developed a planning idea that was supported and publicized by a national citizen interest group, the Royal Dutch Touring Club, and eventually, with a developed and budgeted participatory component, became part of the regular planning system in Holland. My interest will be primarily in why this particular concept has been so successful, and the depth of that success.

History

In the early 1970s, after years of professional discussion about how to design for the auto in residential areas, a group of young planners and urban designers in Delft, together with a creative police chief, developed the *woonerf* concept. A *woonerf* has two parts: pedestrians legally have the right of way, and the street is designed as shared pedestrian-vehicle space with a predominantly residential character. The redesigning of street space was not difficult in Holland since, due to ground settlement, brick paving set in sand is replaced every five years or so. The difficulty was in changing the law. The police chief said he could not enforce traffic rules under existing laws, which made pedestrians as responsible for "accidents" as drivers. So the group began to think of prototype laws. Meanwhile, the Royal Dutch Touring Club became interested in the problem.

The Royal Dutch Touring Club, perhaps the largest citizens' group in Holland with 2.4 million members (out of a population of 14 million), started as a cycling organization and has always concerned itself with traffic safety and control. It is also concerned with those who tour by automobile; however, there also exists a Royal Dutch Automobile Club. The Touring Club has a staff of 2,400, with 600 in Den Haag. Most of these people deal with the touring needs of all kinds of travelers. In the Netherlands, there are frequently separate paths (and stoplights) for motorized vehicles, cycles, and pedestrians. The Touring Club holds annual concurrent conferences in four cities, with 3,000 participants per year. From 1970 on, the *woonerven* were discussed and developed at these conferences. The Touring Club has a traffic department and advises, among other things, on signing of highways. It has supported the *woonerven* and produced important publicity on them (20,000 copies of a brochure designed for international use, a second in 1979, a film, etc.). They have also staged up to 10 one- or two-day tours per year of *woonerven,* taking up to 400 city officials, police, engineers, and others to see a variety of different examples (new and old, large and small, good and not so good).

After 1971, Delft started to design precincts and in 1973 asked the ministry if new laws could be enacted. A ministerial commission developed the new laws, which went into provisional operation in 1976. Once this happened, several other cities picked up the idea. Government subsidies are provided for maintaining roads, which in Holland are mostly of brick and require relaying every five years. Now the program is in full swing, with 224 cities participating. Amsterdam and Rotterdam have between 7 and 14 *woonerven,* but several smaller cities have

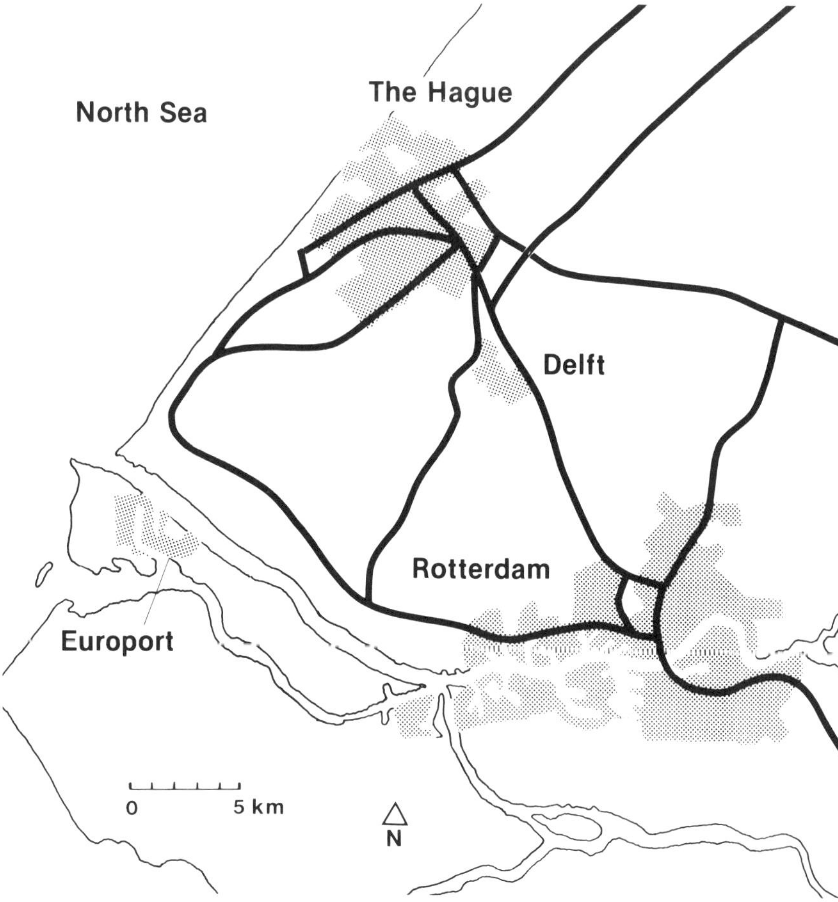

Figure 11. Central-western region of the Netherlands, showing the location of Delft.

larger numbers (up to 50): Delft, Papendrecht, Weert, Ede, Huissen, Eindhoven, Tilburg, Enschede, Alphen, Utrecht, Vlaardingen, Schiedam, Nieuwegein, and Katwijk each have 20 or more. Socialist cities tend to prefer the *woonerven;* conservative local governments see less need for them. However, all parties favor them, although they differ over methods.

There will be two major evaluations of large neighborhoods in the coming years in Eindhoven and Rijswijk, where $10 million has been allocated. The evaluation will be carried out under the direction of J. Kraay, who will take before-and-after safety measures. They will be constructed early in 1980, with three levels of restriction: a complete *woonerf,* a half-complete one, and only a partial one.

Organization of City Government

The city of Delft has a population of 85,000, of whom 10,000 live in the beautiful medieval city of Oude Delft, with its small and large houses, huge churches, narrow streets, canals, and bridges. Niewe Delft is a vast CIAM-style housing project with high-rise slab buildings as well as new houses, apparently the densest project in Europe at 50 to 60 dwellings per hectare, which—with family sizes averaging 2.7 persons—can climb to 400 persons per acre.

The city is run by a council, with a burgermeister appointed by the central government on 5 to 7-year terms. This system of appointment goes back to the Napoleonic era and creates a link between local and central government. Five aldermen are elected at large every four years. The political power in 1979 was divided between a Socialist–Liberal coalition and a Catholic minority. Aldermen are politically responsible for the five main city departments: public works; public health, social welfare, and sports; finance; housing and education; and culture, well-being, and citizen participation *(welsijn)*. The Burgermeister is responsible for public information, police and fire protection, and general affairs.

The Public Works Department (Openbarewerken), mostly responsible for the planning of *woonerven* and working in coordination with the police and fire departments, is divided into four sections; civil engineering (construction maintenance, surveying, and sanitation), parks (design and maintenance), planning (city and traffic planning, historic conservation and building permit review), and building (architectural design of public buildings apart from housing). The planning department is the primary manager of the *woonerf* program.

The Welsijn agency, created about 1969, has managed citizen participation programs for the city, although those connected with the *woonerven* are now run by the planning department. In addition, there is a government-financed advocacy planning organization, Opbouwerk, who have a small office in the city and channel sizable funds to citizen groups to be described in the next section. Finally, there is an ombudsman, who is there to listen to citizens' complaints on major issues such as housing needs, rent controls, and so on.

Citizen Groups

The national government, through the Ministry of Culture, Recreation, and Social Work as well as the Department of the Interior, subsidizes citizen groups in Delft in collaboration with the city. In 1978 the subsidies were approximately $850,000 from the national government and $40,000 from the city. Both youth and neighborhood groups were supported. In addition, the neighborhood coordinating group, Opbouwerk, was supported at the level of two staff members, a coordinator and secretary, with an office shared with the ombudsman. The neighborhood groups meet monthly with the Opbouwerk staff to discuss common problems.

In Delft, 20 neighborhood groups are subsidized. The allocations have until 1978 been made by the national government; from now on they will be made by the local Welsijn office. The funding criteria depend on special problems, especially "poverty pockets," and the size of the neighborhood. They do not cover all the neighborhoods in Delft. Until 1979, neighborhoods in the inner city have been far more active, but now those in the newer parts of the city will receive their share of the funding.

A neighborhood group must form a nonprofit foundation *(stichting)* to be eligible for support. A *stichting* is not legally responsible as a collectivity, like a corporation, but it has a statute requiring a chairman, secretary, treasurer, and committees. These neighborhood groups have enough funding for a staff member or two, a meeting room, and a newsletter. Their interests are reflected in their newsletters, some of which concentrate on neighborhood events and issues while others are more ideological and national in perspective, concerned with the neutron bomb, nuclear power, and so on. Most of them contain protest articles or news on such matters as rent control or the elderly.

Wippolder: A Low-Income Neighborhood

This is a large area in southeast Delft with some very low income pockets. Nearly all the population consists of low- or lower-middle-income groups. For 1979, Wippolder (10,456 people) received 242,000 gulden ($125,000) for social work. This supported four offices, one full-time neighborhood worker *(buurtwerker),* two half-time social workers, and a part-time cleaner *(schoonmaakracht).* There is a general committee with chairman, secretary, and treasurer; this is responsible for all decisions. Several special committees deal with the elderly, neighborhood workshops, recreation, traffic, bathhouses, workers. The leaders include a sizable number of students, several of whom are communists, together with communist workers. The organization therefore has a strong ideological bent, although its members are mixed. The students chair several of the committees, including the general committee. The ideologists who are more active in the two lowest-income parts of the quarter give priority to housing renovation and rent control, issues that involve more direct confrontation with the local government. Downtown pedestrianization schemes are also attractive to those involved in class struggles, since they lead to direct confrontation with business. The January 1979 cover of one of the monthly newsletters, *De Wip,* attacks national agency budgets with a cartoon showing how the rich are subsidized while social benefits decline. Another depicts the "municipal scavengers." Inside, there is a call to demonstrate against the neutron bomb and nuclear arms race. Other pages have announcements from various work groups and about recreational events.

The neighborhood is organized at two levels: the neighborhood committee and a number of street committees. The traffic issue is raised at the street committee level. Residents in general worry about the things closest to home: rent control, renovation, traffic on their street, local recreation, and so on. The most active streets are those that are troubled by through traffic but are not major roads. Residents of quieter streets are more cohesive and know that a solution, *woonerven,* is possible. However, the problem of the traffic is not susceptible to easy solution. Street committees differ on the solution. The young families with children are most in favor. Other groups—those with grown-up children, the merchants, those with new automobiles, those who like things to stay the same—are all opposed in different ways. These internal conflicts are not susceptible to ideological solutions. In fact, the neighborhoods are forced to ask the city, the high level of government, to arbitrate their conflicts. The particular neighborhood problem, therefore, profoundly affects the nature of participation. It also takes many years to resolve problems, and the solutions are very patchy. In the Wippolder neighborhood, there are *woonerven* on some streets but not on others. Traveling through the neighborhood, one passes in and out of *woonerven;* this is a bit confusing for drivers, but more responsive to the needs of residents on different streets.

These street committees are very active. After one *woonerf* was installed, cars making shortcuts continued to use the street, so residents blocked it with cars, forcing the city to redesign the plan so that it would not block through traffic.

It is at the street committee level, at this "lower base," that citizens are more directly involved. As scale goes higher—to neighborhood and city—the professionals take over: social workers, planners, and so on.

Participatory Process

Initiation. *Woonerven* can be initiated by the city or by citizen groups. Now that the *woonerf* is a well-known developed concept, citizen groups frequently demand to have one.

To convert a street or neighborhood into a *woonerf,* a number of preconditions must be fullfilled. They are as follows:

1. No through traffic that cannot be diverted.
2. Adequate parking space for residents within or adjacent to the *woonerf.* Parking is usually reduced for visual reasons. (The *woonerf* should not look like a parking lot.)
3. The majority of citizens must be in favor.
4. Construction must fit in with the street maintenance schedule (repairs every five to ten years), which is a government-subsidized program and limits the city's capacity to install *woonerven.*

Many citizen groups demanded *woonerven.* They are guided to the alderman in charge of the traffic subcommittee of the council. This has eight people, including appointees by other council members. When the *woonerf* does not meet the rules, citizens are frustrated. There is a waiting list of 10 to 20 groups.

Originally, the public works department was quite rigid in its definition of a *woonerf*. They argued this rigidity was necessary in order to establish the concept to avoid confusing residents or drivers whose behavior had to change. This policy has now become too rigid, and they are beginning to loosen it up. People can now have "a bit of a *woonerf*"—such as street bumps without signs. As the result of these preconditions, the city selects about 80% of the *woonerven*, while 20% are the result of resident demand.

Selection. When the public works department decides on a *woonerf*, it makes an announcement, sends out a brochure to all addresses, calls a meeting, explains the concept with slides, and asks for a vote on whether to have one. To go ahead requires a 60% majority. The department is now quite cautious, because in the past unanimous votes in favor have been followed later by unanimous votes against after the design has been developed and completed. So public works requires an individual vote, either at the meeting or through mail-back envelopes.

Design. If the citizens say yes, public works goes ahead, receiving citizens' ideas about the design in later meetings, when the planning department displays its design. Since the residents know the details of how their streets work better than the planners, their ideas play an important role. For instance, they will know where a handicapped person lives.

Implementation. Implementation takes place from three months to one year after a decision to go ahead. One week before implementation, the neighborhood is blanketed with handouts. In nearly all cases, the city carries out all the physical changes, planting, and so on.

Evaluation and Modification. There is no formal evaluation of the effects of *woonerven* in Delft. If residents find problems, they speak up or act. On one street, Frederickstraad in Wippolder, the *woonerf* did not succeed in stopping through traffic. Residents therefore blocked the street with their cars. In a meeting with the planners, they said that they wanted street humps. There was no more money for street humps, but the city will probably put in posts to block the street.

Maintenance and Street Territory. Since the *woonerf* involves turning a street from a channel into a residential precinct with plants and flowers, it opens up the chance for residents to make the street space more their own. In a number of areas, residents begin to plant their own flowers, putting out their own flower boxes and watering the plants that the city has put in. There is, however, no definite program to have residents take on the responsibility for maintenance. In most cases, it appears that the originally active residents will keep up the watering for a year to two; then care will decline as some move away.

In some cases, plots of landscaping in the street have been designed to be

Figure 12. A *woonerf*. (Courtesy of The Royal Dutch Touring Club.)

more attached to particular houses, and residents have been allowed to rent the plots for a nominal 1 gulden a year provided that they maintain it. However, the city retains title and can take it back if necessary. This public ownership of the street may be required for future public flexibility, but it limits the involvement that ownership of the streets would offer.

Trompetstraat

Trompetstraat is a narrow street in Oude Delft, leading from one of the canals to the center. The residents are old working-class and young middle-class professionals; according to the middle-class activists, the composition is quite stable. Rent control keeps 50% of the public housing at 30 gulden ($15) per month. The very small houses have in many cases been converted to enlarge rooms and extend accommodation. They are built right up to the street and have minute backyards.

The Trompetstraat residents are part of the Raam Vlamingstraat e Omgeving neighborhood group, which covers about seven streets. The group formed in 1972 around transport and parking problems. Their main purpose was to improve living conditions, conserve the neighborhood, and participate in social activities. They have a group crèche and bingo evenings. "People acting alone have little power," but getting together in a *stichting* has "forced the police to pay attention."

The neighborhood constituency consists of about 600 families of students, academics, and working-class people. Though the leftists are more active, the group has no connection to political parties. The streets lining the canals are richer; other streets are working-class. Everyone is viewed as a member. There is no subscription. The leaders change, but are usually the young 25- to 35-year-old people of a more or less academic type. There is a chairman, a secretary, and a committee of seven members. It would be preferable for one to come from each street, but not all are active. The most active are on the quieter streets.

The subsidy of 12,000 gulden ($6,000) provides a meeting place (actually the kindergarten building), utilities, events, and a publication which is sent to

Figure 13. A *woonerf*.

every address monthly. There are no paid staff members. Decisions are made informally, with voting only on important issues. There is general agreement on the parking problem and need for renovation. Conflicts occur between residents and those who do not live in the area but have stores there. They are in a minority, however. The organization is well known among neighborhoods and a representative attends the monthly meetings between all neighborhoods at the Opbouwerk. But the most widely read newspaper, *Den Haag,* only has a section on Delft and reports neighborhood news only when there is nothing else.

One of the problems for Trompetstraat residents is the parking of trucks, since small industries exist along the streets. These trucks fill the street, block light to the windows, and sometimes scrape the paint from the walls as the street is narrow. Also, outsiders using the shopping center park illegally on the streets, and it is unsafe for the few children who live there.

The residents (three or four of whom are active) asked for a *woonerf* in 1975. They belong to the neighborhood *stichting,* but have carried on their activities more or less independently. They presented their plan to the city. The city turned them down; downtown businesses need the parking space since they are losing trade to suburban shopping centers. The city would have to find alternative parking spaces if a *woonerf* were to be installed.

The residents became very frustrated, and one Saturday in 1978 they started their own *woonerf* by erecting wooden posts to prevent parking next to their houses. The planning department actually welcomed this protest, because it is too trying to eliminate illegal parking. The citizens could rightfully sue the city for lack of enforcement (though they cannot in fact afford the legal costs). The city ended up ordering the posts out with a promise to build a *woonerf* (though the cost will be 10 times as much as the citizens' posts) provided residents were willing to park their cars in a neighborhood parking lot at about $15 per year. The residents agreed. The posts are now being used as street planters, and one resident is constructing benches and lamps for the street. They wait expectantly for their *woonerf,* after four years of effort.

The residents have tried to exert most pressure on the traffic subcommittee of the council, whose alderman chairman is politically powerful and has a city-wide perspective. They feel that the *woonerven* are beginning to look all the same. They have "no fantasy" about them; they are "too definite," "too complete," leaving no creation for residents. "You have to give people a handle." The city says the residents can do what they like behind the new posts, but they must leave room for strollers, so there is little scope for creativity. The residents think the *woonerven* could be cheaper. They only need boards and posts, but they are being offered more than they need; they have to have the whole package. There are also disagreements about the number of acceptable parking spaces on the street. The residents conducted an inventory of parking spaces, but the council did not accept it and argued only to use legal spaces. Since there are 35 car owners in the street, 25 will have to find places elsewhere if a *woonerf* is installed.

Effectiveness of the *Woonerf*

Woonerven offer many advantages. They offer a clear promise of improving the residential environment, making it safer for children, and providing a more pleasant ambience. Since the *woonerf* is a clear concept, now written into the law, people know what they can get; they imagine a clear and positive solution that they can demand. This is an improvement on the normal complaints about traffic. The clarity of the concept and codification of procedures help to make the planning agency's work more effective. Standard procedures economize on staff time. At the same time, the concept is flexible in interpretation; it can be applied to streets and areas of different sizes and character, and the traffic-control devices can vary in character with the situation. *Woonerven* are also fairly easily achievable; they are cheap by comparison with major public works or buildings, and they can be part of normal maintenance procedures. *Woonerven* extend citizens' rights over their home territorial space, perhaps the most important part of the urban environment outside the home. Finally, the *woonerf* concept has benefited large numbers of people: 800 *woonerven* would include at least a million people.

But *woonerven* are not without their critics. Traffic is not always slowed. Street humps and severe width restrictions are not always installed, so the streets look more beautiful but remain dangerous. Indeed, *woonerven* can encourage a false sense of security. This seems to happen when designers, forgetting that the fundamental need is to control traffic speed, concentrate only on beautification. When *woonerven* do not actually control traffic, they become tokens, appearing to solve a problem but not actually doing so. Moreover, *woonerven* do not help the parking problem. For visual reasons, parking often must be reduced to create a *woonerf*. This is objected to by certain residents, especially those, usually low-income car owners, who like to keep their cars outside their houses, and also by merchants who want as much parking available for shoppers as possible.

The codification of the concept and accompanying procedural roles have tended to become too rigid and elaborate, closing the system off from citizen involvement and strengthening the hand of the bureaucracy. The standards preclude simpler, easier solutions in some cases; the procedures can take a long time, and residents can become very frustrated. According to some, it is becoming more and more difficult to get a *woonerf* as the rules become stricter. Also, the designs of *woonerven* are becoming more predictable and less imaginative.

Woonerven could be used to encourage more permanent citizen control over their own streets (cultivating and maintaining plants, monitoring parking, etc.), but this has not been done. On the contrary, some *woonerven* have led to interstreet and interneighborhood conflicts, which then have to be resolved at the city council level. This is an unattractive facet of *woonerven* for those on the left who want to unify the working classes. Fighting for rent control or downtown pedestrianization is more fruitful terrain for struggles against landlords or business. Rent control and maintenance of public housing are also seen as the basic issues in a city where some claim that 10,000 people are looking for new houses and only 70 houses become available each month.

Borough of Camden: Residential Neighborhoods

Camden is one of London's more progressive boroughs, with substantial experience in efforts to have citizens involved in the planning process. They have also had a long history of planning for traffic in neighborhoods and appear to have developed a participatory program that works to the satisfaction of a majority of citizens.

History

Camden tried to develop a comprehensive traffic plan in the early 1970s, which would redirect traffic out of neighborhoods and onto arterial streets. They held six public meetings where a few thousand people turned out in opposition to the plan, mostly because they lived on the main streets that would receive more traffic. Later the planning and communications department carried out an attitudinal survey showing that, in fact, a majority were in favor of the plan.

The plan was scrapped because of the opposition, and the department lowered its sights to neighborhood improvement schemes. Whereas in the early schemes technical measures were taken to predict changes in traffic, noise, pollution, and pedestrian delay levels, they now concentrated more on attitudinal surveys in parallel with community meetings. This section describes what happened in Lamb's Conduit.

Description of Area

Lamb's Conduit, or Central Holborn, is the name given to an area 3/8 mile by 1/4 mile, about a mile north of the Thames, 1/2 mile northeast of Covent Garden. The surrounding major traffic arterials to east, south, and west are predominantly commercial, with hotels, office blocks, and shops. The district, designated as an "environmental area,"[2] contains four large hospitals and two types of residential development: Georgian terrace houses sometimes converted to offices and new terraced blocks of flats. There are no continuous east–west routes through the interior of the area, and only two north–south routes. One of

[2]"Environmental areas," originally proposed in the Buchanan Report, 1963, are areas where traffic is limited and where the residential environment is of dominant concern. They are similar to traffic-controlled precincts.

Figure 14. Lamb's Conduit: The pedestrian mall through which trucks and service vehicles are allowed to run.

these, Lamb's Conduit Street, partly a shopping street, had an average daily traffic volume (adt) of 7,000 to 8,000 while John Street, mostly offices, had 2,000 adt.

Jurisdiction

The Borough of Camden has jurisdiction over the area, but the Greater London Council has say in all transportation issues. The Borough of Camden has a council that is predominantly Labour/Liberal. The planning and communications department combines transportation, land-use, and environmental issues under one agency.

The Process

A Camden councilor on behalf of a resident group initially asked the planning department to develop a plan for turning the district into an "environmental area," that is, reducing through traffic.

The First Stage: Meetings and Surveys. The planning process took place partly in accordance with the Town and Country Planning Act of 1971, which stipulates the need for some form of public participation and, if there are objections, a public inquiry.

Meetings were held with local traders, hospital officials, and local associations. At the first public meeting in February 1974, the planning department explained their scheme for closing Lamb's Conduit and John Street with some alternatives. The residents at the meeting wanted more closure; the merchants were against any closure at all. Having received these responses, the planning department prepared a 3-page illustrated leaflet explaining the reasons for the scheme, two maps showing the likely changes in traffic flow on each street, with decreases internally and "some increases" in two peripheral roads. They then explained how an experimental scheme would be carried out on a three- to six-month basis to determine the effects of the temporary closures. Resident, business, and other views were solicited on a simple stiff card sheet containing 12 questions with space for responses, which could be torn off and mailed back without cost. This was distributed in the autumn of 1975, and by December an 18-page mimeographed report on the findings, with an equally long appendix of tables, was available for internal circulation.

The questionnaire aimed to discover the views of three distinct groups; residents, shopkeepers most affected, and shoppers and visitors. The brochure and questionnaire were handed out to all residential units except hotels and motels and to all retail shops, pubs, and restaurants within the area but not on the peripheral streets. After two to three weeks, reminder letters were sent. In addition, three staff members distributed the materials randomly to pedestrians in Lamb's Conduit Street. Of 1,250 residential forms, 462 (36.7%) were returned. The relatively low response rates were attributed to lack of concern, predominance of other problems, or, in the case of merchants, other ways of expressing opposition.

The questions asked for opinions not only about the proposal, but also about general attitudes toward traffic regulation and specifically which roads the respondents thought should be main roads and which should be closed.

An overwhelming majority thought traffic should be reduced and confined to main roads while pedestrian facilities should be improved. Some 70% of responding residents thought the proposals would improve the area; 18% did not. In analyzing subsamples, it was found that car owners were less in favor of traffic control; those on the trafficked peripheral streets wanted more distribution of traffic. The resident majority contrasted with the vociferous opposition of the merchants at the public meeting.

The Experimental Installation. The Camden Council applied to the GLC for an order to implement the scheme. It took two years to receive the go-ahead and actually proceed with an experimental closure. Meanwhile, a Conservative had been elected as the local councilor.

The Second Stage: Meetings and Surveys. Eight months after the scheme was implemented, a second questionnaire was sent out, and five months later a report was produced. Again a simple sheet headed "Success or Failure? Camden wants *your* views," with two maps showing physical changes and alterations in traffic flow, were distributed with a mail-back questionnaire. Separate questionnaires were designed for shoppers and traders.

Every *other* residence in the area was given a questionnaire, and every residence on the peripheral roads was given one. The former group, only 50% sampled, were weighted by a factor of 2 in reporting the final results. Response rates were between 50 and 59%. Interpretation of the results was complicated, because many respondents wanted slight changes. On the final count, 60% appeared to favor a permanent scheme; 33% wanted out. Some 53% of shoppers thought it more pleasant, 31% less pleasant. Of 33 traders (59%) responding, 24 were against the scheme, 21 reported a loss in trade, and 14 had delivery and access difficulties. Significantly, very few of the traders lived in the area.

Camden Council decided to go ahead with the permanent scheme, at which point opponents demanded a public inquiry from the secretary for the environment (the ultimate appeal), which was duly held 14 months after the experiment was first installed.

The Public Inquiry. The report of the public inquiry is an extremely detailed 120-page mimeographed document, written by J. H. Ross, an inspector, planner, and surveyor, to the Secretaries of State for Transport and for the Environment, who had appointed him. During six days of hearings in Camden Town Hall, the inspector heard representatives from all sides. The opposition came from the Lamb's Conduit Street Traders' Association, the London Licensed Taxi Drivers' Association, and the London Fire Brigade, together with 60 individual objectors, mostly from representatives of businesses.

After listing the objectors, the case for the Camden Council was made. The main objectives of the scheme were stated, restricting through traffic as far as possible to the peripheral roads, providing safer pedestrian movement in the area and safer access to the new school and the shopping center on Lamb's Conduit, all without unduly affecting access to the hospitals, shops, businesses, and other activities in the area. This scheme was in accordance with GLC policy and with the borough's district plan to "promote schemes to protect those living or working in areas . . . from the adverse environmental effects of large flows of traffic."

Traffic flow changes and noise levels were reported. While internal roads showed significant reductions, the peripheral roads showed only minor increases except for two streets whose capacity was able to absorb them without "substantial increase in congestion." Reports here came from police and London Transport.

Accident levels had declined, though this was not necessarily attributable to the scheme.

The council conceded that servicing of shops was more difficult but claimed that the permanent scheme would be more convenient than the present one. They could not come to an agreement with the merchants to audit their accounts, so the council staff counted people calling at a sample of shops before and after the experiment, showing only small changes. The merchants' unaudited indexes showed decreases of up to 10 and 15%, but the council questioned whether these were caused by the experiment.

In support of the scheme were the Holborn Society, registered with the Civic Trust; the headmaster of the school, who noted that 119 pupils had a safer journey to school; a resident association, which mentioned the 5,000 or 6,000 people living in the area and produced 250 signatures; the Camden Community Center; and a local playground group.

The four hospitals did not raise major objections, although they expressed concern about access. At a reported meeting, the council stated that the only dissenters were businesses and some commuters. The council calculated that about 60% of residents would be satisfied with the scheme if made permanent, and just over 30% would not. Over 50% of shoppers thought the temporary pedestrianization was more pleasant, and 30% thought it was not. A larger number thought permanent pedestrianization would improve matters.

The objection of the Fire Brigade was to the temporary scheme, a problem that the council maintained would be resolved when the permanent scheme was installed. As for the objections of the traders, the council quoted other pedestrianization schemes to show that traders in the long run gain rather than lose from such changes.

The various traders recounted their losses in some detail, to be followed by the other objectors. The council was then allowed to rebut these arguments before the inquiry concluded. The 65-paragraph conclusion by the inspector carefully summarized all the pros and cons of the scheme once again, commenting on inconsistencies, and ended up recommending that the scheme be made permanent while opening up the part that was on neighboring John Street.

Summer 1979. A visit to Lamb's Conduit in the summer of 1979 found it to be a pleasant, pedestrianized shopping street, quite busy with people, though blocked most of the time by a delivery truck.

Lessons of Camden

On the whole it can be said that the Lamb's Conduit scheme was a success, for a majority of the residents of the area were satisfied with it. The delicate balancing of those who gained and those who lost was brought out through the several surveys that were made—of traffic flows, store turnover, accident rates, and resident opinions—as well as the evidence from letters and public hearings.

No doubt it was the public inquiry that forced the accumulation of this degree of detail, but it was also the long experience of the Camden Planning Department in assessing traffic schemes that determined the quality and relevance of the information collected. The mixture of quantitative material and strongly felt human response for and against the scheme make the minutes of the public inquiry a fascinating and human document, though its meticulous detail may seem tedious to some.

The incremental nature of the scheme, which consisted of several components on different streets and which commenced with an experiment before final permanence, allowed intervention by the public at different stages and in different parts of the project. This accessibility of the scheme to intervention allowed a high degree of participation. Camden planners, it seems, are more flexible than those in Delft. They responded to resident demand not with a standard solution but with one that was particular to the place.

Concluding Note

A brief concluding history of citizen participation in Britain might be helpful in identifying what could be useful for participatory programs in the United States.

After World War II, the British government enacted a series of town and country planning acts that created a planning authority at the local and national government levels. This authority had extensive powers over land development and was able to create whole new towns, often over the opposition of local residents. The original planning acts did have some provisions for public participation. They consisted of public notice columns in local newspapers concerning plans and applications, with the public having a right to complain; mass publicity programs with exhibits to inform the public of new plans; an open register of all planning applications; and, most importantly, a public inquiry system.

The Public Inquiry

A public inquiry is a public hearing into a particular proposal or decision from a central or local government body. There are two types, those that are appeals against decisions such as refusal of a planning application and those that are concerned with objections to a particular proposal. A public inquiry is not a court of law, and *any person affected has a right to attend and give evidence* without being legally qualified. Inquiries are chaired by an inspector appointed by the Secretary of State for the Environment. He presents his report to the secretary or minister who can accept, modify, or reject it. Thus, the secretary is both judge and jury. This is characteristic of the British planning system, in which appeals are made more to the administration than to the courts. A public inquiry must be held if there is any group that wishes to appeal for one.

Public inquiries have several advantages. They are accessible to the public. Anyone can object without the need for legal training, although larger inquiries now restrict evidence. The inspector is bound to hear all evidence and report it in detail to the secretary. Inquiries such as that of Covent Garden have been instrumental in preventing large-scale developments that might destroy local neighborhoods. These inquiries are frequently well reported by the press. The series of motorway inquiries which John Tyme (1978) disrupted in a long campaign of opposition and civil disobedience received nationwide attention.

But there are also objections to the public inquiry. They come late in the planning process. They are only initiated when a group objects to a plan. By this time the proponents have a plan well developed, substantial money and time have been invested, and the opposition is put in the position of obstruction. Moreover, the secretary or minister is both judge and jury of schemes that are often proposed by his own civil servants or those of a local authority. In rare cases, appeals are made to the High Court. The inspector is appointed by the secretary or minister, who makes a decision independently of the inspector's report. Inspectors have often restricted the scope of inquiries, disallowing discussion of the need for a project or its relation to a national policy. This was the main objection of John Tyme in his book *Motorways versus Democracy,* Finally, information at the public inquiry is mostly in the hands of the government agencies, and its diffusion can be limited at the discretion of the inspector. For these reasons, opposition at public inquiries has usually been forced to seek support from layers and "experts" to counter the official case. *Community Action,* one of the most useful nationwide citizen action journals, has published an *Action Guide to Public Inquiries* that, in 40 pages, explains to community groups how these inquiries work and how to approach them.

More progressive local governments now hold hearings less formally at earlier stages of the planning process. For instance in Warwick, panel hearings are held "at which local politicians informally listen to, consider, and evaluate public comments." Nevertheless, it seems that the public inquiry continues to perform an important ultimate function when disputes are otherwise unresolvable. However, as Tyme says, larger issues of policy must also be open to discussion or be clearly resolved in some other form in which public representatives can participate. In the Motorways program, as with the Covent Garden plan, key debates and decisions were mostly confined within the bureaucracies.

Public inquiries continue to be a focus for citizen participation. Although the more famous ones, such as the Roskill Commission's investigation of alternatives for the third London Airport and the Greater London Development Plan Inquiry, involved "a massive volume of paperwork and a bewildering array of expert talent" (Fagence, 1977) and took a period of years to complete, most public inquiries are of shorter duration and complexity, taking only a few days, as in the Camden case. Nevertheless, the scope of public inquiries has changed. Whereas originally they were intended to involve only the parties directly interested, they have been extended to include third parties and all others affected. They have also increasingly become an examination of policies, as in the case of the motorways inquiries. And finally, the public has become much more active and organized. In the original inquiries related to the new towns around London, the opposition of local residents was not seen as sufficient to obstruct the plan to bring inner-city Londoners out to the country. Today, according to Fagence, the inquisition has become increasingly by the public rather than on behalf of the minister. The more complex public inquries are run by a panel or commission and the numbers of participants have become restricted in order to streamline procedures, though the selection of participants is still meant to cover "the spectrum of opinion" pertinent to the issues being examined (Fagence, 1977). But public inquiries are now seen as dependent upon a meaningful discussion of issues before the inquiry is held.

The Skeffington Report

In 1969, the Skeffington Committee (headed by Lord Skeffington) was appointed by the Minister of Housing and Local Government to consider the best methods for securing public participation in the formative stage of development planning. This was an official acknowledgment of the inadequacy of the current planning process. The report, although couched in rather general terms, did come up with a number of interesting recommendations. Besides emphasizing the importance of improving the information about plans to the general public and generally to improve public education about planning, a diversity of modes of participation was endorsed. Two methods were however emphasized, one directed toward *the actives,* those citizens who take part in influencing community affairs, and *the passives,* "who, although deeply affected by decisions, do not make themselves heard because of diffidence, apathy or ignorance of what is going on." For the "actives," the committee suggested the creation of a *community forum* to promote useful discussion between the local authorities and identifiable groups. This was the instrument set up in Covent Garden. For the "passives," the report recommended the engagement of a *community development officer* to act as a catalyst for expressions of local opinion. These recommendations were not mandatory, however, and have only been adopted where local authorities saw fit or were specifically required by the minister, as in the case of Covent Garden.

Participation in Britain, 1979

Numerous experiments in participation by local authorities have taken place in Britain in recent years, and there has been some systematic research on their effectiveness, notably the work of Peter Stringer, a social psychologist from the University of Surrey, and William Hampton, of the University of Sheffield, which has concentrated mostly on participation in structure planning (i.e., major land-use and transportation plans at subregional scale). Other work has been carried out by social science research consultants such as SCPR, led by Gerald Hoinville and Roger Jowell. Some of the innovations being suggested include the following.

Focused Public Meetings. The findings of Stringer and associates show that the numbers that come to public meetings are a very small percentage of the community. For structure plans which are admittedly of less immediate interest to residents than neighborhood plans, between 0.04 and 3.4% of the potential audiences turned up in the cases studied. The benefits of the public hearing, the educational value of seeing councilors and planners under questioning, are set off by the small number of people who actually speak up. Local issue-oriented public meetings are likely to be more successful than those dealing with larger scale problems. Politicians and activists like them, but their lack of use to the general public who do not turn up demands that they only be part of a broader program.

Neighborhood Councils. Community forums suggested in the Skeffington Report and adopted in Covent Garden as well as other places are supported by the Department of the Environment. A study of the community forum in part of Chelsea and Kensington, an upper-class community, showed that, like Covent Garden, the forum had been appointed by the local council, had a planning officer as its secretary, ran orderly meetings, and was closely involved with community decisions. However, it was not an elected body, like that in Covent Garden, and its representativeness was questioned. The problem of cooption so clearly pointed out by the Covent Garden Community Association is the cost of such an intimate relationship.

Neighborhood councils have been discussed by various communities, sometimes under the name of parish councils, which have a long history in rural Britain. The Blythe neighborhood council in Hammersmith was compared to the Chelsea community forum. The neighborhood is much more mixed, but the neighborhood council is more representative than the forum. It has little connection with the council, however, and seems to be as isolated as the Covent Garden Community Association. Meetings are unorganized and cover many more issues than the forum. The council is, however, renovating a hall, providing play facilities for children, organizing an annual fair, and planning to start a newspaper and events for the elderly and handicapped. In 1979 it had a community worker employed under the Job Creation Program and an Urban Aid Grant. It therefore seems to be a more viable social group, though its wide-ranging concerns allow it less focus on planning issues.

Approaching the "Passives." While the Skeffington Report suggested the use of community development officers, a rather elitist designation, Hampton and Belle report on an effort in South Yorkshire to use community workers to approach nonparticipant groups. Since nearly two-thirds of the electorate belong to one group or another, the technique of reaching out to groups that do not normally participate could be a productive one. The program was related to participation in a structure plan, the type of plan for which it is most difficult to arouse public interest. Of over 600 groups, 370 were given kits and 160 made response. The program focused on "planning kits," which explained the plans and asked for group ranking and preferences concerning a number of issues. The community workers contacted the groups, explained the kits, and often acted as secretaries. The technique was seen as successful though impermanent. Similar ways of involving small groups in the course of a plan have been proposed by SCPR. Members of these community panels are selected by the consultants. Groups of six or seven meet to discuss their needs informally over a number of meetings.

Tony Gibson at the University of Nottingham has developed a number of model kits ("Neighborhood Action Packs") which can be used by residents and schoolchildren to develop plans for their own communities. These models are made of cardboard and are apparently generalizable for any community. "All parts of the model (8 feet by 8 feet) were easily detachable and replacement pieces were provided—new houses, pedestrian crossings, demolition blanks, street closure kits, and so on." People in the Raleigh Street neighborhood, with students from the local planning school, came up with their own plans for the neighborhood by using these kits.

A more advanced media project can be found at the new town of Milton Keynes, where a package called Viewdata seeks to use television for locally interactive discussion of planning issues.

Evaluations of typical planning publicity materials and exhibits and the role of the press have been made by Peter Stringer and are reported on with many practical suggestions in documents from the Linked Research Project (e.g., Stringer, 1978; Stringer & Plumridge, 1975).

ACKNOWLEDGMENTS

I would like to thank the following people for giving me their time, insights, and material: Tish Frances, Martin Coomes, and Ian McNicol of the Covent Garden Community Association; Rene Schoonbrodt, Paul Vermeylen, and Maurice Culot of Inter-Environment and ARAU; Bob Lane and Bruno Schlaffenberg of the borough of Camden; Peter Jonquiere and M. Hakkenbroek of the city of Delft; and Peter Stringer and David Uzzel of the Program in Environmental Psychology, University of Surrey.

Bibliography

Borough of Camden, Department of Planning and Communications. *Central Holborn Environmental Area: Lamb's Conduit Street Improvements*. PC7506 (6), London, December 1975.

Christensen, Terry. *Neighborhood Survival: The Struggle for Covent Garden's Future*. London, Prism Press, 1979.

Community Action. *Investigator's Handbook: A Guide for Tenants, Workers, and Action Groups on How to Investigate Companies, Organizations and Individuals*. London, undated.

Community Action. *Public Inquiries: Action Guide*. London, undated.

Convent Garden Community Association. *Keeping the Elephants out of Covent Garden*. London, 30 Sept. 1976.

Covent Garden Community Association. *Newsletter*. London (monthly).

Culot, Maurice. *The Reconstruction of the European City*. Brussels, 1979.

Fagence, Michael, *Citizen Participation in the Planning Process*. London, Pergamon Press, 1977.

Greater London Council. *Covent Garden Action Area Plan*. May 1978.

Greater London Council. *Covent Garden: The Next Step: Revised Plan for the Proposed Comprehensive Redevelopment Area*. London, Covent Garden Joint Development Committee, 1971.

Hampton, William. "Adult Education and the Urban Programme," in *Bulletin of Environmental Education*, London, 1979.

Hampton, William and W. Beale. *Methods of Approaching Groups in South Yorkshire*. Linked Research Project LRP, No. 11, London, 1976.

Inter-Environment Bruxelles. *La ville et l'habitant*. Brussels (monthly).

Jonquiere, Peter. *Woonerf: An Environment for Man and Transport*. Delft, Department of City Planning, 1979.

Langley, P. *Public Participation in Local Planning: A Discussion Paper*. London, March 1979.

Mazis, Aristidis. "The Covent Garden Planning Participation Case Study." MSc thesis, University of Surrey, Department of Psychology, 1975.

Monaham, Jim. "Up Against the Planners in Covent Garden," in *Community Politics*, edited by Peter Hain. London, John Calder, 1976.

Royal Dutch Touring Club, ANWB. *Woonerf*. The Hague, May 1978.

Schoonbrodt, Rene. "ARAU: Balance and Prospects after Five Years Struggle," in *The Conservation of European Cities*, edited by D. Appleyard, Cambridge, Mass., M.I.T. Press, 1979.

Secretariat d'Etat a la Region Bruxelloise. *Plan de secteur de l'agglomeration Bruxelloise*. Brussels, 1980.

Stringer, Peter. "Participation." Department of Psychology, University of Surrey, 1975.

Stringer, Peter. *Tuning in to the Public: Survey Before Participation*. LRP, No. 14, London, 1978.

Stringer, Peter, and S. Ewens. *Participation through Public Meetings: The case in Northeast Lancashire*. Linked Research Project, LRP, No. 2, London, 1974

Stringer, Peter, and G. Plumridge, *Publicity Guidelines*. Linked Research Project LRP, No. 7, London, 1975.

Town and Country Planning Act 1971, Section 217. *Proposals by the Secretary of State for Transport. The Conversion of Highways into Footpaths or Bridleways (London Borough of Camden) Lamb's Conduit Street*. Ref LH # 38/5008/9/010, London, 1976.

Tyme, John. *Motorways versus Democracy: Public Inquiries into Motorway Proposals and Their Political Significance*. London, The Macmillan Press, 1978.

Uzell, David. "Participation in Friary Ward." PhD dissertation, Department of Psychology, University of Surrey 1979.

Citizen Action in Brussels: Some Comments

JEAN-FRANÇOIS LEJEUNE AND RENÉ SCHOONBRODT
Associate and Director of Inter-Environment, Brussels

Professor Appleyard's paper leads us to offer the following reflections about citizen action in Brussels and its success.

Similarities and Differences between Brussels and Other Cities

We think that it is important to define the sociopolitical situation in Brussels by making comparisons with London, Paris, and American cities. In Paris we are witnessing a process of citywide gentrification: the population of the working-class neighborhoods is being progressively replaced by upper-middle-class and upper-class inhabitants. In London, Brussels, and, it seems to us, the majority of American cities, the processes of urban transformation are leading to the exclusion—over the long term—of *all inhabitants* from the urban centers (including the lower-middle and middle classes that have always lived in the city). Only zones occupied on a very temporary basis by marginal groups will be left.

The difference between Paris and the other cities of which we are speaking can be explained only by the behavior of the upper strata of the population with regard to housing. The effect of this behavior is reinforced when these upper strata also control the economy. Their behavior—which itself has historic, economic, and cultural origins—becomes symbolic for the rest of the population: in Paris the decision of the upper class to live in the city leads to imitative behavior by the groups that follow it in the social hierarchy. In the other cities and especially Brussels, the fact that all strata of the population have chosen suburban living has its origins in the behavior of the upper class.

The City Gives Power to Its Inhabitants

The research of Henri Lefebvre and many others—Castells, Remy, Harvey—has shown that the city is the spatial structure that permits the reinforcement of the dominant power over the society as a whole. This dominant power is economic, administrative, financial, industrial, and cultural in nature. Brussels is an example of the seizure of ancient (1,000-year-old) structures for the benefit of NATO, the EEC, and the multinational corporations. ARAU, relying on the theoretical research of contemporary economists and sociologists, has been able to demonstrate the true nature of the urban transformations in Brussels. The debate has been seen in its true terms and has gone beyond secondary subjects such as urban activity, respect for the historic patrimony, urban esthetics and even public nuisances (in the strict sense of the term). For ARAU, the effect of urban transformations is to diminish the power of the population and to increase the power of the socioeconomic structures.

Within the framework of this analysis, ARAU has progressively—that is, in almost 11 years of activity—developed a clear image of the city, an image that has been accepted by the population. Briefly stated, ARAU's point of view is that the city is the seat of power of the population: it gives power to those who inhabit it. It would take too long to give a more precise analysis. Let us simply say that the urban image advocated by ARAU includes the following points: The city should be inhabited. It should be organized in mixed neighborhoods: planning in these neighborhoods should permit the development of economic activities related to the capacities of the population. Driving should be limited rather than facilitated. Urbanization should favor the creation of public and collective spaces and restore the streets and public squares. This is our theory for reconstructing the European city. In this perspective, the urban action program of ARAU is part of a project of democratization of the entire society in all its political, economic, and cultural aspects.

Why Does ARAU Undertake This Effort?

Especially in the area of urban planning, neither the public authorities nor the political parties that share or fight for power have other plans for the city than those directly tied to the development of the industrial and multinational corporations. This complete absence of plans has led various people on the left from different (and sometimes opposing) professional and political milieus to work together to formulate a plan and defend it in the political arena. Professionally, the members of ARAU are architects, economists, sociologists, lawyers, and philosophers—all involved in militant action within different economic and political structures. The professional qualifications of the members of ARAU have not, however, evoked accusations of technocracy, given that ARAU's mode of action is to subject all its theoretical analyses to direct public criticism—from the authorities, and, above all, from local urban action groups.

How ARAU Makes Its Projects Known

The population of Brussels and sometimes even certain public authorities have given rapid support to the ideas of ARAU. This is the case with programs for the rehabilitation of older housing, for the area plan, and for the means of informing the public.

The ideas of ARAU are spread by the communication system provided by Inter-Environment Brussels, an organization serving as a common meeting ground for the local working-class communities and middle-class groups.

ARAU's ideas are also spread by the media and by the various cultural activities directly sponsored by ARAU: guided visits, lectures, and an annual "urban school" consisting of a week of evening workshops.

Reasons for the Success of ARAU and Other Urban Advocacy Groups in Brussels

We first cite the evidence of the plan for Brussels. ARAU's plan is based on a clear analysis of the real reasons for the transformation of urban spaces. The question that motivates ARAU is: Who profits from changes in the urban structure—changes that some people try to justify on the basis of the necessity and desirability of progress? ARAU's approach is to consider the various problems in their economic context; that is, to look at the economic causes of urban transformation and also to formulate economic objectives to be pursued in order to respond to the needs of the current population of existing neighborhoods. It is on this basis that ARAU has developed a capacity for architectural and urbanistic analysis and the preparation of counterproposals. While the images presented in ARAU counterproposals have been criticized as being too conservative, too "nineteenth-century," ARAU's position is that only the traditional city composed of streets and public meeting spaces allows the creation or recreation of a spatial structure that gives power to the inhabitants, economic power included. All other forms are repressive because they cannot be built except by dominant financial groups and their public and private allies. This position of ARAU and the connections of ARAU's members with social movements and labor organizations enable ARAU to make its ideas known to various groups less strictly oriented toward urban problems.

We next examine our rules of action. Experience provides several lessons. Initial action must be sustained. The process of decision making that influences the built environment is slow. In order to defend a given position, it is necessary to sustain action over a very long period as choices pass by different actors in the decision making process. This demands an organization structured for continuing action.

Action must be public. All action must be made public. The temptation to negotiate secretly with public authorities must be resisted. To this end it is useful to make good contacts with the media, to have your own publications, and to place the emphasis on systematically keeping the public informed.

Trained leaders are needed. The success of the action requires that each group have a core of persons who are experienced in dealing with urban problems and who have a vision of the society as a whole. The emphasis should be placed on ideological and technical training.

Negotiation is necessary. To refuse to negotiate is to cling to a revolutionary utopia, an ideal that does not correspond to the desires of the population of Brussels. Negotiation requires those human qualities that lead to victory and also permit positive lessons to be learned from defeat.

4

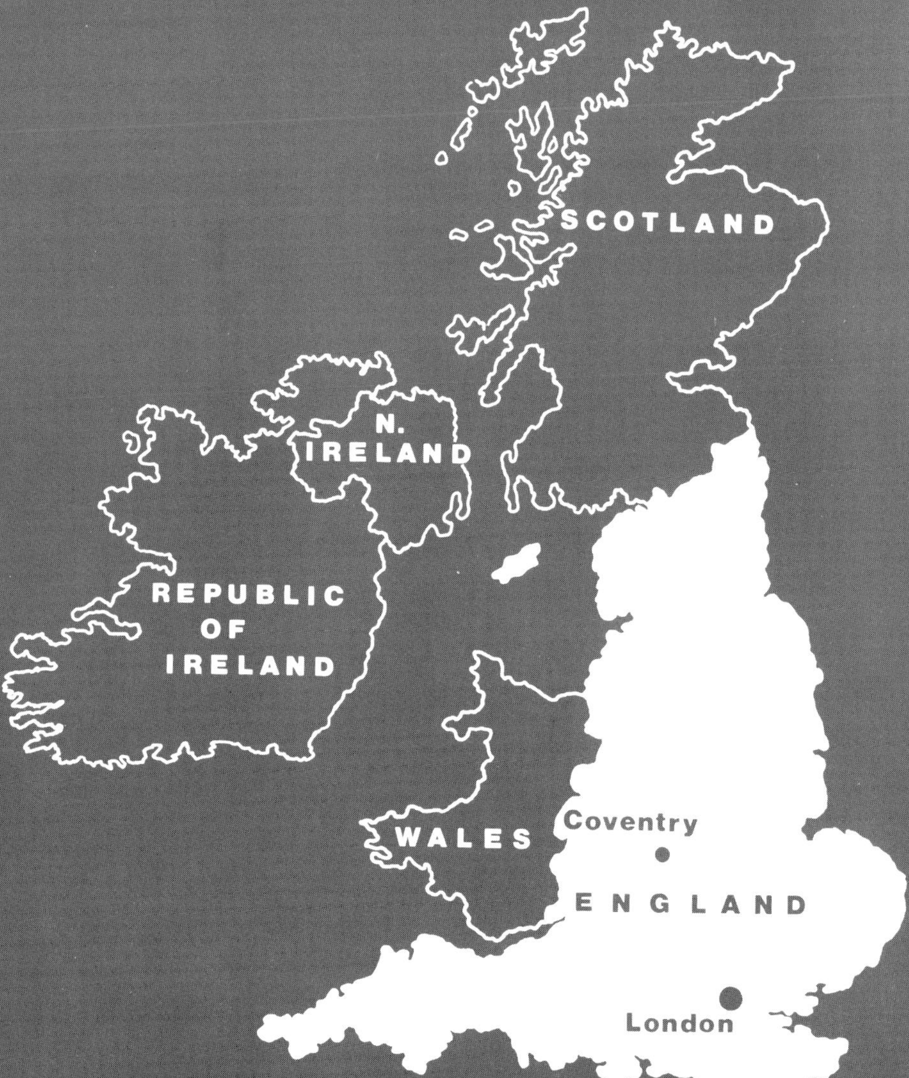
SCOTLAND
N. IRELAND
REPUBLIC OF IRELAND
WALES
Coventry
ENGLAND
London

4

DOCKLANDS AND COVENTRY

Two Citizen Action Groups in Britain's Economically Declining Areas

HANS SPIEGEL AND JANICE PERLMAN

Introduction

Both the following cases describe mature, militant community action and technical assistance organizations utilizing research as a principal instrument of social action. The emphasis on in-house research (and its corrolary of long-range planning) seemed highly suggestive for American community-based groups. This method of fostering social action was reason enough to engage in a study of the British groups. In addition, we chose to look into these groups because they dealt with similar issues as those faced by many American community organizations, especially those that combat gentrification and try to protect the interests of working-class and poor constituencies in a threatening economic climate. And as in our country, both the British groups were finding it difficult to fight basic structural and technological issues with limited weapons that can win some battles but are not powerful enough to win the war. There is a bittersweet quality to both of these British cases, as the successes—but also the limitations—of this sophisticated and depth-probing style of citizen action are detailed; it is a story of struggles with mixed results that also emerges, we might quickly add, when analyzing various U.S. grass-roots groups, as we have done in previous papers.[1]

[1] Janice Perlman, "Grassrooting the System," *Social Policy,* September–October 1976; Janice Perlman, "Grassroots Empowerment and Government Response," *Social Policy,* September–October 1979; Janice Perlman, "Grassroots Participation from Neighborhood to Nation," in S. Langton, ed., *Citizen Participation in America* (Lexington, Mass.: Heath, 1978); Hans Spiegel, "Citizen Participation in Federal Programs: A Review," *Journal of Voluntary Action Research Monograph No. 1,* 1971, pp. 20–22; Hans Spiegel, "From Protest to Program: Three Grassroots Coalitions in Their Formative Stages," Graduate Program in Urban Affairs, Hunter College, 1978; Stephen D. Mittenthal and Hans Spiegel, *Urban Confrontation: City versus Neighborhood in the Model Cities Planning Process* (New York: Columbia University, Institute of Urban Environment, 1970), pp. 395–470.

In both Britain and the United States, the central governments have launched wars on poverty and initiated new policies regarding citizen participation in community development. Our colleague Don Appleyard reviews a number of these policies in his case; we shall confine ourselves to a brief note about the changing political patterns in which these policies are now caught.

The Labour government's position on central-city development was most clearly stated in the 1977 white paper *Policy for the Inner Cities*. This policy statement, part of which was later implemented through the Inner Urban Areas Act of 1978, declares that "the time has now come to give the inner areas an explicit priority in social and economic policy, even at a time of particular stringency in public resources." There follows a series of proposals for assistance to municipalities to tackle their inner-city problems. Reminiscent of the U.S. Model Cities legislation (1966), the white paper states that "comprehensive action is needed" and chooses a limited number of inner-city areas for "special partnerships" between central and municipal governments. Citizen participation requirements in the white paper and other programs are limited to the admonition that "Public authorities need to draw on the ideas of local residents, to discover their priorities and enable them to play a practical part in reviving their areas. Self-help is important, and so is community effort." Advisory letters to municipalities and ministerial speeches give what one observer has called "gentle encouragement to consult residents."

At the time of our investigations, both the Docklands area and Coventry operated under the bygone administration's mandate to upgrade central cities. Docklands, being the largest of the seven designated partnership areas, had more leverage in obtaining central government funding, but both it and Coventry were eligible for categorical assistance for housing, open space, social services, and so on. In 1979 the political sands had just shifted and the Tories were in power. The new Conservative government made a number of proposals to change existing policies for housing and urban development. Americans have probably been most aware of the policy shift that would empower local authorities (as municipal governments are called in Britain) to sell public housing units, at a discount, to individual tenants. In addition, the Thatcher government is advocating the establisment of urban development corporations for the London Docklands and Merseyside. These corporations are modeled after the new town development corporations and supposedly would facilitate large-scale land assembly, planning, and development management of these tracts. This proposal received some support but also considerable opposition from within the Docklands area.

Finally, the Tories introduced a measure that may be copied by the Reagan administration, the enterprise zones, or "EZs," as the British press likes to call them. Dubbed "a bold experiment in unfettered capitalism,"[2] these zones are to attract business enterprises into depressed urban areas by exempting them from

[2] British Information Service, News Division Release P107/80, 19 Sept. 1980.

Figure 1. Southern England, showing the location of the Docklands and Coventry.

various taxes and regulations. The Isle of Dogs within the Docklands has been mentioned as an enterprise zone, together with six others spread across the British Isles. (The proposed urban development corporation would administer the Isle of Dogs Enterprise Zone.) Reaction to the EZs from affected local government bodies was initially warm because of anticipated employment gains, but lately some questions have been raised whether newly established or relocated enterprises within the zones would not actually drain off already hard-pressed business elsewhere in the central cities. A number of citizen groups, including the one in Docklands we are about to describe, have additionally argued that anticipated gains in economic activities within the zones do not justify the loss of government revenues and the relaxation of planning and environmental controls. The British EZs were initiated in 1981; if they are to serve as a model for U.S. economic development of depressed areas, they deserve close scrutiny from Americans.

Joint Docklands Action Group

The advertisements placed in the slick business magazines by the developers of the East London's Docklands area declare in bold headlines, "London Docklands Over £1,500,000,000 Will Make Sure It's Not Just a Planner's Dream." The text continues,

> Situated in the heart of the city alongside the River Thames, London Docklands is the largest area for development in the world. All the dreaming and a lot of planning have been done. And now we are getting on with it. . . . It is also a great opportunity for business. . . . As well as providing for people presently living in Docklands, we are hoping to attract a wider cross section of new residents to the area.

Many of Docklands' long-time residents read danger signs in such pronouncements. They are far from convinced about the projected opportunities for people now living in the area. They fear for their jobs and their neighborhoods. One of the organizations that articulates their fears and frustrations is the Joint Docklands Action Group (JDAG), a militant and brassy organization that does

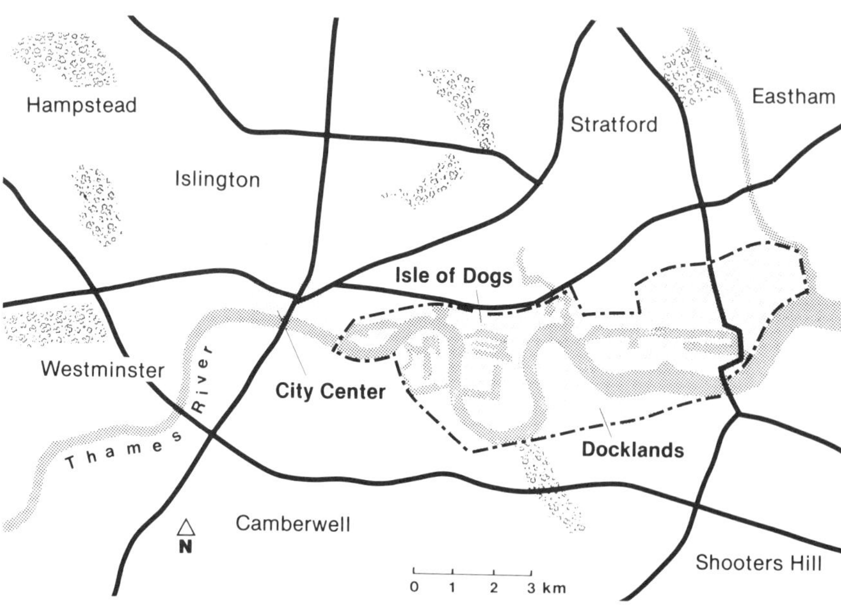

Figure 2. Central London, with 5,500 acres of Docklands stretched out along the Thames River.

its homework about issues and coordinates community action against many of the plans. However, JDAG is not only in the business of being opposed to plans presented by higher authorities; it also proposes alternate ways to improve the area. JDAG frequently opposes the views of the people planning the huge development, but it earns their admiration, perhaps grudgingly, for being a tough, skilled, and creative adversary and sometimes a collaborator, too. JDAG has thus become an economic and neighborhood development institution whose views carry weight.

The Setting

The Tower of London represents the westernmost border of the Docklands area. From there, the Docklands area stretches some 8 miles east as the crow flies and many more miles as the majestic Thames curves its way downstream. The Docklands land area covers 5,500 acres within five London boroughs. Only 585 acres (slightly more than 10%) are utilized for residential housing; the rest are used for docking facilities, industry, institutions, roads, and so on, while 1,440 acres are vacant land. Some 56,000 persons now live in the area; government plans call for a population of 82,000 to 94,000 by 1986 and over 100,000 by 1997.

The sheer size of the anticipated development operation is rather staggering. The Docklands Joint Committee, which is the intergovernmental planning agency overseeing the renewal project (not to be confused with the Joint Docklands Action Group, which is the indigenous federation of resident groups), asserts that Docklands "is the largest single area of urban redevelopment in Europe today, and the largest that has risen in London since the Great Fire in 1666." A recent government white paper stated that "The redevelopment of the London Docklands is of its kind, the greatest challenge of our time."

The Docklands area gives the visual impression of strength and activity. The eyes are assaulted by cranes and gas storage towers, huge docks and warehouses, river traffic of every description, and the architectural gem of Christopher Wren's Royal Naval College just across the Thames from the Docklands area. Indeed, many years ago the area was called "the hub of the Empire," since most trade routes found their nodal point there. In this setting the casual visitor is tempted to overlook the houses that are tucked between these rather overpowering facilities. But the people have become visible through a variety of activities undertaken to draw attention to their plight.

The people and the houses are principally in the moderate- and low-income category. On the average, the housing conditions rank considerably below the median for London. Depending on the borough, from 10–25% of the housing stock is in "poor" or "unfit" condition. Only a tiny portion of the houses are owner-occupied, the overwhelming majority of the people being renters. In several of the boroughs that contribute to the Docklands area, public housing

Figure 3. The Docklands area, with its huge docks, was once considered the hub of the British Empire.

(called "council housing" in Britain) constitutes more than half of the housing stock, compared with about 25% in all of London.

Two of the boroughs, Tower Hamlets and Southwark, represent the second- and third-poorest boroughs in London; every one of the three remaining boroughs shows average household income considerably below the London norm. The labor force in the area comprises mostly manual workers; in the case of Tower Hamlets, three-fourths are so classified.

World War II brought considerable destruction into the Docklands area, and economic trouble for the area was further speeded up by the postwar movement of industries away from central cities into industrial areas at the periphery. For example, between 1961 and 1971, the southeast London boroughs (of which the Docklands area is a substantial part) lost 96,700 manufacturing jobs. East London increasingly felt the impact of London's changing role away from manufacturing and toward services. Most immediately threatening to the economic viability of the area was the decision, in the mid-1960s, to move the docks downstream. These new docks would be especially appropriate for handling containerized cargoes. Thus, a technological change considerably contributed to threatening the viability of the century-old docking facilities. In addition to the industrial retrenchment, urban renewal activities eliminated some of the deteriorated housing stock, new roads and a tunnel directly affected the area, and large tracts of the land became vacant.

As in America, such a massive exodus of facilities and jobs from prime land invites politicians and planners to fill the vacuum. And just as in our country, the vacuum is often filled with glistening visions of sparkling projects, such as the idea to use a goodly portion of the Docklands for business and professional activities, exhibition halls, malls, and offices. One thought was to replicate London's West End, which was then being renewed, with a number of office buildings going up and quickly being rented. In 1971 a plan was commissioned to study redevelopment possibilities for the area by the then Tory government's Department of the Environment and the Greater London Council. When the plan was completed in 1973, it unleashed strong opposition in each of the five boroughs of the Docklands, which had remained strongly Labour throughout. The plan, which gave five options for development, was opposed because the residents and their borough governments saw "their" land cavalierly being used for purposes that would change the character of their neighborhoods and that did not appear to benefit them substantially. One of the options proposed, for example, was a new-town-in-town. This was seen by many residents as an attempt to colonize the area by changing its population, its economic style, and even its governing structure. They would have none of it.

The various organized groups in the area protested vigorously, and marches and demonstrations were held. The time was ripe for tenants' organizations, trade unions, and ad hoc action groups to join into a coalition that took on the name of the Joint Docklands Action Group (JDAG). The hostile reception to the plan made at least two things abundantly clear to the Docklands residents and the government and private developers of the area: (1) citizen participation in the planning process needed to be considerably strengthened, and (2) an alternative plan for the development of the Docklands was necessary. The economic decline was continuing and needed to be reversed; in the words of a recent draft JDAG paper, "neither the residents nor the politicians and planners in the Docklands could have plausibly pressed for the simple maintenance of the status quo. . . . The dilemma for Docklands was not whether the area should be redeveloped or left undisturbed. Rather the key question was and continues to be in whose interests is change to be effected?"

Thus JDAG originally came into being as a general reaction against the decision of the Port of London Authority to move the docking facilities 20 miles downstream and, more specifically, as a form of profound disaffection from the 1973 plan—in other words, as a weapon of protest. But soon JDAG also developed its own positive programs and alternative plans; therefore it can serve as a good example of the move "from protest to program," which also occurred among a number of American groups in the 1970s.

But JDAG has never relegated its protest function to the sidelines. Indeed, the Tory government's intent to create an urban development corporation has given new impetus to protest activities. JDAG was quick to remind local residents of the decade-old proposal to create a new-town-in-town and the dangers that such a proposal would hold for consequential participation in planning by resident individuals and organizations. In late 1979, a JDAG release conjectured that an urban development corporation would mean that the Docklands were being "handed over to developers, landowners and companies whose interest will be in making money out of Docklands rather than meeting the needs of the community." Public meetings were held, petitions circulated, and representations made to the House of Lords Committee which will have to deal with the issue. Opposition to the urban development corporation led to the formation of two new citizen action groups in Southwark and Wapping.

JDAG's Function and Structure

JDAG was organized in 1973—in the words of one of its papers—"to ensure that any redevelopment in the Docklands meets the needs, first and foremost, of local people in East London." It was a loose coalition of trade union groups, tenant organizations, and so on. A management committee of five members (one from each of the Docklands boroughs) oversees the work of JDAG, and an annual general meeting is the ultimate organizational authority. There appear to be no standing committees, but ad hoc projects are undertaken by staff, student volunteers, and indigenous leaders around a variety of issues. While there are no regular meetings of the total membership except for the annual meeting, there is frequent contact within the leadership cadre of the organization. The two-floor storefront office of JDAG is the heart of organizational activity; area residents drop in for information, student volunteers write up their reports there, announcements of meetings come off the busy mimeograph presses, a row of filing cabinets contain considerable documentation and plans, and the phone lines are constantly in use.

At the center of this activity is a full-time staff of three. Nick Sharman was, until recently, the staff head of JDAG. Young, energetic, and articulate, Nick appears the very model of a modern working-class intellectual community organizer. He is fully conversant with the history of the area and with the organizational strategies of unionization and antifascist drives in the Docklands area of bygone years. "The Docklands have always been Labour party all the way," he says: "Single-issue politics don't work here. A Ralph Nader could not be sustained here. You must develop a broader political base." Building this broader political base has been the boiler plate of JDAG's strategy—at least of its leadership.

One of the objects of this strategy is to get labor organizations more affirmatively into the JDAG coalition, especially the union of the dockers. The dockers, traditionally proud and independent, have surely seen the writing on the wall: their jobs have been threatened and the fate of the dockers is inextricably linked with that of the community. (Indeed, it was reported to us that every job on the docks helps to maintain three jobs in the community.) There was a common enemy that could best be fought collaboratively: the government and industrial institutions that would make workers and residents redundant. Perhaps the participation of the dockers union in JDAG is still not as fervent as Nick Sharman had hoped, but the union seems in JDAG to stay, joining demonstrations, signing declarations, and making financial contributions.

On a day-to-day operational basis, however, the closest link seems to be between JDAG and the local action groups, especially the two JDAG-affiliated community resource centers. (The latter are neighborhood information and referral centers funded by government grants.) The action groups appear to be relatively informal organizations made up of working-class residents, sympathetic planners, and related professionals whose efforts are coordinated by JDAG. The action groups focus on a number of local issues and frequently cosponsor political action or studies with JDAG.

Practically all JDAG funds come through the Docklands Joint Committee (the agency responsible for the planning and coordination of Docklands development activities), which, in turn, received its money from the government's Urban Programme. The Docklands represent the largest of Britain's special, noncategorical urban assistance programs serving seven large "partnership areas" and 15 smaller "programme authorities." We were told that 17% of these national funds had been allocated to the Docklands. Clearly, the Docklands project is nationally recognized as having high priority.

JDAG receives approximately £25,000 (or $60,000) per year from the Docklands Joint Committee for services that are being rendered to the population of the area. Only a small fraction of JDAG's budget comes from voluntary contributions and private donations. JDAG's funding seems to be assured until at least March 1981, the date by which the joint committee itself may be replaced by an urban development corporation.

JDAG organizes and coordinates groups within the Docklands, but it also represents this federation of groups to external agencies. The first of these functions involves JDAG in rendering technical assistance to its constituent organizations and launching joint ventures for the entire Docklands area that include political action, study groups, research enterprise, the publication of pamphlets and position papers, and the distribution of its own newsletter. The second function is discharged by JDAG's representation of its constituency toward the larger government departments, labor groups, and related publics "on the outside." This second category of functions includes the demanding task of influencing the planning effort now going forward for the Docklands Joint Committee (which is in charge of the actual planning), the Greater London Council, the London Port Authority, and similar bodies.

JDAG is thus an interesting organizational hybrid when compared with American models of citizen action. On the one hand, it resembles some features of Alinksy-type mass-based coalitions, with their emphasis on creating "organizations of organizations" that exercise collective political power. On the other hand, JDAG also resembles American "intermediary" community research and training organizations, such as the Center for Community Change, which is one step removed from but strongly supportive of grass-roots groups.

As JDAG relates to planning agencies, it has to contend with government-initiated programs of citizen participation. This structure requires a short explanation.

As has been mentioned, the official planning and development organization for the area is the Docklands Joint Committee (DJC). It is a creature of the Greater London Council, which appoints eight members to the committee, and of the five Docklands boroughs which also appoint eight members. These 16 members appoint an additional eight members representing the Port of London Authority; banking, industrial, planning, and labor interests; plus a member representing the Docklands Forum. This last group is one of the principal means of gaining citizen participation in the planning and implementation of Docklands development.

The Docklands Forum is a broad-based participation group, staffed and funded by DJC, composed of not more than 45 official representatives. They come from labor unions, labor federations, employer groups, tenant associations, churches, educational institutions, and so on. The Forum sponsors community meetings and information programs and is asked to react to planning proposals. JDAG is a member organization of the Docklands Forum. The Forum is a place for debate and exchange of views intended to assist the development organization. However, the Forum is no action group, and its activities are confined by its limited mandate. (Interestingly enough, in late 1980 negotiations were under way about the possibility of JDAG assuming staff responsibility for the Docklands Forum.)

The Forum is not the only avenue used by DJC to reach out to the people of the area. DJC's staff (which recently has been renamed the Docklands Development Team) has taken its strategy plan (1976) to the community in an effort to obtain public reactions beyond those offered by the Forum. Vehicle trailers equipped with maps, photographs, and information booklets were parked in convenient locations throughout the Docklands area, public meetings with slides were held, a popularized version of the plan was inserted in every mailbox in the area, interviews surveyed sample groups among residents, and others sought response from employers. DJC claims that approximately 70% of the affected population in the Docklands, according to one of its survey results, "were aware of the DJC's efforts."

When the development coordinator of DJC was asked why, in view of all these avenues of citizen participation, government funds were utilized to support JDAG, he responded that an action group has a vital part to play in helping people to articulate their independent views, even if these views are in opposition to the plan. JDAG helps to focus the debate on real issues, he said, rather than on emotional or impossible schemes. He saw the approximately £25,000 yearly appropriation for the resource and information services as a good investment. DJC exercises no control over JDAG, he pointed out, except for an audit of books, which were being kept impeccably.

An official of the central government who deals with central-city areas acknowledged that some voices have been raised among municipal officials about the appropriateness of the national government funding municipal government's potential opposition. There is an effective counterargument to such worries, he felt, and that was to remind local officials that it is far more effective and convenient to deal with only one action group, even if it is in opposition, than with several groups representing divergent and often superficial and contradictory views.

Issues and Activities

The issues that propel much of JDAG's actions are the familiar ones for advocacy groups in an economically declining area: jobs for people now in the Docklands, housing, transportation for the area, public amenities, social services, and so on. The strategic plan, published by the Docklands Joint Committee in 1976, provides opportunities for periodic changes and alterations. JDAG judiciously alternates between supporting and attacking certain features of the plan. In the words of a recent flyer,

JDAG . . . attempted both to improve the plan and to ensure it is actually implemented. This has meant campaigns to:

1. Expand public spending to rebuild the area's infrastructure and to meet current shortfalls in social services.
2. Resist the closure of firms and the general process of withdrawal of industrial capital (the campaign has concentrated particularly on the retention of two remaining dock systems).
3. Resist developments harmful to the area (e.g., offices, luxury housing).
4. Expand the role of the public sector in creating jobs, especially in industry.
5. Expand public participation in decision making at all levels.

Even though JDAG's efforts go primarily into issues that are adversarial in character, there is also an element of service provision in its work. Advice is freely given to residents who come to JDAG's storefront office, and there is ready encouragement for self-help projects that lead to beautification of the area, or the establishment of cooperative child care among mothers, or the street art that can be seen on formerly empty walls.

A unique feature of JDAG activities and strategies are the alternative plans that it proposes as substitutes for the "official" plans drawn up by the Docklands Joint Committee. These alternatives are offered in 9- to 60-page mimeographed and attractively bound pamphlets that are frequently cosponsored by one of the local affiliates. Thus JDAG has prepared a document, together with the Southwark Community Development Project, entitled *Alternative Forms of Tenure: Preferences and Costs*. Another, prepared in conjunction with the Tower Hamlets Action Committee on Jobs, is called *London's Docks: An Alternative Strategy*. These are carefully researched and closely reasoned tracts that criticize existing plans or policies and then propose alternative solutions.

An old-time resident and JDAG leader was asked what he thought about the organization spending so much time in preparing alternative plans. He commented that the action group could, of course, concentrate only on arguing against existing plans, "and we would be successful at frustrating these plans" he added. "But then," he went on, "we would have nothing in their stead. We needed to switch from negative thinking to positive thinking about planning. The easiest thing is to shoot proposals down. But you need to have something that you are positively for and to put it into action."

For example, one of the major issues that concerns JDAG is that the Docklands area will be gentrified to an extent that will force existing residents out of the area. A number of rehabilitated structures in some of the more desirable locations (such as along the Thames) underline such fears; spanking clean flats have attracted affluent households, and the hotel–office–tourist complex to the east of the Tower of London was certainly not constructed in the image of the average east Londoner. JDAG objects, therefore, to the proposal that 40% of the area's new housing be for "equity sharing" between occupants and public housing authority, since "this form of tenure will deny access to most Docklands residents and lead to imports of the more affluent." Instead, JDAG proposes housing upgrading and rehabilitation for present occupants, in order to protest the development of luxury apartments instead of housing for the dockers and other area residents.

Another unique aspect of JDAG's activities is its sophisticated approach to long-range "macro" issues as well as the immediate here-and-now concerns. The pamphlet called *The Engineering Industry in Docklands,* for example, carefully analyzes the tendency of engineering jobs to leave the area and looks for explanations in the international marketplace as well as in private investment goals and public policies. At one point the pamphlet points to an apparent contradiction in government policies affecting the engineering industry, namely, on the one hand,

> to raise the productivity and hence the profitability of industrial investment, thereby stimulating a further cycle of renewed investment and profitability. On the other hand, Government is committed to encouraging jobs, especially in areas of high unemployment. Yet in an industry like engineering, while output remains static, any attempt to raise productivity must necessarily place more jobs in jeopardy.

The JDAG pamphlet advocates the need for determined government intervention in order to keep engineering jobs in east London. It recommends specific government supports such as grants, loans, and public sector contracts to achieve this objective. The report is replete with manpower statistics, industrial profit statements, and analyses of historical trends. JDAG and the people who helped to put together a proposed strategy of revitalizing a whole industrial sector in Docklands had done their homework. This point was not lost on the policymakers at whom the statement was directed.

Both adversarial and self-help activities are sponsored by JDAG. The range of programs engaged in can be illustrated by some of the projects undertaken by the local action group and the JDAG-affiliated Resource Center on the Isle of Dogs.

1. A "No Road Rally" designed to stop proposed roadways through the neighborhood.
2. A protest against an American community development organization that was viewed as a smokescreen for its sponsor, an American Insurance corporation, to increase its sales and profitability in east London (the newsletter states, "You were not invited to our Island, we don't want you to stay. You have lied your way in—now get yourself out!").
3. An action that preceded the present organization but that is still relished by people with a decade-long memory, namely the Isle of Dogs Universal Declaration of Independence in 1970, which sparked a day-long citizen closing of the main arteries leading into the island and brought a great

deal of publicity to the blighted conditions of the "seceding" neighborhood.

The self-help action, on the other hand, deals with the creation of an urban farm. The "Mud Chute" is the last available large open space on the Isle of Dogs; the chute was created by an excavation made many years ago in the course of building the docks and looked not unlike an area that has undergone strip-mining. The Mud Chute's 40 acres were slated for the development of 3,000 new dwelling units, which caused considerable anxiety among many residents. Instead, the action group decided unilaterally to establish "an urban farm" on the land. Today there are garden allotments for people desiring to grow vegetables, horses and goats are lovingly tended by children, as are geese, community gardens, a playground, and open fields used for community recreation. (One of these recreational activities is "Wellie chucking," which consists in throwing a size 11 Wellington rubber boot as far as possible; the local champion holds the

Figure 4. In the midst of the neighborhood, the bookstore of the Joint Docklands Action Group.

record with an impressive toss of 97 feet.) The farm is a well-supervised facility, greatly used by young and old. A visitor was recently told with obvious pride (which appears to be shared by many of the 16,000 inhabitants of the Isle of Dogs) that 37 different species of wild flowers have been identified on the Mud Chute.

Who initiates the various activities of JDAG? The answer appears to point either to the local neighborhood constituency group (such as the above example of the Isle of Dogs Resource Center, which originated the projects) or to the central JDAG staff which, together with volunteer students and scholars, initiates a project such as a research report. In either case, however, the other party is involved, too. The local neighborhood group is likely to receive some help from JDAG's central office on Cable Street, just as a research project will likely receive the cosponsorship and active collaboration of the affected local groups. The ebb and flow between JDAG's grass roots and its organizational center is one of the most intriguing features of this coalition. This pattern of mutual reenforcement requires delicate skills on the part of JDAG leadership and staff as they consult with constituencies before launching what otherwise might be seen as trigger-happy action.

An Assessment of Achievements

At a distance, it is difficult to make an accurate judgment of the achievements of JDAG as an instrument of citizen involvement in reviving an economically declining area. When asked to cite their achievements as an organization, some JDAG leaders mentioned the fact that two of the docks scheduled for closing have been kept operating, that the 1973 plans were discredited, and that the new planning procedures make serious efforts to involve residents as a key constituency. It might be added that JDAG and the local action groups have managed to make themselves heard in the halls of local and metropolitan government, but especially with the JDC, by utilizing a variety of approaches that include demonstrations, careful research resulting in alternative proposals, and continuing and steady pressure upon decision makers.

Whether JDAG's voice is loud enough to alter fundamental planning priorities is, of course, another matter. The Tory government may turn a deaf ear to the pleadings against an urban development corporation for the area. Indeed, such a Corporation may stop the financial support of JDAG altogether. It appears unlikely that even under friendlier political circumstances the present plans for expansion of office facilities can be completely traded off for new opportunities for manufacturing jobs. Or, for another example of JDAG's still incremental decision making power for its area, it is generally agreed that the plans for the Southern Relief Road were scrapped because of lack of money (the road's price tag had been estimated at $500 million) and only partially because of JDAG's opposition.

Figure 5. Coventry. The fight against demolition and for on-site renovation by existing residents is highlighted by a simple mural.

The fate of JDAG is perhaps typical of multiissue grassroots coalitions on both sides of the Atlantic. Coalitions attempt to pull together existing organizations into a well-oiled fighting machinery. That objective may work for one or two highly visible issues (the defeat of the 1973 plan and the proposal of alternatives), but the organizational going may become tougher once the issues become larger, more subtle, and less comprehensible. Once the primary, superordinate issue that unites constituent groups becomes faded, there seems to be a tendency to retreat to the original "primary" organizations. And the Docklands are full of organizations that can satisfy limited needs of their membership, including trade unions, churches, the Labour party and various splinters on the political left, tenant organizations, and many others.

JDAG's financial dependence on governmental funds—especially now, when those funds must flow through the Thatcher administration—makes the organization's future less than secure. The budget seems assured at least through March 1981; by that time, it is likely that the Docklands Joint Committee will have been disbanded and a new urban development corporation put in its place to manage the planning and development process. The enterprise zone for the Isle of Dogs may also be authorized by that time. In such an event, JDAG must choose from among a number of policies and strategies for its own continuating; the following two seem to be two polarized alternatives: (1) JDAG could assume leadership in a citizen participation effort that collaborates with and is funded by the urban development corporation. Such a government–citizen partnership endeavor would preserve citizen input to the plans and provide an arena for mutual discussions, limited though these activities may be. (2) JDAG could deliberately divorce itself from involvement with the new corporation, intensify political and educational efforts against it, and hope that after suffering through some lean financial months the Labour party may gain some victories (such as taking over the Greater London Council) that might eventually lead to renewed governmental funding for some JDAG projects.

Of course, JDAG might seek a third method to deal with its financial uncertainties, and that is to follow the lead of the Coventry Workshop, to which we now turn, which deliberately avoids governmental funding.

The Coventry Workshop

The Coventry Workshop bills itself as "a local research and advisory unit for trade unionists and community organizations." Its activities cover a number of areas of vital interest to workers and community activists, including deindustrialization, unemployment, new technology, industrial democracy, occupational and environmental health, public services, and housing. It is entirely beyond the scope of the present case description to do justice to all these facets. Therefore, we have concentrated our attention on describing the workshop generally and specifically looking at its work with the people who are on the bottom of the housing pile. As was the case with the London–Docklands case, the Coventry Workshop illustrates how research can be integrally tied to community action and the building of a coalition between trade union and resident groups to help victims of changing neighborhoods and changing economic conditions.

The Setting of the Coventry Workshop

The workshop is located in one of a number of unimposing row houses that front Binley Road. The first room that the visitor enters on the ground floor is the library, which is crammed with books, reports, periodicals, newsletters, photographs, maps, and newspaper clippings. Bulletin boards display posters and graphics that were used in community meetings. A student volunteer is cataloging new items. A sheet of newsprint tacked to the wall contains the workshop's current budget, with contributing foundations and organizations listed by name. Back copies of the bimonthly *Coventry Workshop Bulletin* are available;

leafing through the latest issue, one notices brief reviews of three newly acquired books: *The New Technology, Jobs and Community Action,* and *Demystifying Social Statistics*.

The visitor is ushered into the large front room on the second floor. Instead of one staff member as expected, there are three, aged approximately 25 to 40. All three consider the questions posed; observations during a subsequent staff meeting reinforce the first impression of shared leadership and mutual trust in the staff. Conditions of the poor in Coventry are described with low-profile intensity—a combination of controlled anger and unpretentious authority.

Coventry itself has long been a major center of the metal-based engineering industry in Britain. It can be compared with Detroit, since both cities are heavily engaged in automobile and related manufacturing enterprises and both are strong union towns. Because of Coventry's key industrial role, it was a target for bombings during World War II.

A raid on November 14, 1940, destroyed much of the city, including its historic cathedral. The effect on the housing stock was severe. But like the cathedral that literally was rebuilt out of—and around—the ruins of the past, the houses and the industries were reconstructed in a major program of municipal planning and renewal.

In the mid-1960s another crisis hit Coventry: manufacturing jobs decreased, and they continue to decline through the 1970s. With 60% of the city's jobs in the manufacturing sector and 75% of these controlled by only 14 giant firms, the city was especially vulnerable to corporations—often transnational ones—that merge, move, introduce labor-displacing new technology, or otherwise reduce their work forces. From 1965 to 1975, jobs in the automobile and other vehicular industries in Coventry dropped from 226,000 to 188,000; in electrical engineering, from 122,000 to 104,000; and in mechanical engineering, from 144,000 to 129,000. While Coventry's unemployment rate a decade or more ago was far less than Britain's average, this favored position was no longer evident in the 1970s. By 1980, unemployment had risen to a postwar high of 8%.

Along with unemployment, housing for moderate- and low-income families became increasingly problematic. Although Britain did not impose a Nixon-type housing assistance moratorium, government assistance for housing was cut in the mid-1970s. The workshop's progress report for the years 1976 to 1977 states that "the impact of this . . . on the housing situation of working people has been dramatic—with rising costs, deteriorating conditions, and diminishing choices."

This is the context of economic decline that is facing the workshop. The auto workers and machinists in Coventry, like their American colleagues in Detroit, appeared to prosper with the postwar boom in the auto industry; but the precariousness of their position has now become clear, as firms attempt to protect their profits by cutting the costs of labor and increasing their control over production. There was little to cheer about among the unionists and poor people during the time of our visit in the summer of 1979.

Workshop Origins and Structure

Just as many American neighborhood and community development groups can trace their roots to the antipoverty and Model Cities programs, so the Coventry Workshop's origins go back to the British equivalent of these U.S. programs. In 1968 the British set up their Urban Aid Programme, and in 1969 they started the national Community Development Project (CDP) as a "neighborhood-based experiment aimed at finding new ways of meeting the needs of people living in areas of high social deprivation." A Coventry neighborhood was chosen as a pilot CDP and staff was employed in 1970 to engage residents in community improvement activities.

From 1970 until 1975 the CDP staff worked under the directorship of John Benington to bring about changes for the disadvantaged in Coventry. They started by utilizing the "traditional" strategy of antipoverty efforts (similar to and in some respects copied from the U.S. War on Poverty) of improving service delivery mechanisms, helping to bring about better communications between indigenous people and decision makers, fostering self-help efforts, and so forth. They worked with the people themselves, the service providers, and the government on wide-ranging issues of income maintainance, housing, adult education, "the transition from school life to work life," the problems of the elderly, social planning, and so on. However, engaging in this antipoverty litany brought no salvation. The staff discovered, for example, that efforts at "improving communications" between target-area residents and the decision makers frequently failed because the messages were being misread or simply ignored.

John Benington and the staff tried to find the reasons for such failures. In 1975 they produced the CDP Final Report, in which they argued that first of all, the very definition and analysis of a poverty area is usually faulty. Poor neighborhoods, they said, should not be viewed as receptacles of pathologies that might be improved through individually oriented services nor even as manifestations of maldistribution of resources. Instead, they insisted that the "underdeveloped area is functionally related to the political economy of Coventry as a whole." It was reasoned that such areas serve as a "buffer against market uncertainties" by providing "a reservoir of labour for industry, a valve responding to fluctuations in public investment, a fluid pool of land." Thus the prosperity of some areas and classes was bought directly at the expense of these poorer areas, which therefore performed "necessary" functions within the city. "The problems of poverty neighborhoods cannot be dealt with by tinkering with technical procedures; they demand new kinds of political intervention." And where might the most strategic leverage point be for such intervention? The CDP Report continues at a later point, "The greatest potential for change may lie in new initiatives which create alliances across the neighborhood, the factory floor and the local political parties." Such a coalition might be able to convince the Coventry City Council to use its "power to intervene to protect the rights of . . . citizens, and . . . to act politically to challenge other institutions and forces affecting the welfare of the local community."

In 1975, at the end of the 5-year government project, John Benington joined with two other community workers to found the Coventry Workshop. Their experiences convinced them that only an independent and autonomous organization, not tied organizationally to national funds or to local government, could fundamentally address the issues of poverty and deprivation in Coventry. They wanted to bring into being an organization that would collaborate closely with both resident groups and labor groups; a coalition between the two was and remains one of their firm objectives.

Thus the Coventry Workshop and its predecessor organization have had the advantage of working under the leadership of an unbroken chain of key staff members for 10 years. Moreover, workshop operations appear to continue in a nondoctrinaire and open fashion. The workshop engages in many and long discussions about causes of underdevelopment, goals for workshop operations, and appropriate strategies to be adopted. The organization tries with rare persistence and thoroughness to learn from its past experiences. It possesses what might be called a reflective organizational style that does not dull its action capabilities but tempers "ad hocism" or attempts at seeking premature refuge in doctrinaire solutions. This may be one of the ways in which the workshop differs from the short-range pragmatism of a number of American community action organizations and even some technical assistance groups.

For the first four years of its existence, the workshop was essentially run as an independent collective of five full-time staff members. These five people deliberately, earnestly, and with obvious devotion to their task hammered out strategies and programs. They made themselves accountable to the groups of workers and residents with whom they worked on specific campaigns but decided that during this pilot stage they as staff had to take responsibility for the overall structure and financing of the organization. Their aim was to avoid token structures of community control until sufficient groups had concrete experience in workshop endeavors to enable them to assume *effective* control of the organization. After four years, the staff invited a wide spectrum of individuals from local labor and community groups to act as representatives who would make the transition to full community control. The first Annual General Meeting was held in mid-1979 and a constitution was adopted. This constitution proclaims the aims of the Workshop as follows:

> To support workers, the unemployed, tenants and residents, and their organizations, in their efforts to gain control collectively over their lives, and to understand the forces which deny them this control.

> To assist in examining issues faced by labor and community organizations in the workplace, at home and in the community; to investigate the relationship between these issues, and to help overcome existing divisions between trade union, labor, and community organizations.
>
> To make relevant information and skills available to workers' and community organizations in ways which will be understandable and useful to their membership as a whole.

Six weeks after adopting the formalized structure, the workshop had 30 organizational members (annual dues £5 to £50 "depending on size and financial status"). Individuals—including skilled workers, professionals, and academicians—can also join as associate members. The budget of the workshop is still modest, most of the £28,000 receipts coming from foundations. The five "core" staff members each receive the identical salary of £3,504, the six other staff members being supported by temporary CETA-type governmental funds. (A $7,000 salary does not buy much more in Britain than it does in the United States, and professional work at the workshop is obviously not entered into for financial gain.)

Issues and Activities

The Coventry Workshop conducts its work in a number of ways:

1. Research and investigation is first on its list. In the words of a workshop publication, "[the workshop] collects information, and analyzes trends to help workers' and community groups to investigate the facts about their industry, workplace, or neighborhood; and how economic and political decisions may affect their jobs, housing and the whole country." This function of data collection and analysis is apparent in all phases of the organizational life: the workshop's dozens of publications have dissected such issues as *Chrysler's Crises: The Workers' Answer, Unemployment and the Multi-Nationals in Coventry,* and *Making Urban Renewal Decisions*. Charts and land-use plans hang on office walls; projects are carefully planned and documented.
2. Technical advice and assistance to community and labor groups are, of course, key activities. This includes help from volunteer lawyers, accountants, planners, public health specialists, and so on. The workshop takes pains to point out that such advisors "do not come in as 'outside experts' but . . . as committed advisors with the groups concerned."
3. Organization and action would include assistance to groups in organizing and programming their activities.
4. Informal education, which may mean leading discussions, giving short courses or other forms of adult education.

5. Library and information services.
6. Bimonthly bulletins.
7. Duplicating, typing, and printing services for affiliated groups.

The issues with which the workshop concerns itself are thus not only the here-and-now manifestations of a social problem but also the more permanent solutions to it. This approach to issues is well illustrated when one looks at the workshop's efforts in the field of housing. On the one hand, the workshop continues to work with separate tenant groups and even individuals who present problems that have to be immediately addressed. For example, the workshop has been actively working with a resident organization fighting to save their homes from demolition, helping them to draw up alternative plans for the area. It has also helped residents take the city and private landlords to court for failure to observe health codes and assisted a neighborhood group in the time-consuming steps in rehabilitating their row houses, including helping to design the project, steering residents to appropriate city authorities, examining funding issues, strengthening the resident organization itself, and so on.

On the other hand, the workshop now deliberately emphasizes another, less parochial, and more permanent approach to improving the status of temporary tenants. The workshop decided to go beyond giving assistance to the separate neighborhood and tenant groups and to "build a constituency not just from a series of street-based groups with parallel campaigns, but around a city-wide organization which can reflect the housing interests of working-class people as a whole in Coventry" and to "put the housing issue on the agenda not only of the local Council, but also of the local trade union movement." The thrust of this strategy, obviously, is to better marshall the political muscle to change citywide (and eventually perhaps even nationwide) housing policies and practices. The workshop now tries to help the groups it deals with to change from being *reactive* to narrowly perceived threats to becoming *proactive* in a larger campaign.

One of Coventry's most flagrant housing problems is represented by the 130 families who were homeless and often in rent arrears who are occupying "temporary council housing." This housing stock is of very poor quality, as even the local officials will admit. But, the municipal authorities will point out, such structures are only temporary, short-life properties, serving as a stop gap until standard housing can be located. However, as frequently happens, the "temporary" accommodations become permanent when there is a shortage of alternative housing. Tenants have resided there for weeks, months, and even years.

The Coventry Workshop investigated the situation and decided to help organize a Coventry Temporary Council Tenants Association. It was not an easy organizing task, since these tenants lived in a number of scattered sites and often felt themselves to be vulnerable to official retaliation because several had not been able to pay their rents. The Coventry Temporary Council Tenants Association organized meetings where tenants could share complaints and agree on action strategies; they submitted written grievances; they organized collective actions such as sit-ins at the office of housing authority; they demonstrated at a city council meeting; and they collected signatures for a petition. With the workshop's help, the association went on the offensive. The voice of the new organization in meetings and in their newsletter was strong and assertive. An Alinsky-type manner of poking fun at the opposition can be discerned in items such as the following song, reproduced from the mimeographed newsletter and sung to the tune of "Daisy, Daisy":

> Sawdon, Sawdon, give us your answer do,
> We're not fooling, we can see through you.
> Injunction you can issue,
> We'll use them for bum-tissue.
> And you'd look sweet, upon the seat,
> of a typical Part III loo.

(Sawdon refers to the chairman of the city council's housing committee; Part III is a section of the National Assistance Act of 1948 which provides temporary accommodations for homeless persons).

One of the workshop's major achievements was to bring organized labor into the struggle for upgrading temporary council housing. This was accomplished in 1978 when, following discussions within thc Coventry Trades Council (the equivalent of American central labor councils), a housing subcommittee was created. Perhaps the time of the late 1970s was particularly propitious for the labor movement to recognize its vital interests in the community, beyond the ever-present concerns of the shop; the Labour party's hold over the electorate appeared to be waning, jobs were down, and inflationary pressures were pinching the working person's pocketbook. But in addition to these external factors, the Coventry Workshop clearly influenced the trades council from within, as workshop-affiliated delegates played substantial roles in coming to this decision.

The housing subcommittee was composed of representatives of various trade unions and of tenant organizations. The relationship between the two groups on the subcommittee was to work together, on an equal-status basis, to solve housing problems. "It is not, I must emphasize," said the trades council's president, "a one-way process: the Trades Council coming down from on high, with the assistance to tenants." Whether the commitment of all the unions in the city to the subcommittee is really enthusiastic can be debated; but the structure for beginning collaboration is in place.

One of the subcommittee's initial activities was to find out first hand what the temporary council houses looked like in one site. One of the trade union visitors described her reaction as follows:

> The first home I visited I could only describe as disgusting. It was infested with rats, it stunk of dampness, the wallpaper was coming away from the walls and the plaster, the wiring was dangerous and the windows and doors were neither safe nor weatherproof.

These visits made a deep impression on the participating union members. The subcommittee issued a report and urged the abolition of this type of council housing. The city council and the local press questioned the validity of the observations, whereupon the subcommittee decided to launch a survey of all temporary housing tenants, with a group of 50 volunteers interviewing over 100 persons. A majority of the dwellings examined in this survey, not unexpectedly, were reported to have "extensive defects and lacked basic amenities such as bath, hot water, or inside toilet." Aside from the well-publicized report, the effect of the survey tended to underline the severity of the problem for a number of trade unionists who now were able to personalize the issue and also helped the temporary tenants to strengthen their determination to organize and change their own conditions.

There followed considerable public information about the housing issue (including photographs taken by a number of interviewers), petitions, resolutions, demonstrations, and related political activity. The trades council and the Coventry Temporary Council Tenants Association cosponsored a number of these events. One of the more immediate results of labor's concern with the issue is that the building trades members who were recently asked to do some minor repair work in temporary council housing complained to the housing management that their assignment was to perform a substandard and inferior job. No outright refusal to undertake such substandard work has, however, yet been reported. The city council, for its part, has to date authorized the above repairs and has responded with a survey of its own.

In these activities concerning temporary council housing, the workshop played a key role. Staff members assisted in calling meetings, helped organize visits, provided information, made suggestions for strategy, and provided meeting space. This assistance, however, was never up front; the workshop was careful to play an enabling role in which it shoved here and tugged there and made sure that the credit—and the final responsibility—belonged to either the organized tenants and/or the housing subcommittee.

During a recent meeting of the subcommittee, for example, over 30 persons were seated in the second floor room of the workshop's Binley Road center. The staff member who had been the principal contact with the subcommittee sat on a table against a side wall. Tenants and union members discussed a list of actions and demands that were being contemplated. A written draft statement covered issues from "new houses, repairs, sale/purchase of land, sale of Council houses, finance, rents, arrears" to "eviction and public spending cuts." The staff member had an important hand in culling these points from the previous meeting and in helping the subcommittee's leadership commit them to paper. Vigorous discussions ensued in which a temporary housing lady tenant was probably the most persuasive voice. During the two-hour-long meeting, the workshop staff member spoke only infrequently, usually to offer some clarification or to make a suggestion as to possible strategies.

An Assessment of Achievements

One way to approach the question of achievements of the Coventry Workshop is to speculate what would have happened in the city if there were *no* workshop. Perhaps there would not have been any major differences in the politics of the city council or in overall unemployment figures or even in the conditions of most public housing. Principally missing would be (1) a new sense of awareness about the conditions in Coventry that is backed by research data and that appears to be shared by still small pockets in the left-of-center community leadership and (2) the beginnings of an organizational infrastructure to address some of these issues from an alternative perspective. The temporary tenants, for example, would probably not be organized, the labor unions would not be as aware of community-related issues, the city council members and the city bureaucracy would be less constrained to conduct "business as usual," the bulldozer would have flattened more houses that instead are now being rehabilitated, and a heterogeneous cadre of people—trade unionists, students, some professionals, and some intellectuals—might be off on their separate agendas. And, who knows, perhaps the lack of a workshop, with its infusion of research and deliberate rationality, might have resulted in more municipal strife and vindictiveness.

To draw a balance sheet about positive achievements and observable failures is difficult. In addition to the above, here are some first impressions. On the credit side of such a balance sheet must be the research and technical support given by the workshop to trade unions in their efforts to question official redundancy figures and to save jobs in local industries, the delineation of alternative strategies to save future jobs, and the bringing together of coalitions of people and interests around common issues. Surely the attempts to link labor and community groups have succeeded in a few limited dimensions, principally around the housing issue. But in other ways, too, the workshop seems to have achieved a measure of legitimacy with organized labor. "Shop stewards work jointly with us now on long-term projects, and some of the full-time trade union officers come to obtain the answers to some technical questions," is the comment of one workshop staffer. At the very least, the workshop has impressed upon the trade union movement and community groups that the struggles in which they are engaged often have the same root causes and can be considerably aided by information unearthed by research.

Turning to housing specifically, workshop actions have probably saved some residences from premature destruction, helped organize not only the temporary tenants association but also neighborhood residents, and brought about the collaboration of an important element in the labor movement. The workshop has also helped to etch bad housing conditions indelibly in the minds of organized labor, the bureaucracy of the city, and possibly even of the general public. A middle-management city official said that most people at his level of municipal government believe the workshop to be a valuable community resource. "From a purely bureaucratic point of view, it's much easier to deal with one group that has professional support than with dozens," he said. But, he added, this view is not shared by the higher-ups in the council, who tend to see the workshop as an irritant.

Among the problems, shortcomings, and failures of the workshop, the following appear to be noteworthy: one persistent problem is money. At the beginning, the workshop was reluctant to accept government funding that might compromise its independent advocacy role. As a result, the workshop has been operating with meager funds and still depends a great deal on foundation grants. Workshop leaders hope that alternative funding might take the form of, say, a penny a week from every member of a trade union serviced. This is still an unrealized goal as far as the factories are concerned, but one of the tenants groups (a co-op) has committed all its members to a weekly donation to the workshop, collected automatically with the rent. In the meantime, the workshop is vulnerable financially, a condition that it shares with a number of American advocacy groups which are loathe to accept funds from the public purse. The workshop is not opposed to government funding in principle but has never been able to obtain government grants because of its uncompromising commitment to work from the bottom up, with working-class people exclusively.

A second problem is the workshop's still tenuous relationship with some reluctant elements of the labor movement. Although we have noted the links of the workshop with labor, especially at the level of the shop, workshop leaders acknowledge that a number of union officials are less than enthusiastic about the workshop and its services. A beachhead of acceptance has been established, but there appear considerable territories to be won over.

Third, despite the demonstrations, petitions, and presentations to the higher reaches of Coventry officialdom, the impact of the workshop on city policies and practices in such fields as housing is still minor. Another way of putting this is to say that the workshop has not yet generated enough political muscle to make a telling imprint on municipal policy decisions. In terms of the workshop's fundamental concern with employment and the quality of life in Coventry, the verdict must remain "small, incremental achievements, but no cigar!" No one, of course, can expect the Coventry Workshop to be the principal mover in turning around these fundamental conditions. Just influencing the level of debate about them must be viewed as a victory; it is a significant victory which the workshop has already won.

A Few Observations

Some concluding remarks, rather briefly and randomly noted, may serve to underline a number of recurring themes in these cases that have implications for citizen action in America.

1. In the face of problems that are embedded in the basic sociological,

economic, and political structures of the society, even fine-tuned community education and action by two sophisticated and dedicated groups resulted in only incremental victories. The most impressive victories were probably those that prevented the problems from becoming as bad as they might have been is unchecked. The worst options for the affected populations were eliminated. Of course, some important groundwork *has* been done that might pave the way for some significant future changes that can affect the Docklands and Coventry in a more fundamental way. But for now, these are hopes and expectations.[3] In sum, one of the bitter lessons we learned anew in Docklands and Coventry is the limitation of neighborhood citizen action in altering well-entrenched societal patterns.

2. Both cases seem to illustrate the progression in the early phases of citizen action from (a) general discontent, to (b) catalyzing and organizing this discontent into targeted protest, to (c) the creation of deliberate programs to achieve first stopgap objectives and then longer-range goals. The two groups were especially active in the second and third of these phases. In the second stage, a great deal of work went into focusing the discontent by identifying and polarizing both victim and villain. This was done through surveys, research, discussions, protest newsletters, handouts, and even songs. The targets were being frozen. (The similarity with Alinsky's methods, deliberate or not, is striking.) In the third phase, a series of planned, sequential activities took place to achieve short-range objectives. Thus one of the planned highways through parts of Docklands was stopped and alternatives proposed, and in Coventry the temporary council housing tenants were organized and achieved some concessions. The move from the second to the third stage can be called "from protest to program."

The sticky organizational wicket is to move beyond the short-range achievements and limited objectives to the larger issues. Protests and modest projects can be accomplished by a well-organized neighborhood group. However, a strategy of dealing with an entire industry (like to docking facilities or the large engineering firms) requires a broader coalition that includes, besides the indigenous residents, the active collaboration of power centers such as organized labor and also, some would argue, government.

3. Some municipal adminstrators actually seem to welcome a strong, unified citizen group that articulates its agenda and can make concrete alternative proposals. This may be true because it is easier for administrators to deal with one group that has professional competence than with many groups that may make "irrational" demands. (This very congruence of style and language between municipal and community group leadership may lead, some observers would point out, to subtle co-optation.) Another reason why public administrators may

[3] Since this case was written, there has been the encouraging development of a network of seven research-based community action groups throughout Britain. One of the network's functions is the establishment of training programs for community organizers and related personnel.

appreciate the strong indigenous organization is that it may rectify the lack of competence of some lower-level bureaucrats and thus shorten response time that might otherwise hold up government-initiated projects.

There seems to be a tendency among a number of public administrators in America to prefer citizen participation of the "single-table model" where, quite literally, governmental officials and citizen representatives exchange views and make decisions around a single conference table. The "dual-table model," on the other hand, calls on the two parties to meet separately and, once their own agendas are formulated, to gather and negotiate. If our observations are correct, the British administrators appear to be more reconciled to the dual-table model than do ours. Perhaps our elaborate mechanisms for single-table participation (via boards, groups of representatives, advisory councils, and consumer participants) is, in a number of circumstances, less productive for creating mutually beneficial plans and projects than the negotiated settlements between two distinct parties whose boundaries remain intact.

4. The leadership of both Coventry and Docklands groups were meticulously careful not to move too far beyond their participatory base. They constantly checked with their grassroots constituencies both about their full understanding of and active collective control over all aspects of the campaigns. There was no evidence of a small cadre of activists undertaking unilateral "exemplary actions" (as we noted in the Copenhagen case). Indeed, the Coventry group views itself primarily as a facilitator of action on the part of others and as collaborator in joint ventures and not so much as a first party intervenor.

The implications of this approach in various European and American contexts might well be further explored. There are, of course, a number of styles of leadership in social action, and we really know rather little about their consequences in various settings.

5. U.S. local citizen groups might well ponder the operational style of their British colleagues, especially in regard to conducting action research and making alternative plans (short-range as well as long-range) that become valuable instruments in negotiating with private and government bodies. (It is recognized that a number of American technical assistance and consumer groups have undertaken such research tasks—as, for example, the public interest research groups, Nader-type organizations, and national bodies such as the National Center for Urban Ethnic Affairs. But there are relatively few local and neighborhood groups that incorporate research, technical assistance, strategy formulation, and action under one roof.)

6. U.S. citizen action might also consider the efforts of action groups in Docklands and Coventry to fashion a coalition of like-minded groups, especially by actively attempting to create alliances with organized labor at the local union and shop levels.

7. The financial maintenance of community action groups is, as we have noted in both cases, a perplexing problem in Britain, as it is on this side of the

Atlantic. It would be a major breakthrough if the Coventry Workshop could convince trade unions to make or negotiate for contributions on a regular worker/day check-off basis. Organizational self-sufficiency will depend, of course, on reliable sources of funds with as few strings attached as possible.

In March 1975 the founders of the Coventry Workshop wrote, "The greatest potential for change may lie in new initiatives which create alliances across the neighborhood, the factory floor and the local political parties. This would demand radical changes in stance in all these areas." Both organizations we have described are struggling to implement this goal. In the process, their own organizational autonomy is being severely tested and they are confronting, as is to be expected, overwhelming obstacles.

ACKNOWLEDGMENTS

We gratefully acknowledge the assistance of many people who helped gather information in Britain. Foremost among them are the staff members of the Joint Docklands Action Group—especially Nick Sharman, its former staff director, and Heather Newham—and the Coventry Workshop, especially John Benington, Slim Hallett, and Paul Skelton.

We talked with many residents, community leaders, government officials, planners, and academics; we thank them all for answering our questions so patiently and for giving so generously of their time. We wish to acknowledge especially the following: B. T. Buckle of the Docklands Development Organization and Nigel Beard who formerly directed the Docklands Development Team, Richard Kirwan and Peter Willmott of the Centre for Environmental Studies, Ted Johns of the Association of Island Communities, Timothy Hornsby and John Peterson of the Department of the Environment, John Haw of the Coventry Department of Architecture and Planning, and Elizabeth George and Foster Murphy of the National Council of Social Service.

5

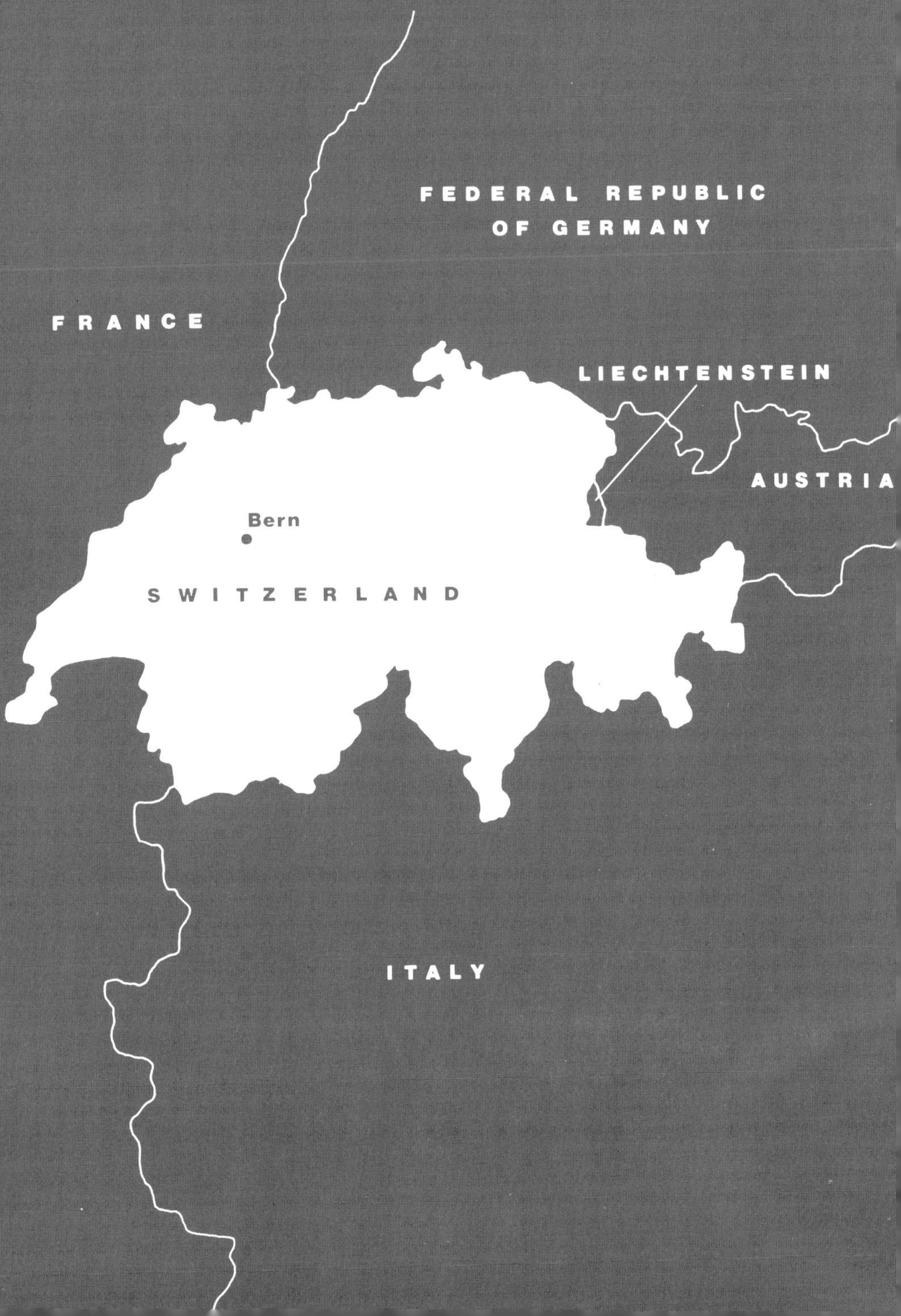
FEDERAL REPUBLIC OF GERMANY
FRANCE
LIECHTENSTEIN
AUSTRIA
Bern
SWITZERLAND
ITALY

5

THE VOTER INITIATIVE AS A FORM OF CITIZEN PARTICIPATION IN SWISS TRANSPORTATION POLICY

CHESTER HARTMAN

With few exceptions (notably bond issues to finance capital expenditures for public transit), transportation questions almost never come before the electorate in the United States. Yet the potential for a significant form of broad-based, innovative citizen participation in transportation policy—namely, the voter initiative—exists in a substantial portion of the United States. Thirty-two states (plus Guam) currently have initiative provisions for local and/or state levels of government.[1] While specific provisions vary from state to state, in general signatures of anywhere from 3 to 5% of the electorate voting in the previous election (sometimes with geographic distribution requirements) can place most any matter on the ballot, which then becomes law if passed by the voters.

In Switzerland, where techniques of direct democracy in the electoral arena are well developed, the initiative process is often used at the national, cantonal, and local levels to deal with transportation issues. Some of the outstanding recent examples include the following:

[1] States that have initiative provisions for state as well as local governments are Alaska, Arizona, Arkansas, California, Colorado, Idaho, Maine, Massachusetts, Michigan, Missouri, Montana, Nebraska, Nevada, North Dakota, Ohio, Oklahoma, Oregon, South Dakota, Utah, Washington, and Wyoming. States in which the initiative provision is available only to local governments are Georgia, Kentucky, Louisiana, Minnesota, New Jersey, Pennsylvania, South Carolina, Texas, Vermont, Virginia, and West Virginia. [*Book of the States, 1978–1979* (Lexington, Ky.: Council of State Governments, 1978), p. 243.] These 32 states contain over three-fifths of the country's total population. While, unlike Switzerland, the United States has no provision for a national initiative, it is important to remember that the entire Swiss nation is only about one-third the size of Pennsylvania and has a population roughly equal to that of Massachusetts.

1. The Albatross initiative, mounted by environmental groups, which sought to set stricter emission standards (at roughly the level now in effect in the United States, Japan, and Sweden) for automobiles in order to deal with the country's growing air pollution problem. (It lost in the September 1977 national elections by a 39% to 61% margin.)
2. The initiative for democracy in the construction of the national highways, led by a Nader-type environmentalist named Franz Weber, which sought to reconsider Switzerland's 20-year-old autobahn plan and subject still unbuilt segments to voter approval. (It lost in the February 1978 national elections by a 39% to 61% margin.)
3. The so-called Burgdorf initiative (named after the technical college whose students originated and carried out the campaign), which sought to ban private automobile, boat, and plane traffic on one Sunday each month. (It lost in the May 1978 elections by a 36% to 64% margin.)
4. The Gratistram initiative in the city of Basel, which was developed by a far left political party and sought to institute free public transit. (It lost in the June 1972 municipal elections by a 7 to 1 margin.)

While on the surface not exactly a string of success stories, these instances of voter involvement in the development of transportation policy suggest, upon closer examination, an important means of advancing citizen participation in a critical area of urban services.

Swiss Use of the Initiative

According to one student of the Swiss use of the initiative, it is a device for consciously politicizing a problem that has been neglected by the political system.[2] In recent times Switzerland has seen an increase in the number of voter initiatives, which have been introduced by underrepresented groups in Swiss society as an expression of unresolved social problems and political tensions. During the 1960s and 1970s, a period of rapid economic expansion in Switzerland, the initiative was used primarily by ad hoc groups to deal with problems generated by growth: social security, infrastructure, environment. The initiative is a way of making the political system take note and deal with problems. By and large, the groups that have been most inclined to turn to the initiative have been those that are too nebulous, large, and heterogeneous to develop into formal organizations in the highly organized country of Switzerland. Most notable examples of such interest groups are consumers, renters, and users of public services—all interest groups outside the production system. The initiative serves to push issues into the forefront of the political arena, often by dramatically putting forth a nonincremental proposal. It is relatively easy to get an initiative on the ballot; the signature requirement for a national initiative until recently

[2] Hans Werder, *Die Bedeutung der Volksinitiative in der Nachkriegszeit* (Bern: Francke Verlag, 1978), and interview with Hans Werder, September 1979.

was 50,000, but the increase in such initiatives has now led to doubling of the signature requirements; cantonal and local requirements are roughly proportional. Since the Swiss system provides a three- to four-year period for government reaction between the time the initiative qualifies and the time it must go on the ballot, an issue is forcefully thrown into the limelight for a protracted time. During this period, government response,[3] publicity, and organizing both within and outside the extensive Swiss party system all occur.

The Burgdorf Initiative

The initiative this case study focuses on is that which proposed "auto-free Sundays" in 1978. Its immediate origins were the three auto-free Sundays of November and December 1973, immediately following the Arabian oil crisis, in Switzerland and other western European countries. Its general context was the increasing public awareness of what unfettered use of the automobile was doing to Switzerland. The center of the effort was a group of students (originally about a dozen in number) at the Technikum Burgdorf, an engineering college. (That the idea and leadership throughout the campaign came from an ad hoc group totally independent of any of the political parties is typical of recent environment-oriented initiatives in Switzerland.)

The initiative was very simple. The 65-word text read roughly as follows:

1. On the second Sunday of each month throughout Switzerland all private motorized traffic is prohibited, on land, water and in the air, from Sunday 3 A.M. until Monday 3 A.M.
2. The Bundesrat [federal council] shall develop rules and regulations providing for exceptions to this prohibition.

Other than necessary exceptions (which might be for emergency vehicles, transport, border and other kinds of workers who must work on Sundays, farmers feeding their stock etc.[4]), all movement would have to be by nonmotorized

[3] Under the Swiss system, the government can respond in one of several ways: take part-way actions in the direction of satisfying the goals of the initiative, thereby undercutting or co-opting the effort and often leading to a withdrawal of the initiative after it has qualified; it can put on the ballot a formal counterproposal with similar co-optative aims, or, more cynically, merely to undercut the initiative's chances of getting the majority vote it needs to pass (since voters are then faced with three choices: for the initiative, for the counterproposal, or rejection of both); or it can make no response to the initiative and merely seek to defeat it.

[4] Bundesrat regulations governing the three auto-free Sundays in 1973 made the following exceptions: (1) total exemption for most taxi service, emergency vehicles, invalids and people traveling to and from work; (2) nonpostponable official and professional work of the army, police, customs officials, doctors, veterinarians and other medical personnel, ministers, press and other media, farmers, private guards, funeral officials, and certain functions related to repairs and the care and distribution of essential goods. Needless to say, the question of allowable exceptions to the general ban is key, not only to political support and workability but to the overall impact of the ban in producing its desired effects of carlessness.

means (walking, bicycling, skiing, riding) or by public transportation. It should be noted that Switzerland has a highly reliable and extensive (although not inexpensive) national public transportation network in its railroad and bus system and, within urban areas, in its trams and buses. It should also be noted that the Swiss like to walk and that the bicycle is a standard means by which adults and children get around: 1.7 million people own bikes (compared with 2.2 million cars) in a country of 6.3 million people.

Motivations and aims for the dramatic monthly car ban were several. Most prominent were (in random order): General environmental and health grounds (less air pollution, noise, traffic accidents, and other environmental degradations); energy savings (particularly in a country with no domestic oil deposits); reduction of reliance on the automobile, increasing reliance on public transportation; *Umdenken*—an opportunity to rethink things: what one does with free time; how one gets around; to experience a life and environment independent of the private automobile and gain a new kind of freedom in abstaining from something that has been over-identified with freedom. Such social experimentation was, of course, related to the energy crisis and the possibility of necessary limits on free use of the automobile in the future.

Campaign Issues

The signature campaign to put the auto-free Sundays initiative on the ballot began in February 1974. The outreach campaign was extremely simple: small ads, asking people to sign the initiative and help circulate petitions, were placed in papers around the country. With neither a preexisting organizational base nor a tie-in to any other organizational network, signatures were collected in 10 months; by June 1975, the 115,000 signatures were verified, qualifying the initiative for the ballot.

Interestingly, most polls in 1976 and 1977 showed the initiative winning by a fair margin—58 to 61% for it, 26 to 38% against it—in three polls done between September 1976 and April 1977. Support was greater among women, lower-income people, older people, and, of course, among those who did not own autos (although in all three polls a small majority of auto owners favored the initiative). In general, those least reliant on the auto or most oppressed by it were more favorably inclined. But in a country with 72% auto ownership, such widespread support was remarkable.

Opposition—which included the government, most newspapers, and those well-organized economic interests associated with auto use—concentrated on several dominant themes (also in random order): Violation of personal freedom by government prohibition (in a country where "freedom" is something of a national battle cry); negative impact on tourism—an activity central to the Swiss economy (images of border signs saying "Sorry, Switzerland is closed on Sun-

Figure 1. Illustrations from pro-Burgdorf Initiative literature.

day" were evoked in the anticampaign[5])—with accompanying warnings of possible retaliatory measures by bordering countries; other, less drastic solutions exist to deal with the problems caused by the auto (emission controls, mandatory usage of seat belts, lower speed limits, improved public transportation, and so on).

The level of debate on the issue was intense, touching on philosophical, economic, and even religious issues (the "pro" side, in mailings signed by ministers, urged keeping the sabbath holy; the "con" side predicted that church attendance would drop if people could not drive on Sunday mornings). Effects on family life were debated: on the one side it was claimed that carless Sundays would mean time for restoration and home life; on the other side were claims that, without cars, weekends would become boring, leading to family tensions. A feminist perspective on the plan was put forth in a piece of campaign literature produced by the Burgdorf group. It held that the automobile is essentially a man's vehicle in terms of who owns and drives cars: because cars make streets dangerous play spaces for children, women must devote more of their time and attention to child care; the woman in the family generally sits in the "death seat" while the man drives; the stress of keeping the children pacified during an auto trip falls on the woman.

The economic arguments against the initiative related not only to tourism but to loss of government revenues from taxes derived from auto use. Proponents of the initiative countered that arguments of that sort are like supporting wars to keep soldiers employed, and that Switzerland was in danger of ruining its attractiveness to tourists through an excess of tourism. They also downplayed the negative effects on tourism and suggested that, in fact, the novelty of an auto-free environment, rather than discouraging tourism, might make Switzerland an even more attractive tourist spot, especially to visitors from cities and countries where the effects of the automobile are even more deleterious. Proponents also pushed the image of Switzerland as a pioneer in social innovation (rather than the laughingstock of Europe, as opponents suggested, with their "closed on Sundays" images[6]), and argued that auto-borne tourists who merely wanted to pass through Switzerland on Sundays could use auto trains.

Another concern was expressed for those living in rural and mountain areas, who would be far more isolated and disadvantaged by such a ban than urban dwellers because of poorer public transportation services and fewer social and cultural activities nearby (and who at the same time might suffer most from any loss of tourist business).

The issue of social class and privilege was raised: if special permits were issued to exempt people from the ban, those with influence would find a way

[5] The various arguments against the initiative lent themselves to easy parody. Figures 2 and 3 show a sampling of the cartoons used in the opponents' literature.

[6] An article in the May 5, 1978, *International Herald-Tribune* quotes a spokesman for Switzerland's largest automobile association as warning that if the Burgdorf initiative passed, the Swiss "would become the village idiots of Europe."

to get them; the poor needed escape from the city more than the rich; the mobility of the rich would be unhampered, since they could better rearrange work schedules to avoid private auto travel during the interdicted hours. Proponents of the initiative commented that this concern for the poor on the part of those making such arguments was outright cynical, since the economic and political interests putting forth these reasons rarely exhibited such sentiments at other times.

The role of public transportation in Swiss life also came sharply to the fore. During the three consecutive carless Sundays in 1973, the public transportation system performed well: surveys showed that the two- to fourfold patronage increase in highly settled areas was easily handled by increasing frequency of service; in rural areas, there was an insignificant increase in use of public transportation. The government argued, however, that the 1973 experience was of limited relevance, since it came at a time of the year (late fall) when people travel less than during the summer months. Proponents of the Burgdorf initiative themselves surveyed public transportation officials of the national and local systems and reported uniformly optimistic response about the systems' capacity to handle extra traffic on carless Sundays; in fact, many public transport officials were eager to have the new business on days when equipment needed for weekday rush-hour traffic normally is idle, thereby reducing the system's deficit. Burgdorf initiative leaders also suggested that, as a way of improving public transportation services in rural areas, various privately owned communal transportation systems—used primarily to get people to and from work—could be incorporated into the public system on carless Sundays to provide scheduled routes in poorly served parts of the country. The general argument was also put forth that an increased demand on the public transportation system was welcome, since it would create a need for greater investment in improving that service and maintaining its currently high standards.

But of all issues raised by the Burgdorf initiative, the social–philosophical one of *Umdenken* was probably the crux. The "leap of faith" into imagined possibilities and benefits was what was most attractive to proponents, while to opponents it was anxiety-provoking or unconvincing. There were eloquent statements that "abstinence can lead to greater freedom" and that "a freedom which hangs on four gas-powered wheels is a questionable freedom." It was argued that the automobile had total freedom 365 days of the year; all that was sought was to make the score 353–12. Proponents quoted from the *Neue Zürcher Zeitung* (Switzerland's *New York Times*), following the three carless Sundays of 1973, which described "almost a type of national euphoria" as bicycles took over the autobahns and horses came trotting through downtown Zurich. Proponents pointed out that government-"imposed" limits are not imposed if the people democratically vote for such limits. Moreover, the plan not only had historic precedent[7]

[7] In addition to the 1973 precedent, proponents pointed to a 1936 day-long national "auto-strike" (organized by car owners' associations to protest an increase in the gasoline tax) and four carless Sundays in Switzerland in the late 1950s as a result of the Suez Canal crisis, the latter reportedly resulting in a 10 to 15% reduction in gasoline consumption.

Autofreier Sonntag: Schildbürger unter sich
Dimanches sans voitures: Le calme et la tranquillité...

Auto-free Sunday: Average citizens among themselves; Peace and tranquility...

Figure 2. Illustrations from anti-Burgdorf Initiative literature.

but also paralleled existing prohibitions on Sunday activity: Swiss law forbids Sunday "work" that most U.S. suburbanites regard as routine weekend chores, such as mowing the lawn and beating rugs. In the transportation area, the country already has a Sunday and nighttime ban on heavy truck traffic. Moreover, until the 1950s, by universally obeyed custom, there was one auto-free Sunday annually, the September holiday known as *Bettag* (prayer day).

The argument was posed by proponents as freedom versus freedom: while the private automobile has brought great freedom, new possibilities of contacts, trips to formerly remote places, a wider geographic choice of workplace, homesites, and feelings of adventure, it has also brought unrest, deracination, concrete deserts, pollution, noise, loss of contact, destruction of nature and cities. Some balance was being sought by regaining the things that were lost because of the auto.

The gap between individual behavior and universal social behavior was hard to overcome. People could imagine, on a personal basis, being carless (indeed, most people with cars experience short periods when they are carless because of needed repairs or other reasons), and if they so chose could even experience the personal effects of abstaining from the use of a car on one Sunday a month. The social and individual impact that would result if virtually no one used private cars on Sunday was another matter.

Another argument against the initiative was the need it might create for an elaborate policing mechanism to ensure that only those in excepted categories actually used their cars. Proponents pointed to the fact that the 1973 carless Sundays presented no policing problems; the counterargument was that the 1973 situation was different, since it was a "national emergency" of sorts and neighboring countries had similar bans. There were also arguments about the extent to which social controls and pressures would provide the necessary policing mechanism.

Yet another argument used by opponents was that a monthly ban on Sunday driving would not lessen movement at all but merely shift it either to Saturdays and Mondays or to other Sundays, thereby undercutting arguments about fuel savings, fewer accidents, and so on. Proponents argued that there was little evidence supporting such a conclusion. They held that if it was true, it undercut other economic arguments being put forth by opponents (deleterious effects on tourism, government revenue, etc.). Finally, they argued that even if there were only a shift in travel patterns, the social, health, and psychological benefits of a day without auto traffic would still be paramount. In truth, the impact on travel patterns could not be firmly predicted by either side, giving something of a tension or contradiction to the major strands of the proponents' vision: Sundays for rest and restoration, free of the need to "go someplace," versus relying on public rather than private modes of travel (i.e., the argument that people would not necessarily travel less but merely in different ways).

The campaign served to provide great debate within Switzerland about the

Figure 3. More illustrations from anti-Burgdorf Initiative literature.

role of the automobile. Newspaper coverage was extensive in a country with a great many papers, high readership, and a style of reporting more subjective and analytical than is usual in U.S. papers. Letters to the editor columns were chock full of pro and con arguments. (In the Swiss press, such forms of reader communication are generally far longer and more intelligent than in U.S. papers: a 500- to 700-word letter to the editor is not at all uncommon.)

Following the mandatory public hearings on the initiative, the Bundesrat produced its (negative) 30-page report and recommendations, giving proponents the opportunity to prepare and submit their rebuttal report.[8] A member of the Nationalrat (lower house of parliament) in 1977 submitted a proposed law, signed by 31 of his colleagues, asking the parliament to mandate at least eight auto-free Sundays. The parliament rejected this compromise proposal and decided to let the matter go on the ballot without a counterproposal.

The campaign itself was carried out very competently, a considerable achievement in light of the total lack of previous political experience on the part of the central Burgdorf group. Regional and local organizations, established originally through the signature drive, were set up throughout the country to raise money, generate publicity, and make contacts with other groups (particularly among the elderly, youth groups, environmental organizations, and other students). The usual paraphernalia of stickers, posters, and leaflets was produced; the basic campaign leaflet (in German, French, and Italian versions for the different regions of the country) was distributed to virtually every household in Switzerland—some 1 million by mail to outlying areas and 2 million hand-delivered by volunteers. About 10,000 people were active in some phase of the campaign, some 500 of these were "hard core" volunteers. The entire campaign was carried out without a paid staff. Endorsement lists were put together—in part to appeal to special groups (doctors, ministers, etc.), in part to dispel the image opponents were trying to create of naive, young, idealistic students and ecology freaks.

The Election Results

Despite poll results (from 1976 and 1977), the outcome of the May 1978 election was no surprise. The government, almost all political parties, powerful private interests, and the press as well, waged a strong campaign against the initiative. A postelection survey of initiative results, regularly undertaken by the University of Bern,[9] showed that the three dominant reasons among the nearly

[8] See *Botschaft über die Volksinitiative "für 12 motorfahrzeugfreie und motorflugzeugfreie Sonntage pro Jahr* (Bern: Mai 25, 1977) and *Bericht zur Botschaft des Bundesrates über die Volksinitiative für 12 motorfahrzeugfreie and motorflugzeugfreie Sonntage pro Jahr (Burgdorfer Initiative)*, (Burgdorf: August 1977).

[9] *Analyse der eidgenössischen Abstimmung vom 28, Mai 1978* (Schweizerische Gesellschaft für praktische Sozialforschung, Forschungszentrum für schweizerische Politik, Universität Bern).

two-thirds who voted against the initiative were (1) opposition to government decrees and the shrinking of personal freedom (31%), (2) the possible deleterious effects on tourism (22%), and (3) the feeling that better, less drastic means exist to deal with the problems of the automobile (8%). The three leading reasons among those who voted for the initiative were (1) environmental protection (63%), (2) establishing other priorities/*Umdenken* (11%), and (3) limiting private modes of transportation (4%). The initiative did worst in rural places (those most dependent on tourism and least well served by public transportation).

Voter turnout was 49%, about average for Swiss elections on issues (elections for candidates, which are held separately, average 60%). The generally low turnout in Swiss elections is deemed to be in part a response to too many elections (the Swiss vote several times a year on candidates and issues) and to the nature of the Swiss governing system, which embodies procedures for preparliamentary negotiations and review by dominant and well-organized economic interest groups (the Konkordanzprinzip). It also has a stable ruling coalition of the four major political parties (the Proporzprinzip); both of these systems tend to detract from the importance of formal democratic mechanisms such as the initiative.[10] Organized interest groups in all parts of the economy (the so-called Wirtschaftsverbände) have access to the government decision making process during the larger and more important phases of preconsultation before a law is passed and during the postpassage phase; formal democratic tools tend to seem less influential to the average Swiss voter, which leads to low turnout at the polls.

Studies of the phenomenon of voter apathy in Switzerland indicate the following leading reasons for failure to vote (in descending order of importance): (1) lack of interest in politics, (2) sense of political powerlessness, (3) lack of competence to judge issues, (4) indifference, (5) privatistic orientation, and (6) faith in the authorities.[11] Women vote less than men (it was just 12 years ago that women got the vote in Switzerland); younger people vote less than older people; and high voter turnout is correlated with high income, educational, and occupational levels. City dwellers vote less frequently than those who live in the countryside (where voting is regarded as a more traditional social obligation and where farmers, whose income is highly dependent on state policy, are very well organized).

Such voter turnout patterns clearly disfavored the Burgdorf initiative. The very people who, in the polls, showed the greatest support for the initiative were those least likely to demonstrate that support by voting.

In many ways it is remarkable that so drastic a proposal as enforced monthly abstinence from the automobile, that symbol of individualistic freedom and

[10] See Wolf Linder, Beat Hotz, and Hans Werder, *Planung in der schweizerischen Demokratie* (Bern: Verlag Paul Haupt, 1978).

[11] Leonhard Neidhart and Jean-Pierre Hobby, *Die Ursachen der gegenwärtigen Stimmabstinenz in der Schweiz*, cited in Linder *et al.*

prestige, was supported by 36% of the voters. Possibly it would have done better had fewer Sundays been involved; the Burgdorf students considered figures such as eight, six, two, and one per year but decided that the regularity and normality of the monthly rhythm was an important part of the *Umdenken* process. The campaign itself has produced—beyond the extensive public discussion and thought given to important policy issues such as greater reliance on public transportation—serious government consideration of a proposal to have one carless Sunday a year *(Bettag);* this is regarded as having some chance of passage. A further result is the establishment of a new kind of "transport club," the Verkehrs-Club der Schweiz, as a direct competitor and challenger to the two principal Swiss automobile clubs. As part of a complete account of the effects of the Burgdorf initiative, it is important to describe this new institution.

The Verkehrs-Club der Schweiz

As noted above, a major political and economic force in defeating the Burgdorf initiative and, possibly more important, in defeating the Albatross initiative (setting emission standards for cars) was the organized "automobile lobby."[12] Following the defeat of the three national transportation initiatives, a group of 10 people—architects, planners, doctors, journalists, and economists representing a wide range on the political spectrum (and most of whom, interestingly, were not centrally involved in any of the three campaigns)—began meeting to discuss development of an alternative to the major automobile clubs in Switzerland, the Touring-Club der Schweiz (TCS) and Automobil-Club der Schweiz (ACS). With 925,000 and nearly 100,000 members respectively, the two clubs are an extremely powerful element within the Swiss auto lobby. Their combined membership of over 1 million, in a country with 6.3 million, makes them a vastly stronger force, comparatively, than the AAA in the United States, which has membership (adding together all its affiliates) of about 16 million (7% of the population, compared with 16% for the two major Swiss clubs[13]). Although automobile owners join these clubs principally if not solely to secure their various services—emergency road service within Switzerland and a *Schutzbrief* (letter of protection) providing for various services in case of mishaps in other European countries (road service, fare to return home, legal and medical assistance, etc.),

[12] For a first-rate case study of the role the auto lobby played in defeating the Albatross initiative and detailed charts showing the components of the lobby and its linkages to all parts of the government, see Paul Romann and Beat Schweingruber, "Die Autolobby Schweiz," *Tages Anzeiger Magazin* (Zurich), 3 March 1979, pp. 14–25.

[13] The smaller ACS has somewhat higher membership fees than TCS and is considered a more elite club than its larger counterpart, although the two work closely together. There are also some minor auto clubs in the United States, for example, All-State, Amoco, Exxon, Gulf, Montgomery Ward, which together have only one-quarter of A.A.A.'s membership. By late 1982, three years after the data in the text were gathered, A.A.A. had grown to 21 million members. See Ron Alexander, "Auto Clubs Vie to Aid Motorists in Trouble," *New York Times*, 4 December, 1982.

the clubs also serve as a powerful political voice on behalf of their members (although it is a matter of some dispute whether in fact the positions and actions taken by the clubs represent the views and wishes of their members and whether their members are even aware of the political activities their membership fees support).[14] At any rate, the TCS and ACS vehemently and effectively advocate the rights of automobiles and automobile drivers, and the 10 people who began meeting in 1978 were interested in forming a counterforce that would speak to and for the interests of transportation users in a more holistic and balanced way.

The results of their deliberations and meetings were the Schweizerische Verkehrs-Stiftung (SVS, or Swiss Transportation Foundation), established in January 1979, and the closely related Verkehrs-Club der Schweiz (VCS), established in May of the same year. In the words of the documents establishing these two organizations, VCS is

> a club for people, not for vehicles. At last there is a transportation club, in which one is not a member as auto-driver, cyclist, pedestrian or train user. Most people daily go a short or long stretch by foot, use an auto, a train or the trolley and often also a bicycle. The VCS is not for or against a particular mode of transportation. It represents people in motion. The VCS and SVS build themselves on the basic principle of modes of transport appropriate to people and nature. They will not represent the interests of a particular group of traffic users, but rather place the person, his security, his health, and the maintenance of a natural environment at the center point.[15]

The foundation and club are openly political but also provide services parallel to those traditionally available to TCS and ACS members, and at a somewhat lower cost. VCS, in addition, appropriately makes available services to nonautomobilists, such as a *Schutzbrief,* providing such features as refunds and insurance on canceled vacation tickets, luggage insurance, accident and sickness insurance, and legal services in connection with mishaps while traveling. (VCS at present arranges with insurance underwriters for these services, whereas TCS and ACS are large enough to be self-insurers; one consequence of this has been the active support VCS has gotten from the insurance industry.) Unlike TCS and ACS, membership in VCS, without services, is available (in fact, encouraged) at a reduced rate. This policy is intended to clarify the organization's two functions and its more open division of functions compared with TCS and ACS.

The foundation and club focus on local and national activities:

[14] The American Automobile Club and its local affiliates play important political roles in this country as well. An example is the lawsuit filed in July 1980 by the Automobile Club of New York against the city's plan to ease traffic congestion by barring cars with no passengers from crossing four bridges into Manhattan during the morning rush hours. (See Glenn Fowler, "Auto Club Seeking to Bar Curb on Driver-Only Cars," *New York Times,* 24 July 1980, p. B3).

[15] These quotes are from materials prepared for the May 29, 1979, press conference announcing formation of the VCS and from the basic VCS booklet, *Der Verkehrs-Club der Schweiz (VCS) stellt sich vor: Für eine Verkehrspolitik, die den Menschen in den Mittelpunkt stellt.*

1. Lowering speed limits (the so-called 50:80:100 goal—referring to kilometers per hour maximums sought for the city, highway, and autobahn roadways, respectively). An immediate campaign was to tie the Year of the Child to speed limits, pointing out that while great progress has been made in eliminating diphtheria, scarlet fever, and polio, traffic accidents are now the leading cause of death in Switzerland in those ranging from 1 to 14 years of age and that the rate of traffic-caused deaths among children in Switzerland is 50% higher than in Sweden, which has a 50 kilometers per hour limit in urban areas.
2. Working for emission controls of the kind sought through the Albatross initiative.
3. Increasing the number of pedestrian zones and *Wohnstrassen* (residential streets where auto traffic is sharply limited) in cities.
4. Providing support for public transportation to move both goods and people. Since 1960, a total of 1,014 kilometers have been added to the national highway system in Switzerland, whereas in the 1955–1975 period 100 kilometers of railroad line were removed.
5. Development of bicycle paths and special protected bike lanes.
6. Removal of specific local problems and danger spots.

The club and foundation are located in a small town about halfway between Bern and Zurich.[16] There is a paid staff of three (the executive director of VCS was for eight years public relations director of TCS; the public relations director of VCS was formerly a reporter for the *Berner Zeitung* and public relations director for a public transport agency). The group has close ties with major environmental groups such as the World Wildlife Fund, Schweizerische Bund für Naturschutz, Schweizerische Gesellschaft für Umweltschutz, the Institut Suisse de la Vie, and the Schweizerische Energie-Stiftung, all of which have delegates on the foundation's board of directors, as does the Gruppe Burgdorf, originators of the auto-free Sundays initiative. The foundation board also includes several members of the Nationalrat (lower house of parliament). The VCS–SVS is careful to remain unaffiliated with any particular political party, although it seeks support and representation from a variety of parties.

Membership in the new Verkehrs-Club der Schweiz stood at 12,100 (as of the end of December 1979) seven months after its founding. The initial membership drive was begun with the aid of the World Wildlife Fund, which gave the VCS a starter grant of 100,000 francs and which sent brochures to its 100,000 Swiss members (the original mailing also went to several other lists, such as that of the Burgdorf Initiative). This initial mailing netted 6,000 members. A second-phase membership drive was undertaken in the fall of 1979, via newspaper ads throughout the country; this campaign brought in another 6,000 mem-

[16] For those who wish to communicate with the VCS or the SVS, their address is Bahnhofstrasse 8, 3360 Herzogenbuchsee, Switzerland; telephone 063/61 51 51.

bers. The original goal of 10,000 members by the end of 1979 was reached, and a further goal of 20,000 members by the end of 1980 was set, with a long-term goal of 100,000 members. Out if its early experience with membership drives, the group feels that it will be easier to get new members among those who are already sensitive to environmental problems than from the general public.

Beyond membership figures, VCS has already organized seven sectional groups in other parts of Switzerland. It is also thinking of expanding its member services, such as free bicycle rentals at the bicycle rental depots that already exist at nearly every Swiss railroad station. Ideas for future services will also come from polling members.

The VCS is an intriguing social experiment. It seeks to compete with the major auto clubs in the realm of practical services for members. It understands how political power operates in Switzerland and is openly seeking to develop a countervailing force to the ACS and TCS—to break their power by fighting them on their own level. How well this strategy will work is impossible to say at this early date.

The Campaign in Retrospect

Although it failed to win voter approval, the Burgdorf initiative served to highlight a major problem in Switzerland by focusing public and political attention on a dramatic proposal to begin to deal with that problem in a radical way. It can be characterized as a relatively low-cost form of citizen action in terms of money and time spent proportional to the goals achieved.

The action was undertaken by political neophytes, outside (but by no means totally alienated from) the system. Their goal was not to involve themselves in any detailed way in the making or carrying out of transportation policy, at either the national or local level, but rather to set some directions and parameters for that policy. If successful at the polls, they would have required the government to take the necessary steps to implement the voters' general mandate (details of implementation were intentionally omitted from the question as set before the voters). Even if they lost, their thinking was to set in motion a process of public debate that, in the long run, might move public policy toward their goals.

The fact that the government is seriously considering an annual one-day ban on private automobile use in itself shows how effective the initiative campaign was.[17] The establishment of a popular institution to deal with traffic and

[17] A recent West German attempt to institute a single carless Sunday by government request was a total failure. Accounts seem to suggest that it was as much "a little referendum in favor of not doing what the Man says" as a rejection of "a kind of flashy environmentalism." I would interpret this as showing the importance both of regularizing the practice as well as the need to obtain popular prior assent before such dramatic changes in habits can be institutionalized. (See John Vinocur, "West Germans Vigorously Ignore 'Carless Sunday,'" *New York Times*, 9 June 1980, p. 3.)

people's movement needs in a more holistic and balanced way is yet another indication of how the initiative succeeded. Those who undertook it feel their efforts were clearly worthwhile, even though they have no immediate follow-up plans in terms of direct political action, and they have not remained together as a political grouping. Public opinion, the final vote notwithstanding, was generally favorable to the Burgdorf group, although to a degree they remained stuck with the "naive student" label the opposition sought to place on them.

Similarly, other related initiatives have elicited a partial response from the system: in response to the Albatross initiative, the government has promulgated stricter emission standards (although not as stringent as called for by the initiative); in response to the initiative for democracy in construction of the national highways, a government commission has been established to reconsider six segments of the nation's autobahn plan. While such responses can be and often are little more than sops and public relations ploys, the Swiss experience also shows that meaningful incremental change can be wrought by introduction of electoral campaigns for nonincremental goals.

The potential for using the initiative in the United States for similar ends in the transportation area remains untested. In many states and localities, the initiative power is rarely if ever used by the voters. In others—California, for example—it has been used extensively and effectively in recent years, although almost never with respect to transportation issues.[18,19] Initiatives and referenda

[18] On a state level, of course, Proposition 13 in the June 1978 election originated as an initiative, as did its successor, Proposition 4 (the local government spending limit) in the November 1979 California election. On a municipal level, San Franciscans tend to use the initiative freely. In 1976, voters passed an initiative measure to switch from a citywide to a district-based system of electing supervisors (city councilors), and in an August 1980 special election the initiative process was used to revert to a system of citywide election of supervisors; in the November 1979 city elections, there were no less than four initiatives on the ballot. The role of initiatives in forcing change from the government is illustrated here as well: a unanimous vote by the San Francisco Board of Supervisors in June 1979 to institute rent and eviction controls is generally attributed to their desire to undercut a far stronger housing cost-control measure being placed on the ballot through the initiative process (see Chester Hartman, "Left Initiatives Lose in San Francisco," *In These Times*, 21 Nov.–4 Dec. 1979, p. 6). In recent times there was one instance of a transportation-related initiative in San Francisco: in the 1950s, voters approved an initiative protecting cable car service levels by requiring voter approval for any future reduction in stipulated service levels.

On the growing use of the initiative in the United States, see Michael Nelson, "Power to the People: The Crusade for Direct Democracy," *Saturday Review*, 24 Nov. 1979, pp. 12–17, and Maureen Fitzgerald, "Initiative Fever: Many Try, but Few Reach the Ballot," *California Journal*, December 1979, pp. 433–34.

[19] As noted in the opening sentence of this article, transportation issues do come before the voters with some regularity in the form of bond issues and other funding decisions that require voter approval; in such cases, the matter is placed on the ballot not via voter initiative but by the government. For an interesting discussion of five such transportation-related voter decisions in California, see J. Allen Whitt, "Toward a Class-Dialectical Model of Power: An Empirical Assessment of Three Competing Models of Political Power," *American Sociological Review* 44 (February 1979), 81–100.

in the United States face some of the same difficulties found in Switzerland. Low and biased voter turnout for our elections means that socially and economically progressive proposals may inherently stand a greater chance of defeat. The ability of entrenched and economically powerful interests to "buy" elections in the United States, together with the development of sophisticated, effective polling and campaign techniques, stack the deck considerably. (A "clear air" initiative—to limit smoking in public places—on the June 1978 California ballot was soundly defeated, the direct result of a $6 million campaign funded by the tobacco industry.) Nonetheless, selective, creative use of the initiative process can be a promising avenue for citizen action.

Citizens' groups interested in public transit and more balanced, socially sound transportation policies would do well to investigate the potential of the initiative process as a means of bringing about change and improvement. This can come about directly (should the proposition succeed on the ballot) or indirectly—through the prominence that an electoral campaign gives to a neglected issue and to proposals for reform, through direct ameliorative government response to an issue once it is placed on the ballot, and through the constituency building that political campaigns generate and accelerate.

ACKNOWLEDGMENTS

I am grateful for Eva Eliassen for the useful contacts and general support she provided during my stay in Switzerland. Philippe Lehmann of the Institut de Recherche sur L'Environnement Construit at the École Polytechnique Federale de Lausanne kindly provided me with many helpful documents and contacts. Marcus Nufer graciously opened to me his extensive files and reminiscences regarding the Burgdorf initiative. Tony Lüchinger of the Verkehrs-Club der Schweiz and member of the foundations's board of directors was an excellent source of information about his new organization. The clippings collection at the Forschungszentrum für schweizerische Politik at the University of Bern was a highly useful source. The following also provided much useful information: Hans Werder of the Eidgenössische Technische Hochschule in Zurich; Franz Weber; Urs Zuppinger of the Department of Architecture, École Polytechnique Federale de Lausanne; H. Bernath of the Institute of Transportation (ITEP), École Polytechnique Federale de Lausanne; F. Grandjean; Pierre Meyer, Robert Stussy, and Janine Durussel of CEAT, Ecole Polytechnique Federale de Lausanne; Carl Hidber of the Eidgenössische Technische Hochschule in Zurich; and Fritz Witschi of POCH-Basel.

6

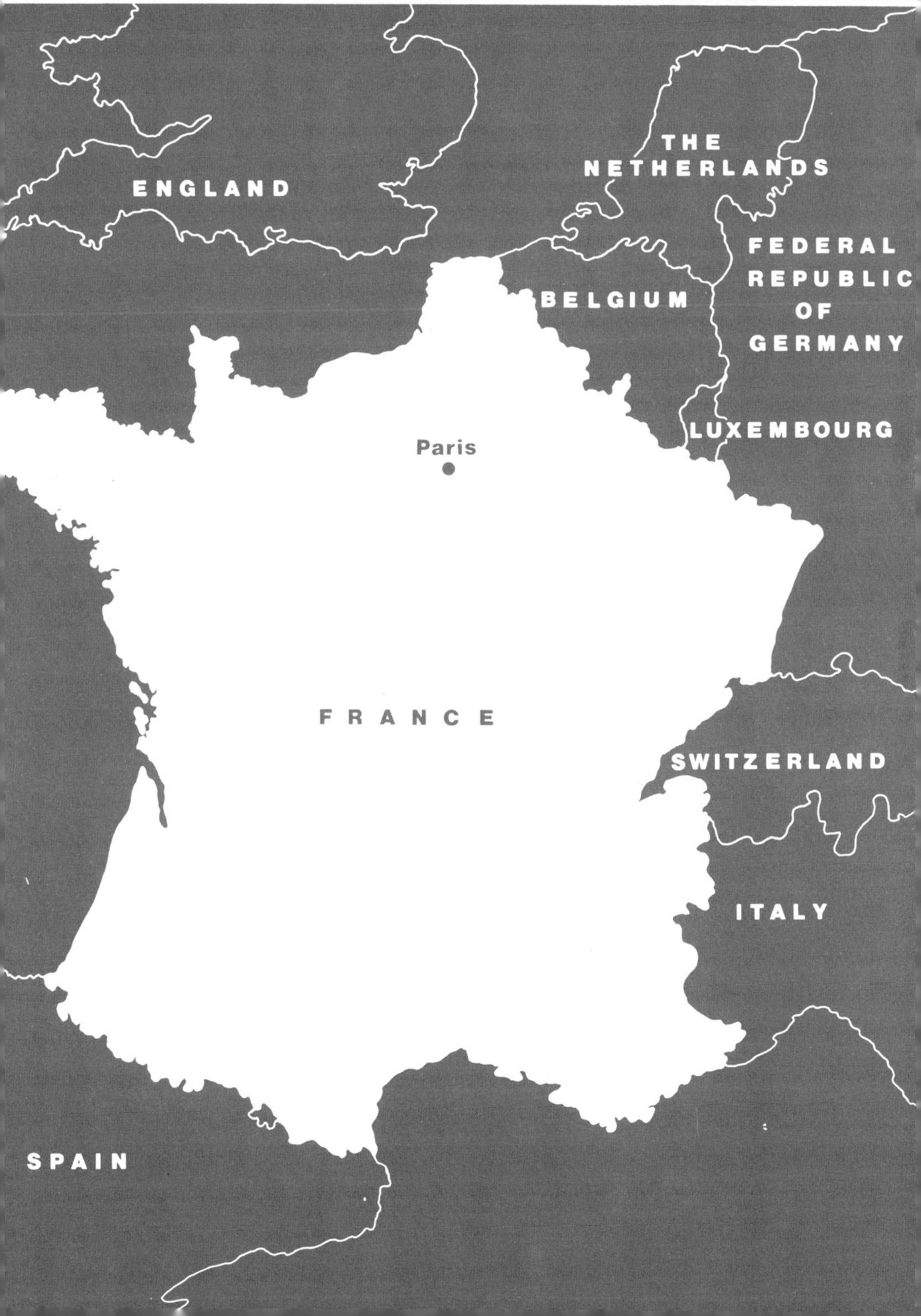
ENGLAND
THE NETHERLANDS
FEDERAL REPUBLIC OF GERMANY
BELGIUM
LUXEMBOURG
Paris
FRANCE
SWITZERLAND
ITALY
SPAIN

6

TRANSPORTATION USERS' MOVEMENTS IN PARIS IN THE 1970s

CHESTER HARTMAN

In comparing France with the United States, one is struck by the paucity of citizens' groups here established to deal with transportation issues. Looking through a list of 180 participants at a 1978 national conference on Transportation's Role in Neighborhood Revitalization (sponsored by the Conservation Foundation for the Urban Mass Transportation Administration) in which a major topic was "citizen involvement" and whose goal was to "seek advice from citizen leaders and state and local officials," just one represented group was clearly a citizens' transportation organization. The other participants were from a range of general-purpose national and local citizens' organizations (47%); public officials from federal, state and local government transportation agencies (29%); private consultants (10%); and the media (3%).[1] The number of local and national citizen-based activist transportation groups in the United States is minuscule. By comparison, a two-week field trip to Paris revealed a plethora of such groups and a long history of successful involvement in influencing various levels of transportation policy and programs.

[1] The Conservation Foundation, *Thinking Small: Transportation's Role in Neighborhood Revitalization,* Report on a conference held Feb. 22–24, 1978, Baltimore, Maryland, for the U.S. Department of Transportation, Urban Mass Transportation Administration (Washington, D.C.: U.S. Government Printing Office, 1979). Quote is from preface by Richard S. Page, Administrator, UMTA. The categorization is approximate and is based solely on assumptions deriving from the name of the organization each participant represented. Obviously, by virtue of their presence at the conference, many general-purpose citizens' groups are interested in transportation questions as part of their overall concerns and activities. But establishment of focused interest groups is a good index of how seriously citizens organize around a problem. The one identifiable citizens' transportation group at the conference was Streets for People/Baltimore Area Bicyclists Association. A second organization, Project Traffic, from Sarasota, Florida, may also belong in this category.

It is not the purpose of this paper to account for these differences in any detailed or comprehensive way. Some factors explanatory of the French case will be offered; with regard to the American scene, the underlying assumption/observation is that there are no insurmountable political, social, or cultural obstacles to the development of a similar range of citizen activist transportation groups here. A further assumption is that creation of such groups would introduce an important element in making our public and private transportation systems more responsive to people's needs and would reduce overall social costs. This paper will offer information about the situation in Paris (and to a minimal extent in other parts of France) in the hope that this exposition and example will provide some basis for triggering similar developments in the United States.

Paris and France

Paris is both a city with one of the premier public transportation systems in the world (80% of all commuting trips and 70% of all trips are made on public transportation) and one whose attractiveness as a place to live, work, and visit is perhaps most threatened by the automobile. As an American observer, now deputy commissioner of transportation for the City of New York, recently wrote of Paris:

> Automobile traffic on the streets of the city is frightening . . . 65 decibels are normal street-level noise readings, 80 each time a horn sounds or a truck accelerates, which is almost always; black diesel smoke, foul-smelling and carcinogenic, hangs in the air of the boulevards; invisible carbon monoxide is trapped in the narrow streets lined with five or six story buildings. Parking garages have been built under most of the monumental squares but there is never enough space to park or drive easily, although only 20% of all trips are made by car . . . . Much of Paris has the feeling of a vast auto race course with pedestrians and old monuments as inconvenient obstacles.[2]

Given the dense settlement of Paris (50% greater than the combined average density of Manhattan and Brooklyn, three times the density of the central boroughs of London[3]) and its street pattern, dominant reliance on public transportation is an absolute necessity. There is no way that easy movement of autos and surface public transportation vehicles can be simultaneously maximized, since, for example, to provide measures favoring the auto is to impede the bus system (which has an average speed of 10 kilometers per hour, compared with 15 kilometers per hour 30 years ago). Similarly, to provide measures favoring the bus system (Paris no longer has trams or trolley buses) is to impede the private automobile. The focus of much citizens' activity in the transportation area has of necessity, therefore, focused to a large extent on the conflict between

[2] David Gurin, "France: Making Ecology Political and Politics Ecological." *Contemporary Crises* 3 (1979), 156, 158.

[3] Ibid.

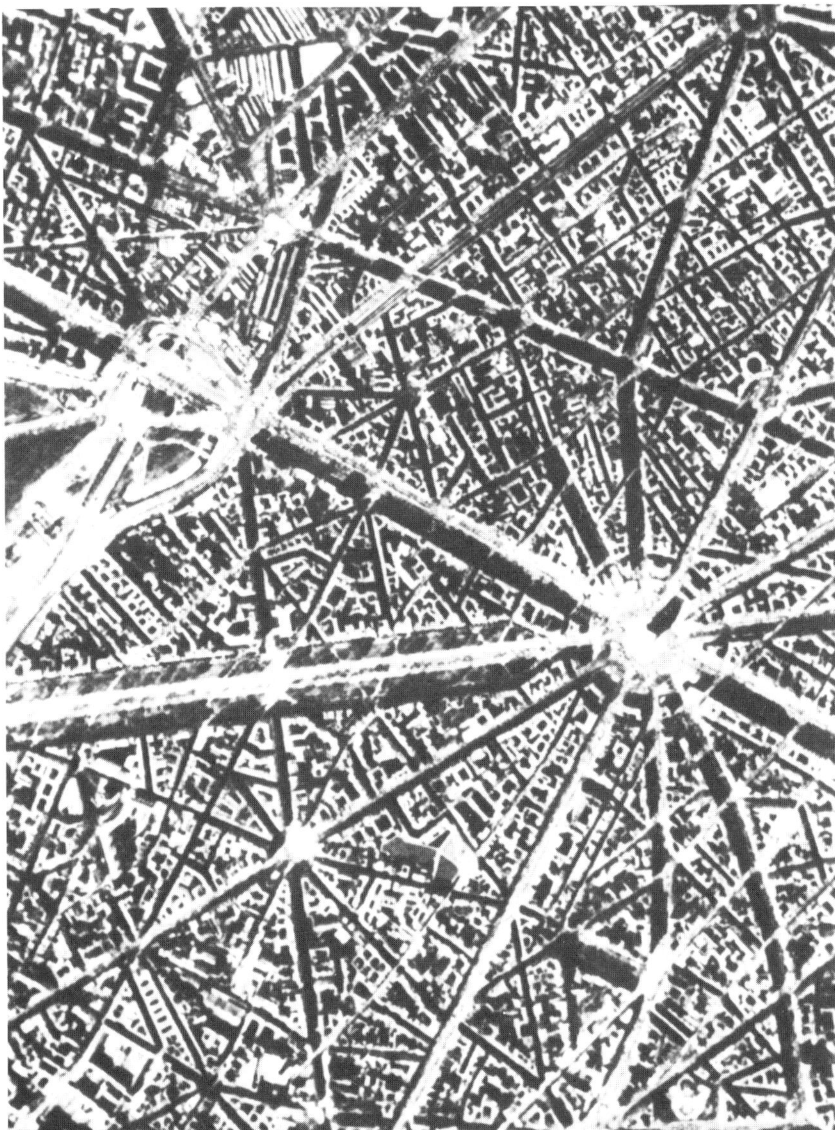

Figure 1. The streets of central Paris, near the Arc de Triomphe. (From the Rotch Library Visual Collection.)

automobiles and other means of transportation (not only buses but also pedestrians and two-wheeled vehicles) as well as on the conditions of the public transit system itself.

Basically, popular activism in the area of transportation services in Paris over the past decade falls into two periods: (1) the early 1970s, which focused in a highly politicized and effective manner on the quality and price of public transportation, and (2) the more recent years—perhaps somewhat less political and effective, although with some impressive successes to be sure—which have seen more of a focus on the automobile–public transit conflict.

France, of course, has a highly centralized government and political system. Some of its institutions—well-developed political parties covering a wide range of the left-right spectrum and powerful trade union confederations linked to the two major left parties—have no real parallel in the United States. It is also a country in which, in a social and political sense, the reality of class is more openly acknowledged and dealt with than is true in the United States. Relatedly, the working class has a far greater sense of its own identity, needs, and potential and has powerful institutions that embody and fight for the class in contest with the state and with institutions representing other classes. The working class and its political representatives hold power at the municipal level in many parts of France, are well represented in the national government, and have come close to taking power at several points in recent history. It should be no surprise, then, that "citizen participation" takes on a quite different form than is true in the United States. In overly simplified terms, the formal government system provides far less structured citizen participation than is true in the United States. Furthermore, the popular activism that occurs around important public issues such as transportation services takes on a more overt political form; it is more a contest for real power that is likely to involve issues and concerns going beyond the immediate and apparent object of struggle.

Early 1970s Activism

In recent times the outstanding period of popular mobilization around the transportation system in Paris occurred in the 1970–71 period, triggered by the announcement in February 1970 of an immediate 16% fare boost (with a further 14% boost announced for the coming July) in the basic ticket used by Parisians at that time: the *carnet* (book of 10 métro ride tickets).[4] The enormous protest

[4] This account of the 1970–71 movement and of the later transit users' movement draws from three excellent sources: Eddy Cherki and Dominque Mehl, *Les nouveaux embarras de Paris* (Paris: Francois Maspero, 1979); G. Ribeill, P. Bertier, F. Lille, and N. May, *Revendications et instances revendicatives en matière de transport urbains* (Paris: Secrétariat d'État auprès du Ministre de l'Équipment [Transports], Prospective et Aménagement, 1978); G. Ribeill and N. May, *Rapports sociaux dans les transport urbains et mouvements revendicatifs transports* (Paris; Secrétariat d'État auprès du Ministre de l'Équipment [Transport]. Prospective et Aménagement, November, 1976). All three works are excellent analyses of the politics and sociology of social movements.

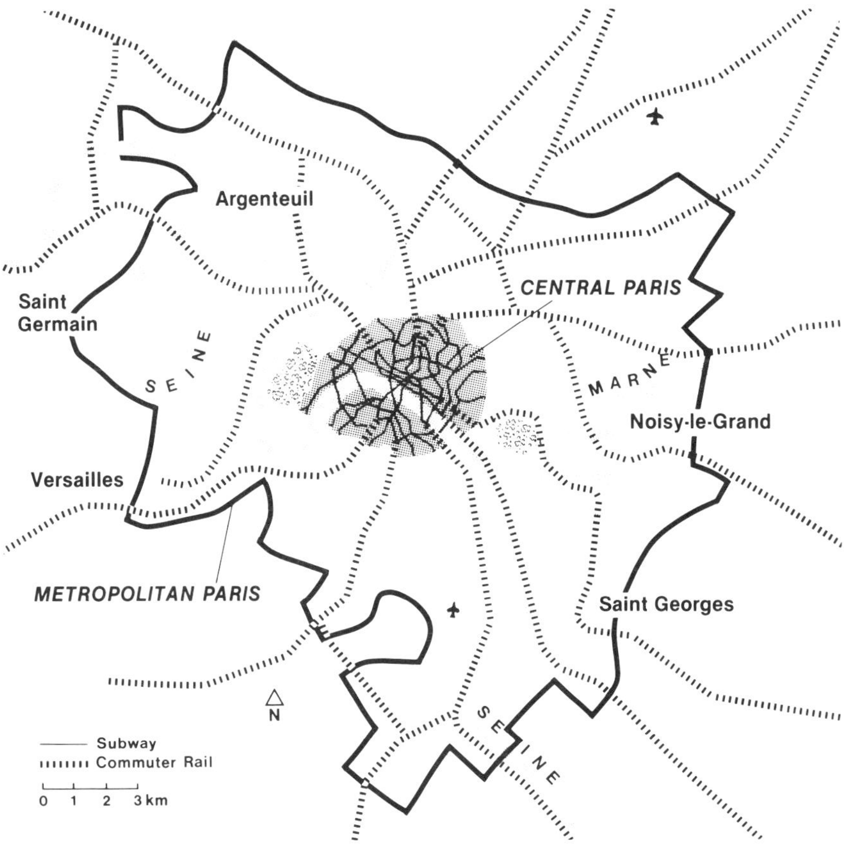

Figure 2. The public transportation system of Paris, showing the subway system located in the core and the commuter rail system that extends out in all directions.

this aroused was, however, more broadly rooted than the question of fares: at issue was the politics of transportation—how the public transportation system functioned in the region. During the 1960–67 period, there had been no fare increase at all in the Paris region; as a consequence, the deficit the system was running had increased markedly. In an effort to overcome this deficit and bring about a fare structure that would provide revenues sufficient to cover costs, the government in 1967 had raised métro fares 60% and bus fares by 100 to 200%. The 1970 raise was an overt attempt to implement the "pay as you go" policy and shift the cost burden from the government to the rider.[5] Unlike the far higher

[5] The deficit of the RATP (the operating agency for Paris public transit) was 8% in 1960, 38% by 1966, dropped to 30% after the 1967 fare increase, and was on the rise again at the time of the February 1970 announcement.

1967 increase, however, it brought about protest and counterforces that were not forthcoming in 1967. One factor may have been the cumulative effect provided by the second increase, which amounted to a 90% fare increase in three years in a service basic to most Parisians. It is likely that the different general political climate in 1970 plus the more open posing of a question of principle were responsible for the more forceful reaction to the second increase. The whole notion of profitability of public enterprises—applying criteria from the private sector—was being raised; with it came a threat to overall levels of service, since "unprofitable" lines and services would gradually be eliminated under such a doctrine. The very nature of public services was being put to question.

The two major trade union confederations, the Confédération Generale du Travail (CGT) and the Confédération Francaise Democratique du Travail (CFDT), had decided at the end of 1969 to mount a joint action on the issue of public transportation. The February 1970 fare increase was the detonator for that plan. The two major demands were (1) the rescinding of the fare increases; (2) the introduction of a *carte unique de transports payée par les patrons* (a single-fare transport pass, regardless of distance traveled or mode of transport, paid for by employers).[6] Elected municipal officials from the left parties strongly protested the increases, and Communist party members organized a massive leaflet distribution in the métro and railroad stations. Huge petitions of protest were presented to the minister of labor. With the July 1970 increase came creation of the Cartel, an umbrella group of eight organizations (the Communist and Socialist parties, three major trade union confederations, and three small left political parties or groupings). When yet another fare increase was announced for January 1971, the Cartel organized a massive demonstration on November 18, 1970, which was preceded by a week of preparatory activities, petitions, and delegations to the minister of transport and the National Assembly. The demonstration itself drew 20,000 to 50,000 people (accounts differed); according to two students of social movements in France who chronicled these events, this was the first time there had ever been a demonstration of this magnitude for an urban demand and the first time the left parties and unions had mobilized to fight around an urban theme.[7] The movement of the trade union confederations into issues of quality of life outside the workplace and defense of workers' buying power in the system of consumption was a new development in France.

The demonstration got a surprisingly sympathetic and important response in the media, which acknowledged that, in Paris, general conditions in the public transportation system as well as traffic conditions were badly in need of im-

[6] The positions and demands among various elements of the left unions and other political groupings differed, but for purposes of this paper these distinctions are not of prime importance. The three works cited in note 4 provide a good and interesting account of these differences and their significance.

[7] Cherki and Mehl, p. 17.

provement.[8] Following the big demonstration, there were smaller ones in métro and railway stations, not only within Paris but in some of the suburbs with Communist local governments; demonstrations were held as well in large workplaces. The demonstrations led the government—with memories of May 1968 quite fresh in everyone's mind—to postpone the third in the series of fare increases, but it then announced that the increase was being put into effect on August 20 (a classic French government move, timing an unpopular event to the summer months when a very large proportion of the population is on holidays). The Cartel went ahead with its planned third major demonstration (the second was held on June 19, with 25,000 to 30,000 people participating); on August 26, 1971, it pulled together 25,000 to 100,000 people (again, estimates in different accounts vary widely) at Gare St. Lazare. This was regarded as an extraordinary turnout, given the fact that so many people were away from Paris.

Result of the Mobilization

The mobilizations by the Cartel did not, in the end, serve to reverse the announced fare increase. They did force the government to back down three times with respect to the timing of the increase, but the fares nonetheless were raised. (The Cartel dissuaded the government from introducing yet another fare increase in 1973 simply by sending a delegation to the minister of transport. It was not until 1975 that the government dared put through another fare increase, and then only because it was accompanied by the offsetting reform of introduction of the monthly pass; see p. 188.) The victories wrought by this popular movement were on two levels: (1) introduction by the government over the following years of major changes in the financing of public transit and the fare structure and (2) the development of local organizations and activities that formed the basis of citizen involvement during the later 1970s.

To deal first with the major policy changes: In May 1971, in direct response to the Cartel's powerful demonstrations, the National Assembly voted to intro-

[8] An earlier version of the Cherki-Mehl study—only the second half of which was published as the book cited in note 4—contains the following description of Paris transportation problems at the time of the 1970–71 movement:

> It's six o'clock in the evening, and drivers are losing patience in the traffic jams, outhonking each other at crossroads. Buses are stuck in their lanes, the subways are overflowing, and the stations are full of hurried passengers pushing their way through to make the next connection. The cars are parked on the sidewalks and on pedestrian crossings and car exhaust is strangling passers-by. The monotony of the dirt, and white tiles and the unending corridors overwhelm the traveler. The magnetic ticket digester has replaced the old wooden benches, and the subway stop at the Louvre has been given special elegant treatment. Despite all that, the Parisian inhabitants must do their daily battle with the ever-present and highly visible symptoms of the transportation crisis. (Translation by Jean Chase.)

duce a payroll tax on employers in the Paris region for purposes of subsidizing public transportation. This tax has become the major subsidy source for public transit and represents a fundamental rejection by the government of the notion that fares should cover revenues. The tax, levied on all businesses located within the Paris region with 10 or more employees, began at 1% for businesses within the city of Paris and 0.5% for businesses in the Paris region but outside the city itself. The taxation rate has been raised several times subsequently and is now 2.0% and 1.2%, respectively. The payroll tax now accounts for about 14% of total operating costs, with fares covering 35% and the remaining costs covered by local government subsidies, advertising, and so on. Twenty-one percent of capital investment costs are also covered by the new tax. Introduction of the tax derived in part from the political perspective and analysis of the 1970–71 demonstrators and their leaders regarding the role the public transportation system plays in the economic system; they held that it is primarily a means of assembling and mobilizing the work force and that therefore those in charge of the economic enterprise ought to bear a large share of the costs of this mobilization.

The second and less direct major policy change resulting from the 1970–71 demonstrations was introduction in July 1975 of the so-called *carte orange,* a monthly pass for unlimited rides on the métro and buses. It was not exactly what the demonstrators had been calling for in that it was not a *carte unique* (that is, a single-price pass). Instead, it is zoned, a higher price being charged for a pass permitting travel to and from more distant zones. Such a system, of course, puts people living in outlying areas at a disadvantage, since they not only must spend more time commuting (three- to four-hour round trip commutes are not uncommon in the Paris region) but must pay more for this travel between home and workplace. Since the Paris region is organized in a way that requires very high proportions of working-class households—in particular, migrant workers—to live in the suburbs and outer segments of the region, the *carte orange* builds in a regressive pricing structure. Nonetheless, it is regarded as a major gain for transit consumers and, along with other improvements to be described below, has had a major impact on the bus system especially. The *carte orange* in 1978 accounted for 43% of all métro trips, 49% of all trips on the RER (the suburban rail system that is linked to the métro), 62.5% of city bus trips, and 51% of suburban bus trips.[9] The price of the pass has been raised several times since its introduction (it was originally 45 francs for the basic two-zone ticket and, as of the latest increase in July 1979, 70 francs—a little under $17, at the prevailing exchange rate). While the various increases (still timed for the summer months) led to some protests (the raise has been considerably higher than general price-

[9] The RATP also has several other reduced fares. A weekly 10-ride trip between two specified stations (the *carte hebdomadaire*) is available, as is reduced fare for the elderly, schoolchildren, and large families; regular fare can be paid by purchasing single-ride tickets or, more economically, by purchasing a *carnet* or book of ten single-ride tickets.

level increases), these protests did not reach anywhere near the scale (nor had they the leadership) of the 1970–71 demonstrations, and they had little impact on government actions. The improved quality of service and changed times politically are important factors explaining this difference.

The combination of the payroll tax and the *carte orange* in fact represent a partial fulfillment of the 1970–71 demand for a *carte unique de transports payée par les patrons,* since the employers' tax covers a large part of the operating deficit of the RATP (the operating agency for Paris public transit), and the *carte orange* represents a lower per-ride fare for transit users, whose fares cover only about a third of the costs of the system. Another victory has been use of the payroll tax to cover operating deficits as well as capital costs; the original notion of the tax restricted it to the latter type of expenditure.

Yet another important result of the 1970–71 demonstrations was to focus attention on the need to improve the quality of the public transit system. While the métro system is quite remarkable in its coverage and frequency/reliability of service, several important weaknesses characterized the Parisian public transportation system as a whole. One was the poor service to and from the suburbs, another was the poor service between suburban areas, and a third was the sad state of the bus system. While the intersuburban public transportation linkages are still poor (and may remain so, given the insufficient densities to support first-rate public transit service), there has been a remarkable improvement in bus service in Paris in the 1970s and development of excellent suburban lines (the RER) linking up directly with the métro. The improvement in equipment on the bus lines (together with such innovations—for Paris—as special bus lanes and bus shelters), combined with the impact of the *carte orange,* have led to a dramatic increase in bus ridership, up 40% in the year following introduction of the monthly pass. Because, unlike the métro, the bus system has a two-zone fare system with no transfers—leading to the possibility of three fares for a relatively short trip—use of the monthly pass eliminated a strong economic disincentive to using buses. (The data above, showing a considerably higher proportion of urban bus trips paid for by the pass compared with the métro, substantiate this point). Much still needs to be done to improve the bus system—as will be described below—but certainly a major gain produced by the 1970–71 movement can be seen in Paris's surface public transit.

The 1970–71 demonstrations were, as noted, centrally directed by the major trade union confederations and left political parties; they represented a centralized battle against the fare policies of the government. Their object was not to develop local actions and organization (except insofar as these were building blocks for the centralized mass demonstrations); the focus was on impressing the government with the opposition and power of working-class transit users. The aim was not to take over provision of this public service but to force the government into making major policy and programmatic changes in how the service was provided. While this mode of activism had its strong points and clear successes, it also had its limits in that, apart from the mass demonstrations and the accompanying delegations to government officials, little was being accomplished in the way of base-building activity to carry out work in between demonstrations and create a permanent political force for implementing transit users' demands once the major demonstrations faded away. The Cartel itself was not truly a specific and permanent transit-based and -oriented political organization.

The 1970–71 movement laid the base for the more specifically transit-based citizens' groups that arose later in the decade, although the role and impact of these groups is now quite different from what they were during the earlier period. As noted, there was much local activity during 1970–71 as building blocks for the centralized mass actions were being laid. These local groups were of greater interest to two of the small parties of the extreme left, the Parti Socialiste Unifíe (PSU) and Lutte Ouvrière (LO)—the former originally a member of the Cartel but an early dropout—which together formed the Fédération des Comités d'Usagers des Transports en Commun de la Région Parisienne (FCUTCRP). The first concrete action of the FCUTCRP was publication of the *Livre noir sur les transports parisiens,* a document analyzing the transit crisis and laying out a set of concrete demands; 30,000 copies of the booklet were sold, and the press gave it substantial coverage. The Fédération comprised 120 to 130 local *comités des usagers* and participated in all the initiatives organized by the Cartel, although as a separate entity with their own banners, slogans, and leadership.[10]

Individual *comités des usagers* put forth their own demands, which can be grouped in four areas:

1. Conditions of transport—frequency of service, transfer-point problems, lack of good service other than daytime on weekdays, need for bus shelters, escalators in the métro, and so on
2. Organization of the transportation network—in particular the lack of linkages among suburbs, lack of service to new suburbs, and so on
3. Tariff—in particular those aspects of the fare system that disadvantaged specific outlying areas (zoned fare, higher fares on the railroad compared with the RATP system, etc.)
4. Improvement of private transportation services that provide communal transportation in outlying areas—including lower fares and better equipment

The PSU also had several goals that it did not put forth as demands, in the belief that the time was not appropriate to achieve them. Among these were free

[10] In style, the FCUTCRP, like its PSU and LO leadership, was a direct descendant of May 1968, which among other things led to sharp divisions between it and the traditional left parties, who disdained the "extreme left." The overall aim of all parties at the time, however, was to create a political mass movement that would make demands on the state well beyond those being put forth at the moment, and in other areas as well.

Figure 3. Interior of the Gare de Nord.

public transit (although they felt the demand for a *carte unique* was a first step toward that goal) and counting travel time as paid work time. The latter was a demand that would totally shift the economic considerations regarding location of economic enterprises and workers' residences, since long travel time would become a production cost instead of being a cost that individual workers would have to absorb.

The Late-1970s Movement

The transit users' movement of the latter part of the decade is to an extent a descendant of the 1970–71 movement (more the FCUTCRP strand than the Cartel strand, however), but it differs in that it has been thoroughly infused with the concerns and organizations of the ecology movement, which in France dates from the 1973–74 period. As such, one can describe a change of underlying focus from concern for the home–workplace trip to concern for the city itself and for quality-of-life considerations. The earlier movement dealt with people as consumers of transportation services, while the new movement deals with people as consumers of urban space; thus it is looking at the city in a more holistic manner. The issue now is more an urban crisis than it is purely a transportation crisis with transportation—in particular the automobile–public transit conflict—at the center.

The Fédération des Usagers des Transport (FUT) still exists. There now is a Fédération National des Associations des Usagers des Transports (FNAUT), and four national conventions of transit users have been held since 1975. Individual groups, such as the Association Combat-Transport, carry out their work in conjunction with larger confederations, but in many ways they are quite autonomous. Many ecology groups have made the transportation issue central to their work. And of course the analysis, set of demands, and mode of action differ substantially from the 1970–71 period.

Probably the central difference revolves about the role of the automobile. Whereas the 1970–71 movement took an ambivalent stance toward the automobile, the current movement is unambiguously opposed to it and, within the bounds of reasonableness, seeks to limit its presence and substitute other means of transportation for it: public transit, bicycling, and walking. The Cartel—the Communist party and CGT in particular—regarded the automobile, for the present at least, as a necessity in French capitalist society. The limitations of the public transportation system and the patterns of urbanization in the Paris region made the automobile something almost essential for a great many working people. While there were, to be sure, class differences in how people traveled, a great many working-class people used the automobile (just as a great many middle- and upper-class people used public transit). Overly harsh measures against the private automobile would hurt too many workers. Demands centered on improving the competitive position of public transit as an alternative to the automobile and shifting public investment somewhat from the private to the public transportation system, but road projects and other investments in infrastructure for private automobile drivers were not totally opposed. By contrast, the FUT, FNAUT, and ecology groups regard the private automobile as anathema to Paris and seek its rapid and total replacement by other means of movement.

Perhaps the easiest way to illustrate this shift is to focus on FNAUT and its October 27–28, 1979 congress at Tours.[11] FNAUT's members (as of July 1979) consisted of 72 local and 5 national associations. It was established in June 1978 as an umbrella organization and includes national organizations dealing with pedestrians' rights (Pour la Cité humaine–Les Droits du Piéton), bicycle and motorcycle riders (Comité Nationale des Usagers des Cycles et Cyclomoteurs), highway safety (Comité Nationale d'Action pour la Sécurité des Usagers de la Route), and the needs of the handicapped (Confédération Générale des Handicapés et Retraités). Local member groups cover the same range of concerns and constituencies. FNAUT's four basic demands are:[12]

1. A guaranteed right of transportation for everyone, with the underlying concept of public transit as a public service not as a profit-making enterprise
2. Outside of urbanized areas, priority of rails over roads for movement of people and goods, thereby lessening energy consumption and social costs
3. Within urban areas, subservience of the automobile to public transportation, two-wheeled vehicles, and pedestrians
4. Reduction of daily transportation needs through coherent land-use and urbanization policies

The notion of a "guaranteed right of transportation" of course embraces both (1) the need for a low-cost (possibly eventually free) public transit system that serves all parts of the Paris region and nation so that, among other things, no one is forced to own an automobile in order to have the right of transportation and (2) the question of accessibility of that system to elderly people, handicapped people, and others shut out from use of the system by its design.

The issue of rails versus road has become a major national question in France. The once excellent national railroad system has been deteriorating rapidly and is becoming ever more expensive to use. Whereas there formerly were 43,000 kilometers of rail track in France, the present figure is 24,000 kilometers, and a recent government report (the Rapport Guillaumat) looks at an eventual cutback to 8,000 kilometers, with service—as on Amtrak—only between large cities. Already there is an entire department (l'Ardeche) without rail service, and if recent trends continue, whole portions of the country will be without trains. Government investment and licensing policies, together with rail cutbacks, have

[11] The Tours meeting was the first national *congress* of transit users; there had been four previous national transit users' *conventions* (the former term is used to refer to a coming together of members of the same association, the latter to a coming together of different associations): in Grenoble (1975), Cannes (early 1976), and Paris (late 1976 and 1977).

[12] See *La F.N.A.U.T.: son organization, ses objectifs* (October, 1979) and *F.N.A.U.T. dossier no. 1: Les conventions nationales des usagers des transports* (n.d. [1978]).

Figure 4. Paris métro station. (From the Rotch Library Visual Collection.)

led to a major shift in the movement of goods; between 1973 and 1978, heavy truck traffic increased by 30%, while movement of goods by rail decreased 16%. The social costs of this shift—in energy use (trains use one-third the energy that trucks use for comparable movements), road accidents, pollution, and other things—are immense.

Within urbanized areas, the range of demands is clearly shown by the six principles of a new circulation plan for Paris contained in the 1977 booklet *Assez roulé comme ça, on réfléchit,* an analysis of Paris' transportation problems, together with some well-illustrated alternative designs, published by Les Amis de la Terre:[13]

1. Paris and its suburbs must be served by a "métro of the surface." The general goal is to make the surface public transportation system as good as the underground system. Absolute priority would be given to public transit vehicles (taxis included); some streets would be limited to public transit traffic, major streets would have reserved bus lanes (ideally, center lanes with islands rather than curb lanes, which endanger and interfere with sidewalk activities, cyclists, turning vehicles, etc.)[14] Public transit vehicles must be accessible to all "transportation handicapped" persons (those with baby carriages and large packages, those with physical disabilities, bicyclists, etc.). More bus shelters and information sources about routes and service must be provided. Particularly for the benefit of suburban areas, bus service must be available seven days a week, with the same hours as the métro (service until 1 A.M.).
2. Bicycling must be made safe and convenient, with special bike lanes, safe parking, and other amenities.
3. Provision must be made for delivery trucks, in terms of special times and areas, so as to minimize interference with other street traffic and ease delivery tasks.
4. Relegating the automobile to low-priority, low-speed status. This will be accomplished by reducing land space available to private autos to one (one-way) parking lane and one (one-way) travel lane per street, prohibiting all parking that is not parallel to the curb, banning cars from sidewalks, pedestrian zones, and all places they should not be (the invasion of the sidewalk by parked autos is one of the most repugnant features of Parisian street life these days), 40–50 kilometers per hour speed limits, more traffic lights and stop signs.
5. Priority to pedestrians: More pedestrian zones (of which Paris has extremely few compared with most European cities), deemphasis of un-

[13] *Assez roulé comme ça, on réfléchit* (Paris: Jean-Jacques Pauvert, Les Amis de la Terre, 1977).

[14] Recently, reserved bus lanes have been introduced in Paris, but they are very narrow (for the most part 3 meters), too few in number, and poorly enforced and respected.

derground passages and other means of separating pedestrian from vehicular traffic which give the pedestrian lower priority, reservation of sidewalks for pedestrians.
6. Development of more green space and easy, safe access to parks, gardens, squares, and so on, plus development of the banks of Seine for people-oriented activities.

Other major demands by transportation users' groups include direct measures to cut down noise and air pollution and further steps to foster bicycling (free railroad shipment of bicycles accompanying passengers; production of a solid, reliable urban bicycle; better safety devices;[15] and free municipal bicycles[16]). Some interesting new organizing techniques have also been introduced, such as attempts to organize riders of individual bus lines into separate users' groups, to deal with the specific problems of that route—hours and frequency of service, obstacles to more rapid coverage of the route, and so on.[17]

Issues and Strategies

Two important developments (and strains) within the present transit users' movement are (1) the degree to which the field of battle should be extended beyond transportation problems and demands and (2) the role of electoral politics.

The issue of how focused and unidimensional transit activity should be in some ways relates to the demands and successes of the earlier period. The several victories of the 1970–71 mobilization—vastly improved bus service, introduction of the *carte orange,* and passage of the payroll tax to subsidize RATP service—were major, and levels of subsequent activism have, to an extent, been affected by these successes. Remaining demands are by no means trivial, but a substantial inroad has been made in meeting the extreme needs of the late 1960s and early 1970s with respect to transit users. While many transit users' groups have incorporated environmental concerns into their work, others have shifted their emphasis to overall environmental work, putting specific transportation-related questions into a lower priority. On the other hand, new groups that have arisen in the past few years around new concerns—such as the environment and the

[15] A very simple, cheap, and effective bicycle safety device, sold by Les Amis de la Terre, is a stiff red plastic flag, about a foot long, which extends to the left of the bike frame and in effect creates a safe space between the bike and passing cars (it can easily be moved 90 degrees to a position parallel with the bike).

[16] The Atlantic coast city of La Rochelle has for several years had a program of free municipal bicycles—several hundred city-owned bikes available for use to anyone within certain boundaries: a person just picks one of the yellow bikes out of the racks or parking spots, rides it to his/her destination, and places it in the nearest rack for the next person's use. See "Le velo municipal a la Rochelle," *Correspondence municipale* 184 (January 1978), 48, 51.

[17] This has been a joint project of Les Amis de la Terre and Combat Transport; some 20 route committees are already in the initial stages of organization.

rights of the handicapped—have incorporated transportation issues into their own sphere. The Groupement pour l'Insertion des Handicapés Physiques, for example, seeks to integrate the handicapped into everyday life and into society's normal services and institutions; access to transportation services has become a major concern of theirs and, in the process, they have sought to make alliances, functionally and conceptually, with other transportation-deprived groups for whom the transit system could be made more accessible: The elderly, children, bicyclists, migrant workers living in outlying areas, and others.

The environmental movement, as part of its rapid growth in the past few years, has forcefully entered the electoral arena. Starting with the running of an ecology candidate for president in 1974 (the internationally renowned agronomist René Dumont)—and with a "green" presidential candidate likely in the 1981 elections as well—and running for local offices, the National Assembly, and the European Assembly, the ecology movement has done quite well, getting 4.4% of the vote in the recent European Assembly elections, and an average of 10% of the vote in the 1977 municipal elections (close to 14% in two Paris arrondissements).[18]

Electoral tactics, however, have produced a substantial split within the transit users' movement, in part around questions of principle and style, in part around personalities. Those presently in the FUT and SOS-Environnement (an electoral coalition formed by FUT and others) have adopted the strategy of running candidates as widely as possible and, in the second round run-off elections, attempting to position the major party candidates to support and work for elements in the "green" platform as a means of competing for SOS-Environnement votes. Other parts of the transit users' movement—notably the Association Combat-Transport (an offshoot of FUT) and the new national federation, FNAUT—have adopted a policy of strict nonalignment with any existing political parties or trade union confederations and reject electoral strategy and tactics altogether. Personal ambitions and conflicts among leaders of individual groups have at times exacerbated this tension. However, beyond the question of election activity and concentration–dispersal of focus (the FUT–SOS Environnement group seems to be placing transit issues in a somewhat secondary position, whereas the Association Combat-Transport–FNAUT group focuses on these issues), there do not appear to be any major substantive differences in analysis and demands regarding transportation.

The question of alliance with political parties is an important one in France. In the 1970–71 period, the left political parties developed an interest in transportation issues not only because the economic and social problems in this area were central to working-class living conditions but also with the aim of party building. The outlook of the major left parties, the Communists and Socialists, is toward eventual taking of power; they believe that demands on the state in

[18] See Gurin.

its present form can bring only limited gains and that fundamental change can only come about with a change in government and the controlling political parties. The outlook of many elements in the present transportation movement and the "extreme left" political groupings out of which those elements emerged is less oriented toward the taking of power. They consequently feel less need to put forth a comprehensive alternative program, and they realize the extent to which large elements of French society, the working class and others, are turned off by the traditional left parties. They are more focused on the achievement of specific demands within the existing state apparatus—on democratizing the government's procedures through regularized consultation, participation, and occasional confrontation between government agencies and citizens' organizations. French government administration is notably impervious and insensitive to outside citizen input (the RATP, SCNF–French railways, and Syndicat des Transports Parisiens—the overall financing and planning agency for public transit in the Paris region—have almost no citizen input except "window dressing" advisory roles that hardly anyone takes seriously); a principal goal of the transit users' movement at present is democratization of that process.

The question of how successful the more recent transit users' movement has been in achieving its goals and demands is open to question. Consciousness of the destructive role played by the automobile has doubtless increased in recent years, as a result of the work of the various users' groups; specific reforms have been introduced albeit at only halfway levels (such as the special bus lanes). There has been some influence of the "green" candidates on the political process (for example, the pressure of the environmental movement in 1974 elections is regarded as having produced President Giscard d'Estaing's decision, shortly after his election, to abandon the controversial Seine left-bank motorway).

But the movement lacks the central direction and campaigns, as well as the ability to turn out masses of people for demonstrations, that characterized the earlier phase. There is a splintery quality to it, with new groupings (often offshoots of existing ones or "second hats" worn by individual leaders) popping up regularly. Clearly the tie-in with the burgeoning environmental movement will be a major source of new strength, although it is unclear how much the "dilution" of transportation issues with other environmental concerns may interfere with the achievement of specific, needed transportation reforms and demands. A further question to be answered is the role of this movement vis-à-vis the existing political parties and trade unions. The environmental movement is less strongly based in the working class than was the earlier transit users' movement. It also has an inherent difficulty in presenting comprehensive programs in the myriad areas that political parties and their platforms must cover. Whether the environment and transit users' movement and the existing left political parties will be able to work together in the coming years cannot at this time be firmly predicted. But it is likely that demands and goals with regard to transportation policy and programs will have only limited success unless they can reach out more broadly, as was the case in the earlier phase. As with the earlier phase, the goal of user groups and actions is not directly to take over provision of public services (except insofar as electoral politics implies a desire to take over the reins of government—not a very likely possibility for the ecology movement, as they realize). The goal is to influence and redirect public policy through political pressure on the existing government and its realistic challengers.

One further issue that must be solved is how transportation users can best be organized. To organize them at the place of employment—traditionally where the French have been best organized—stressing the journey to work, is to compete with more directly job-related issues of salary and work conditions. Furthermore, the variety of means and routes by which workers at a given workplace travel to and from work presents severe organizing difficulties (including facing the reasons the Communist party and CGT in 1970–71 were ambivalent about too sharp a polarity between the private auto and public transportation). Workplace-based organizing also omits the population outside the work force and trips unrelated to the journey to work. To organize users of a particular network or line presents different kinds of problems: people using a specific mode of transportation or given line have very transitory commonalities and contact; a wide range of differences exist among them with respect to their residence and workplace, the places where organization has traditionally and understandably been most successful. To organize on the basis of place of residence involves people with a multiplicity of transportation needs and patterns as well as the obvious imperative to extend residentially based organizations to other aspects of urban life. The French transit users' movement has a good deal to offer as a model to the United States in terms of the range of activities and demands that organized citizen activists can generate. In addition, closer study of the French experience can offer much insight on different modes and possibilities of organizing.[19]

[19] In a letter dated Dec. 15, 1979, commenting on the initial draft of this chapter, Jean Macheras of the Association Combat-Transport offers the following advice (author's translation):

> Based on my own experience, it seems to me that the following conditions are necessary for an effective and durable [transit] users' movement:
>
> To enlarge its sphere of activity to an entire city or even region, and not confine it to a single [transit] line or neighborhood.
>
> To be be independent of all [political] parties and trade unions.
>
> To carry out successive actions toward limited objectives, bringing each to its fulfillment (wanting to accomplish too many objectives simultaneously diminishes the quality of the action).
>
> To federate with other regional associations, and if possible on the national level.
>
> To seek out maximum contacts and common actions with [political] parties, trade unions, and organizations in order to realize one's objectives.
>
> To avoid appearing to compete with the above groups, in order not to alienate them (it is pointless to make other enemies), from which stems our refusal to engage in electoral politics.

It is clear that more and more people will be relying on public transportation in the 1980s and that this will increasingly be a major public issue. It is equally clear that public transportation systems are failing to meet users' needs. In early 1981 the Massachusetts state legislature was forced to create a financial rescue package when the Boston public transit system ran out of money in December and threatened to shut down service entirely; the transit agency there had voted to eliminate Sunday service, cancel bus service for 18,000 schoolchildren, and lay off 400 employees to provide sufficient savings to enable 1981 operations. In Chicago, the regional transit agency raised the basic fare to 80 cents in early 1981 with a $1 basic fare planned by the summer. In New York (and other cities), new buses and subway cars have such serious design and manufacturing defects that they cannot be used. In Birmingham, Alabama's largest city (with 286,000 population), the entire public transit system was shut down in March 1981 for lack of funds. Most alarmingly, the Reagan administration has proposed to cut federal mass transit operating subsidies by $3.4 billion and mass transit equipment and construction subsidies by $4.6 billion over the next five years (compared with the amounts that were available under the Carter administration projections and current legislation).

Organizing this huge constituency of mass transit riders is essential to improving public transit systems.[20,21] One mode of organizing is through election

[20] The *New York Times*' lead editorial of June 10, 1980 (p. 18) was titled "No Voice for Public Transit" and urged formation of a users' lobby, with adequate financial support and strong leadership:

> Mass transit is in deep political trouble and its friends are not doing enough . . . If New Yorkers want to avoid a 75-cent fare this year and $1 fare next year, they need to make the case for transit as strong as automobiles and other groups have made the case against . . . . There should be room for a public transit group that can speak for the need to maintain the system with the fiscal help of all who benefit from affordable fares and good services. These voices could make the case for tax supports, and equally important, act as friendly critics of the Metropolitan Transportation Authority, drawing on independent technical advice.

There are several such citizen groups in New York City (Committee for Better Transit, the Big Screechers, the Permanent Citizens Advisory Committee to the MTA), but they do not reflect the potential political power the 2 1/2 million daily subway and bus riders in that city could wield. (See letter to the editor, June 19, 1980, p. A22, in response to the above editorial, from Arlene L. Bronzaft and Stephen Dobrow, criticizing the media and the MTA for not "want[ing] the grass roots riders to have a real say in transit's destiny."

[21] Air travelers in the United States (a small number within a small stratum, at least) are organized into the Airline Passengers Association, a Dallas-based group claiming 50,000 members, which reflects business travelers (41 flights per year on the average). The association publishes a magazine *(First Class)* and newsletter and offers both services (insurance, accommodations and car rental discounts, assistance in retrieving lost luggage, intervention with the airlines on members' individual complaints) and a variety of more collective political functions—air safety advocacy (a dramatic example was their successful suit to force FAA to ground the DC-10 until design defects had been corrected), and attempts to influence regulations, fares, and airlines' practices in favor of the consumer.

campaigns—particularly those initiated at the grass-roots level, as described in the Swiss case study (Chapter 5), which provides a classic example of political mobilization and organization building. The Paris experience suggests the clear value of broad-based organization of consumers of public transit services and use of a wide variety of strategies, sources of support, and modes of influence.

ACKNOWLEDGMENTS

Jean Chase proved an extremely competent and valuable guide to and translator of French politics, culture, and language; without her assistance, my two weeks in Paris would have been far less productive. Jean-Louis Sarbib kindly referred me to her, as well as to numerous other valuable contacts. To those who gave their time for interviews, I am most grateful: Jean-Claude Delarue of the Fédération des Usagers des Transports; Michael Barbier of Régie Autonome des Transports Parisiens (RATP); Nicole May of the Mission de la Recherche Urbaine, Ministère de la Environnement et du Cadre de Vie; Françoise Guyon and M. Fradin of the Institut d'Aménagement et d'Urbanisme de la Région d'Ile-de-France; Maurice Plantier of the Syndicat National des Usagers des Transports; J. M. Offner of the Institut de Recherche des Transports, Ministère des Transports; Danielle-Anna Sapolsky of the Centre d'Information sur les Innovations Sociales; Rosalie Footnick; Cynthia Ghorra; Odile Hanappe of the Mission de la Recherche, Ministère des Transports; Michele Barriere of Les Amis de la Terre de Paris; Gus Massiah; Jacques Perrier; Paul Josse, Syndicat des Transports Parisiens; Louis Pouey-Mounou of SOS-Environnement; Jean Macheras of the Fédération Nationale des Associations des Usagers des Transports and Combat-Transport; Philippe Saint-Martin of the Groupement pour l'Insertion des Handicapés Physiques; M. Cisnaro of the Groupe d'Étude et d'Action pour la Circulation et les Transports dans la Région Parisienne; Louis Caul-Futy of the Confédération Syndicale du Cadre de Vie; and Roger LaPeyre of La Cité Humaine-Les Droits du Piéton.

APPENDIX

Principal Transportation-Related Citizens Organizations in Paris

Association Combat-Transport (63 rue Raymond Losserand, Paris 75014)

Contact: Jean Macheras. Created in January 1978 as an offshoot of the Fédération des Usagers des Transports, it is one of the most active local groups and a member of the Fédération Nationale des Associations des Usagers des Transports. It focuses on overall reform of the transportation system in the Paris region, advocating strong measures to favor public transit, two-wheeled vehicles, and pedestrians, to downgrade the automobile, and eventually to provide free public transit. Independent of political parties and trade union confederations.

Comité Nationale d'Action pour la Sécurité des Usagers de la Route (31 rue d'Enghien, Paris 75010)

Contact: Roger LePeyre. A federation of 14 groups, it works to lower speed limits, introduce more severe alcohol tests for drivers, improve vehicle inspection procedures and safety devices, eliminate specific danger spots. Member of FNAUT.

Confédération Syndicale du Cadre de Vie (28 Boulevard Sebastopol, Paris 75005)

Contact: Louis Caul-Futy. A general consumer organization with 60,000 members and some 650 member associations, far more working-class in its membership and concerns than parallel groups in the United States. Some focus on transportation issues, but not one of their primary activities.

Fédération des Usagers des Transports (35 rue Bourg-Tibourg, Paris 75004)

Contact: Jean-Claude Delarue. Created in 1970, FUT has been involved in anti-highway fights, defining alternative politics of transportation, improving the public transit system, reopening rail lines. Heavily involved in electoral politics, running candidates on environmental platforms at all levels of government. Regards itself as independent of traditional party politics to the same extent Association Combat-Transport is.

Fédération Nationale des Associations des Usagers des Transports (5 Boulevard Pérerie, Paris 75017)

Contact: Jean Macheras. National federation of transit users' groups, begun in June 1978. Held first national congress in Tours, October 1979. Has over 70 member associations, locally and nationally. Independent of political parties and trade union confederations.

Group d'Étude et d'Action pour la Circulation et les Transports dans la Région Parisienne (5 rue Faraday, Paris 75017)

Contacts: M. Pacary, M. Cisnaro. Created in 1973 (under the slightly different name Groupe d'Étude et de Documentation sur les Transports et la Circulation en Région Parisienne) by the Communist party, it does studies of local and regional transportation problems and issues in order to organize around these. Has individual, union, and local government members, not all of whom are affiliated with the party. Aim is to demystify technical solutions, provide a platform for action, and offer counterproposals.

Groupement pour l'Insertion des Handicapés Physiques (20 rue Paul Appell, Paris 75014)

Contact: Philippe Saint-Martin. Originating out of the needs and work of university students, GIHP has become a 5,000-member national organization and a force advocating the integration of the physically handicapped into French life—rather than establishing separate institutions for work, living, schooling, and movement. In the transportation area, it works to make public transit more accessible and more widely available; for those unable to negotiate public transit it works with RATP and other public transit systems in France to create minibus services. Apart from ideological commitment to integration, GIHP points out that access to public transportation can lead to overall social economies by permitting the handicapped to use regular institutions such as schools and workplaces, rather than having to establish special institutions.

Les Amis de la Terre de Paris (3 rue de la Bûcherie, Paris 75005)

Contact: Michele Barriere. Friends of the Earth is the largest and best-organized ecology group in France, with 170 local groups. Transportation-related activities of the Paris branch have involved opposition to high-speed train lines (based on loss of agricultural land, harm to animals, poor investment priorities, etc.), advocacy of bicycling (via mass bicycle demonstrations—a 1977 Manivelo with 10,000 people, for example), and, with Combat–Transport, helping to organize bus riders into route committees.

Pour la Cité Humaine–Les Droits du Piéton (31 rue d'Engien, Paris 75010)

Contact: Roger LaPeyre. Concentrates on pedestrian safety devices (zebra-stripe crossings, more traffic lights, etc.), antinoise and antipollution measures, more green spaces and rest spots, lower speed limits, accessibility to buses and métro, bus shelters, opposition to parking meters on narrow sidewalks (which impede pedestrian movement), eliminating or at least improving (through better lighting, elimination of stairs, etc.) subterranean pedestrian passageways. Founded in 1959, Les Droits du Piéton has more recently expanded its range of concerns into general environmental issues, altering its name somewhat by the addition of the words *Pour la Cité Humaine*.

SOS-Environnement (31 rue d'Engien, Paris 75010)

Contact: Louis Pouey-Mounou. Formed as the electoral arm of several transportation users' and environmental groups (FUT, Droits du Piéton, etc.). Purpose is to run candidates for local and national office as well as for the European Assembly in order to create pressure on major candidates in run-off elections to adopt SOS positions so as to attract SOS voters. Absolute priority for public transit is a basic element of their platform.

Syndicat National des Usagers des Transport (5 Boulevard Pérerie, Paris 75017)

Contact: Maurice Plantier. Established in 1958 to increase public involvement in transportation planning, it helped to form local groups (Fédération des Comités d'Usagers de Transports en Commun de la Région Parisienne was one of these). Sponsored national conventions in 1975, 1976, and 1977. At 1977 convention, federated structure was recommended, which led to formation of FNAUT. SNUT still remains in existence for membership by individuals living in isolated areas without local groups and by associations not directly connected with transportation (such as organizations of the elderly). Member of FNAUT. In operating style, more "establishment" than other citizens' transportation groups, reflecting its original character as a group of self-appointed, knowledgeable amateurs established to advise transit officials (its first president was a general in the army reserves, formerly director of troop transport).

Transportation Users' Movements in Paris: Some Comments

JEFFREY SUTTER
Community Organizer with the MUNI Coalition, San Francisco

For American transit users and activists, Chester Hartman's account of the Paris users' movements presents both an inspiring array of possibilities and a sobering reminder of how far we have to go before we will have an effective voice in our transit services and can steer U.S. transportation policy away from its current socially and environmentally destructive course. This inventory of diverse groups, techniques, and demands is an important guide to issues we will need to understand and alliances we will need to realize in order to create adequate political support for transportation alternatives. Public transit riders are no more organized than are postal patrons, though the crucial economic, political, and environmental significance of transportation problems make such organization more necessary by the year.

The United States has never had a period like that of 1970–71 in Paris, with its general public demand for improved transit and equitable funding. The politicization of life in France, the relative strength of the left, and the far greater dependence of Parisians upon transit (typical of the older European cities) lend a power and immediacy to organizing around transit issues that do not exist here. The dominance of the automobile and the suburb in American life has left transit as a marginal social service, almost a form of welfare; even in major cities, it has been in decline and eclipsed by political support for the auto. Americans have no awareness of the history of transportation for profit or of the "market" distortions that created the "love affair with the car." In a real sense, the auto age has been a technology force-fed to American society; the resulting problems in the transportation system and the use of resources finally reaching crisis proportions. Few are concerned with the class biases evident in the structure of services and the mechanisms of financing behind our auto, freeway, rail, subway, and bus systems.

Except for the typically isolated citizen advisory panels and other government-mandated "input" mechanisms, few riders' groups exist. Those that exist are mainly small lobbying groups in cities such as New York, Boston, and San Francisco. Some improvements (such as passes and schedules in San Francisco) have been won by individuals or small temporary groups. Few groups have the resources or patience adequately to address the complex planning and fiscal buck-passing scheme built into the multilayered local–regional–state–federal government.

Our transit groups are more of the style typical of the late-1970s Paris movement: smaller, fragmentary, concerned with the urban environment and user amenities. The major immediate task of these groups is to develop politically active "user committees" (usually area- or city-based) to press for basic passenger aids and service improvements (such as information, passes, maps, more regular and reliable service, additional routes, rational fares, shelters, etc). This task is the more difficult in the absence of the foundation laid by a Cartel-style mass mobilization and public debate on the adequacy of public transit services. American transit riders are organizing for the 1980s in the face of a "tax revolt," expansion of military spending at the expense of social needs, and attacks on public-sector employees.

Without large-scale involvement by riders, participation in planning and decisions has depended upon successful use of grass-roots public meetings, constant scrutiny of the city transit authorities, testimony of the "riders' position" at hearings, and wide dissemination of information by newsletter, leaflet, and mass media in order to create the basic preconditions for general rider participation.

The experience of the San Francisco MUNI Coalition has probably been symptomatic of the situation of United States riders' groups. The MUNI Coalition is a loose group of transit advocates and neighborhood activists trying to forge a constituency of riders of the city's Municipal Railway, which enjoys a relatively high level of partronage limited to the core city. A three-year effort to direct city financing policy away from the fare box predictably failed, with little likelihood that more than trivial business and auto-use charges will go into effect to balance a doubling of fares. (Without adequate state and federal funding, public transit in California has been a major casualty of the now infamous Proposition 13 property tax rollback.) However, a citywide redesign of transit routes was successfully opened up to wide public participation when the city became convinced that neighborhood opposition to the imposition of a staff or consultant drawn network would certainly prevent any implementation of plans unless substantial rider participation and support were encouraged. A lengthy series of meetings, workshops, and hearings articulated MUNI riders' needs and drew out some of the latent street-wise expertise of longtime riders and those who wait thoughtfully at bus stops thinking of improvements to "their" system ("The People's Road," as MUNI's motto says!). In this effort, the major focus has been on effectively democratizing the process and widening involvement in the planning and monitoring of the system through a combination of participation and confrontation. Even mutual encouragement and cooperation are sometimes possible: riders have common cause with drivers or with management on some issues, and there are allies within the system. The resulting reforms and improvements have not been very substantial so far; we are (by comparison with the Paris organizations) learning our baby steps.

Major efforts will be required to develop the strengths apparent in the two aspects of the Parisian movement (typified by the Cartel and by FNAUT and Combat–Transport, respectively). First, it is crucial to develop riders' groups to begin the self-organization process; this will be the foundation for any success. It is best done through riders who begin the self-education process regarding transit issues and set limited, achievable goals. (Careful work on specific issues may make the difference between being respected by ignored sideline critics and gaining a real voice.) This process can increase political awareness and foster "doorstep planning" by illuminating for people the "direct route" from angry waits at bus stops to an awareness of the political economy of transportation. It will encourage user involvement in the operation of a system responsive to the mutual needs of riders as people with common working-day conditions, needs for recreation, and mobility. Increases in transit use and dependence due to higher costs of auto ownership and fuel, and the resulting strain on capacity, may help create the basis for future mass demands.

Second, the general ecological critique and multi-issue approach of FNAUT will be absolutely necessary to the long-term restructuring of our transportation systems. Our movement will certainly be independent of political parties: we don't have them in the French sense and they are irrelevant to such policy issues. Comments by Jean Macheras are significant: he stresses this independence, urging a citywide or regional scale of organization (fitting the actual mobility needs of people and the necessary area of reform of the urban structure) with a gregarious approach to all other groups willing to advance transportation demands.

The demands of FNAUT are comprehensive and point in the directions we will need to take and to those we will need to work with in coordination and coalition: first, the right to public transit, inclusive of the demands of *all* the transit-dependent—even those dependent by choice—and, second, a systematic economic and ecological view that favors time-efficient, intensive modes of transport and a safe, human scale of movement. It should be possible to work effectively with environmentalists, alternative-energy advocates, antinuclear people, the antiwar movement, bicyclists—in fact, anyone and everyone who may begin to see and articulate the interrelatedness of these issues. Other urban quality-of-life and neighborhood groups, block clubs, tenants, "advocacy" planners, and economists are natural and necessary allies.

If we are to build sufficiently broad-based support to win any major improvements (and this will require truly independent power) two issues that threaten to marginalize our politics must be addressed. First, transit-priority measures need not be advocated apologetically after decades of slavish auto dominance, but we need not become the attackers of drivers (or of auto workers). Our goal is to make it possible and desirable for drivers to leave their cars yet still be mobile—to lure them, or the resistance will be massive. (Perhaps direct antiauto measures are more acceptable in Europe.) Second, it seems vital to work with the transit workers, whose unions generally support free and expanded transit services (understandably enough). Effective mutual support between riders and drivers may be a precondition for improved funding, adequate wages, and better conditions of service. If some barriers are overcome, such cooperation could have profound rewards for both transit users and workers.

The good example of Paris teaches us that flexible and thoughtful experiments with organizing, advocacy, and coalition building can develop into an effective arsenal of forms of public participation: from making democratic and public the drawing of lines on maps to direct action in the transit ways.

7

FRANCE
PORTUGAL
Madrid
SPAIN
MEDITERRANEAN SEA
ALGERIA
MOROCCO

7

CITIZEN ACTION AND PARTICIPATION IN MADRID

JANICE PERLMAN

Introduction

It is one of history's ironies that Spain, so long the bastion of fascist dictatorship, has today perhaps the most powerful independent citizens' movement in Europe. Almost every Spanish city has neighborhood associations, housewives' organizations, parent–teacher associations, organizations of pensioners and retired workers, and merchants' associations. These are not only relatively more numerous than in other European cities but also, in many cases, more developed in terms of militancy, consciousness, level of organization, and independence.[1] They oppose becoming simply neighborhood governments in a decentralized system or local chapters of the left-wing parties and cling fiercely to the position that their greatest asset is their autonomy.

Because these groups were born and developed illegally under the repressive conditions of the Franco state and because they contribute critically to the general struggle for democracy as well as to the specific struggle for more livable cities and neighborhoods, Spain is an essential component of any study of citizen involvement in Europe.

Within Spain, Madrid's level of citizen action is second only to Barcelona's. (In 1979, Barcelona had 150 neighborhood associations with 150,000 activists, as compared with Madrid's 120 neighborhood associations with 60,000 activists.) Madrid may be more instructive, however, in that (1) the Catalan nationalist nature of the struggle in Barcelona makes it less generalizable; (2) the fact that Madrid is the capital of a highly centralized country and very much under the

[1] See Manuel Castells, "Urban Social Movements and the Struggle for Democracy," *IJURR* II, no. 1 (1978).

Class	Center	Shanty town	Periphery
Working Class		case #1: Orcasitas	
Old Middle Class (elderly, retired, etc.)	case #3 La Corrala		
Mixed: new middle class and working class			case #2 Hortaleza

Figure 1. The three case studies were selected to illustrate the variations in both location and social class found in Madrid.

thumb of the national government permits direct observation of the interaction between the grass-roots movement and the political system; and (3) in metropolitan Madrid, there is now an experimental participatory planning system—Programa de Acciones Immediatas (PAI)—that illustrates the possibilities of moving from paternalism to conflict to coproduction.

A recent study of conflict and participation in Madrid, based on the local news sections of the major daily papers, showed that over the three-year period June 1975 to July 1977 there were 1,445 urban conflicts involving neighborhood associations. This represented 54% of all urban conflicts. The major issues were related to facilities, housing, and infrastructure and the locations spanned the entire metropolitan area, with most concentration on the area just outside the city proper; the actions involved everything from direct contacts with the administration (46%), to propaganda campaigns (30%), to mass actions (17%).[2]

Since almost every neighborhood in Madrid is highly organized and there is a great deal of diversity among them, I have selected three neighborhood case studies that will illustrate variations in both urban situations and class base, as shown in Figure 1.

The Context

During the 36 years of the Franco dictatorship in Spain, decisions concerning urban and regional planning and municipal service delivery—like all others—were made entirely from the top down. In the case of Madrid, this was even

[2] Eduardo Leira, Jesus Leal *et al.*, "La Participation Conflictual en Madrid: Movimentos Urbanos 1975–77," (Madrid: COPLACO, 1979).

more exaggerated, since the national government, through its relevant ministries, in effect ran the city directly, with even less delegation of authority to the local level than in other major cities. The results of this process are evident in the physical appearance of Madrid—much of its charm and comfort from the residents' point of view has been sacrificed to monumental buildings and multilane roadways considered symbols of "Madrid's place in the modern world."

During all those years, two types of indigenous democratic organizations existed: worker commissions in the factories and neighborhood associations in the communities. These groups were careful to observe the limits set by the Francoist law, but both provided an opportunity for experience in discussion of problems, collective decision making, (limited) action, and leadership development.

In Madrid the neighborhood associations were particularly effective. They existed everywhere, from the *chabola* (jerry built) communities on the peripheries of the city, to the new mixed-class housing developments in the inner ring, to the old downtown neighborhoods around the historic center. They helped each other cope with emergencies (as when someone's home was destroyed by fire or flood); they promoted popular culture through plays, picnics, and festivals; and they pressured the authorities to deliver basic services for the welfare of their neighborhoods.

By 1979 there were about 120 such groups in and around Madrid, as mentioned above. Their leaders were articulate and powerful; these groups were a force to be contended with. After Franco's death, they became one of the major building blocks of the new democracy. In fact, the entire metropolitan planning process in Madrid has been designed to utilize their capacity.

In April 1971, the first local elections since 1931 were held in Madrid. A coalition of socialists and communists won, and major positions throughout the city government were occupied by people either highly sympathetic to the neighborhood movement or spawned by it. Tierno Galvan, the mayor, is a socialist and mildly supportive; Tamames, the vice mayor, was part of the Communist party leadership and is an outright advocate; and the Urban Planning Directorate (Genecia de Urbanismo) is run by Eduardo Mangaba, an award-winning architect and Communist party member along with his chief planner, Ignacio Quintana, the former leader of one of the most active neighborhood associations in Madrid—the Union de Hortaleza. In atypical fashion, Quintana still lives in this neighborhood and takes an active role in local community affairs.

At the national level, the Commision de Planeamiento y Coordinacion del Area Metropolitana de Madrid (COPLACO), the planning authority for the Madrid metropolitan area, is headed by Fernando Teran, an architect-planner who took the job in September 1977 with the understanding that he would transform the entire planning process from "top down to bottom up, and from serving the interests of speculators to serving the interests of urban residents." These goals were institutionalized in January 1978 with the approval of the PAI, which

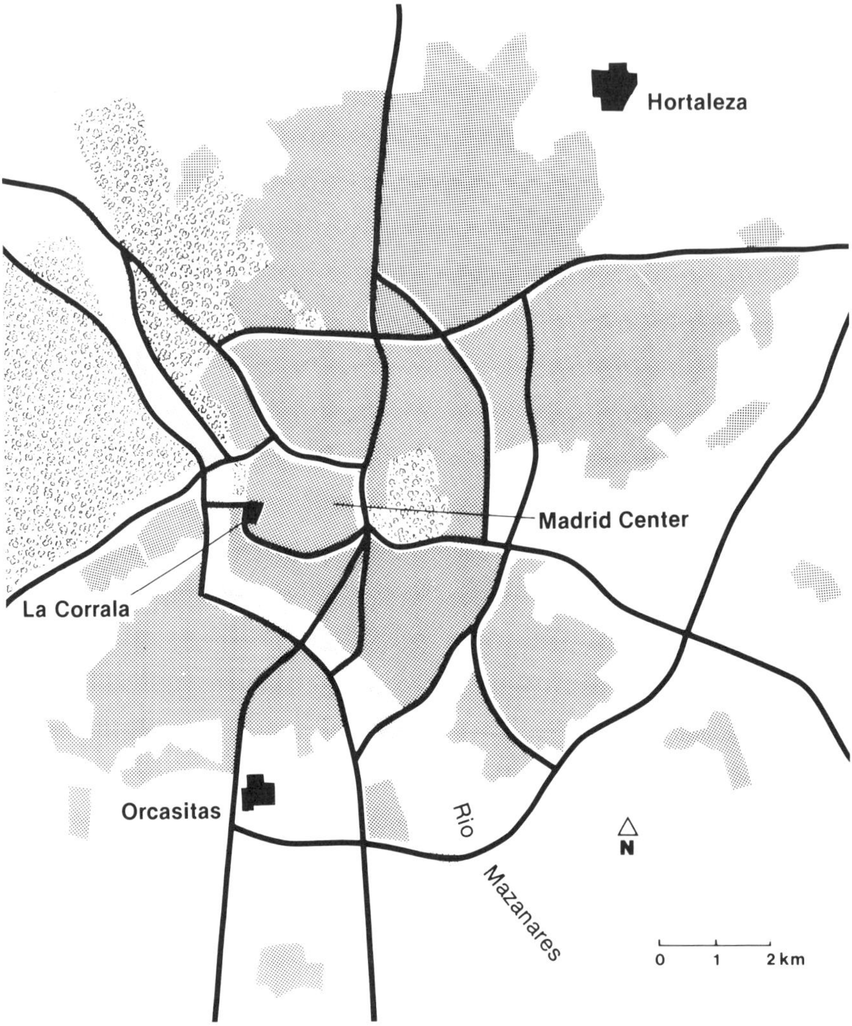

Figure 2. Metropolitan Madrid, showing the location of the three communities discussed in this chapter.

represents a participatory planning process for the entire Madrid metropolitan area.

Thus, in present-day Madrid, the entire logic of urban planning is neighborhood-oriented, citizen-based, and openly participatory. To provide a more precise idea of how this works, we shall describe the PAI and then present three case studies that focus on three distinct areas: Orcasitas, a working-class shantytown; Hortaleza, a mixed-class residential complex on the suburban periphery of the city; and La Corrala, a central-city historic neighborhood occupied mostly by elderly retired people.

Three Citizen Movements

The PAI: Program for Immediate Action

Madrid is a province of 4.5 million inhabitants, of which 4.3 million live within the metropolitan area and 3.3 million in the city.[3] Planning for the metropolitan area is coordinated by COPLACO, a special authority within the Ministry of Public Works and Urbanism. As of January 1979, the planning process for the metropolitan area is officially the PAI, mentioned above, which is premised on "making users the main resource in the planning process."[4] Through the activities of the residents associations, there was widespread public awareness of urban problems in Madrid, and the PAI was designed to utilize that capacity both as a source of information and as a mechanism for setting priorities. The thrust of the PAI is to refute top-down master planning in general and to replace the 1963 Madrid plan with a process that starts with the self-identification of neighborhood needs and builds a broader, coordinated perspective. The operating premise is that planners cannot accurately identify or assess needs without the sort of information aggregated by voluntary associations of all types at the neighborhood level, including traditional cultural and recreational groups and merchants' associations as well as neighborhood associations *per se*.

For the purposes of developing the local PAIs, Madrid is divided into 25 planning districts—17 within the metropolitan area and 8 in the provinces. Each area has a professional team paid by COPLACO (and in some cases by the district) to work with local groups in preparing a detailed inventory of short- and long-range problems. The first stage is to gather information about land use, open space, housing, infrastructure, transportation, and so on through cooperation with the local, "commissions of participation," which are self-organized in each district for this purpose. These commissions can include any association within the area and can designate any form of participatory process—whether through general assemblies, subcommittees, balloting, or other methods. Once the needs are identified and ranked, the group puts forth suggestions and proposals for how they can be met. These are compiled into the Program for Immediate Action with the help of COPLACO's staff. COPLACO publishes and disseminates these local studies and programs in order to permit the broadest possible feedback and to stimulate a secondary participation process. The programs are then coordinated throughout the metropolitan area on a four-year time frame,

[3] The growth of the metropolitan area has been so rapid that it has doubled in just over 15 years; it went from 2 million in 1960 to 4.3 million in 1977. Castells reports that 62% of all buildings in the metropolitan area as of 1979 did not exist in 1960.

[4] Quote from interview with Teran, July 13, 1979.

and an institutionalized commission (composed of representatives from the participating organizations plus the various governmental and private sources of funding and investment) is established to oversee implementation. The goal, according to the director, is to "achieve greater efficiency in the use of our limited resources and to increase public satisfaction with the way these resources are utilized and distributed."

Thus far, the PAI will deal with only existing, precommitted resources, but Teran hopes that more funds will become available and that additional resources will permit a subdivision into smaller PAI districts. He points out that each of the 17 districts in the metropolitan area is larger than most towns and cities in Spain, and hopes for the support and financing to subdivide these and provide the necessary technical assistance teams and central coordinating staff.

The PAI has faced opposition primarily from three directions: (1) political parties of the right for putting too much power in the hands of local groups and raising expectations, (2) planning professional for not giving sufficient weight to their expertise, and (3) large developers and construction companies for exposing their "deals" with the preceding government and limiting their freedom of action through closer monitoring. In fact, the Minister of Public Works and Urbanism, a member of the conservative Union of Democratic Center (UDC) party, which holds a majority in the national parliament, has tried to suppress the PAI and has succeeded in holding it to a much lower profile than COPLACO and Teran would like. The funds and personnel have been limited, and the idea of launching a full-scale national campaign about PAI with film, television, and radio coverage was vetoed as "politically unfeasible."

Only one PAI has been completed at the time of this writing—that of Hortaleza—which was chosen as a pilot project because of its very high level of organization. But as Teran put it, "At least all planning is now coming from dialogue with residents, not from planners playing God. . . . Thus, the results are much better because they take into account symbolic values and subtleties that planners have no way to ascertain."

Orcasitas: Case of a Chabola Neighborhood[5]

Key Actor: Felix Lopez Rey

Chabolas refers to the jerry-built shacks on the outskirts of Madrid. They are typically small, one-story whitewashed dwellings without running water, sewerage, or electricity. They are often built without legal title to the land on

[5] This section is based on interviews with Felix Lopez Rey, Eduardo Leira, and Manuel Castells in the period July 9–16, 1979. Further details on the Orcasitas struggle are described in Manuel Castells, *Ciudad, Democracia y Socialismo* (Madrid: Siglo XXI, 1977), pp. 104–114, and Castells, "La interaccion entre crisis urbana y Movements Social urbano: La experiencia de Madrid," unpublished manuscript, 1979, section 3.1.

which they are constructed. At present they house between 35,000 and 45,000 families.[6] As Madrid has expanded, the peripheral areas they occupy have become increasingly valuable and developers have sought to expel the *chabolistas* with little or no indemnification in order to build for maximum density and profit. The *chabolas* have thus become the focus of a collective struggle against displacement and in favor of highly subsidized on-site upgrading.

Orcasitas is a *chabola* settlement which was built in the 1950s and rapidly grew to about 2,000 families, mostly immigrants from Andalucia, La Mancha, and Extremadura who arrived during the 1950s and early 1960s. Most of the men work in construction. The process they went through in fighting against expulsion and then in favor of direct control over their housing and neighborhood became a prototype for the participatory planning model of the PAI. They established the legitimacy, credibility, and capability of neighborhood associations and they set critical legal precedents. They established the principle that they were entitled not only to remain in their neighborhood but also to receive a high rate of indemnification for their *chabolas*—and to be assigned to one of the new apartments constructed there at highly subsidized rates—on the grounds that they were the ones who had created and constructed that part of the city through their lives and work. In line with this, they also won the right to hire their own architects, planners, and engineers for the development of the new residential blocks and community facilities. These experts were to be selected by them, to be accountable to them, and to be paid for by the government; they were to have equal status with the government technicians. Finally, the residents insisted that no plans by any of the technicians would be effective without approval by the general assembly of the resident's association. "We built this area from nothing, by our own work" they contended; "thus any improvement made here should be acceptable to us and should benefit our people."

The residents' association of Orcasitas began with six friends who met weekly in someone's kitchen to discuss problems of neighborhood improvements. In 1970, they were officially founded to serve as a collective negotiating body to deal with the water company regarding the installation of a system of running water for the neighborhood. They have met every single week since that time, have held yearly elections for their executive council, and have set up an ongoing democratic structure flexible enough to deal with all sorts of problems.

Their original demands concerned the right of their self-built houses to exist and be maintained (they had to do all repair work at night, since it was illegal to upgrade their dwellings) and the struggle for basic urban services such as water, electricity, and sewerage.

In the early 1970s they built their association headquarters, working clan-

[6] This estimate was made by Felix Lopez Rey, president of the Neighborhood Association of Orcasitas. Manuel Castells gives a much lower estimate—6% of the population of metropolitan Madrid. (Interview July 12 and July 15, respectively.)

destinely after dark so they would have a place to hold their meetings. Membership swelled from 100 to 1,400, almost the entire number of households in Orcasitas. The association provided a place to get together for social events as well as meetings, served beer at cost, and provided public toilets and showers which the residents did not have within their homes.

The second stage of demands was an all-out battle against the existing "Local Plan" (created in 1963), which specified the removal of the *chabolistas* and the development of the area into expensive apartment blocks. The Association fought for the right of the residents to remain and demanded, furthermore, that the administration repay the costs of installing urban infrastructure—including water, electricity, roads, and sewers—that the residents had assumed individually and collectively during the previous two decades.

The residents took their cause from the city planning department of Madrid, (Gerencia de Urbanismo), to the metropolitan area authorities (COPLACO), and to the National Housing Ministry. Along with six other *chabola* settlements, they used a clause from the 1969 Local Plan (stating that "decent housing was the right of the residents of an area") to press their case all the way up to the Supreme Court. They waged a massive public opinion campaign through the press and radio, accompanying it with an array of activities ranging from one-on-one negotiations to massive sit-ins and demonstrations. One hundred thousand people were mobilized for a housing demonstration during this period.

The third stage of demands went beyond the right to remain in the area and called for a total upgrading of the housing, infrastructure, and facilities for the benefit of the residents and at a price they could afford. This included parks, schools, medical facilities, and so on.

Again the entire range of tactics was utilized and again the struggle was successful, despite major obstacles. Opposition arose from four sources: government bureaucrats and technicians (opposed to citizen participation in general and local control even more); conservative political parties; some Socialist party members (who felt that residents' associations were redundant and unnecessary since people elected local officials to represent them and defend their interests); and, finally, private construction companies, developers, and financiers who had benefited from the speculation boom of the 1960s and early 1970s and were reluctant to see it curtailed.

The outcome was successful on many levels. In terms of concrete demands, the residents of Orcasitas won (1) the right to remain in their community; (2) the benefit of a very high indemnification for their *chabolas*—enough to cover the down payment on a new flat; (3) the right to one of the 2,500 new flats being built in the area; (4) an ongoing rent subsidy for the new flat; and (5) an array of community facilities including a community center, a school and staff (at a ratio of 35 children per teacher), a 70-hectare public park, a health center, and facilities for the care of children and the elderly. The housing blocks are being constructed in three stages, the first of which was completed in 1975, the

Figure 3. Orcasitas. *Chabolas* and grazing goats.

second in 1979, and the third in 1981. The units cost $2 million pesetas each ($33,000), of which the residents will pay less than half and they will be allowed to spread this payment out over a 30-year period.

A group of sympathetic young architects and planners (directed by Eduardo Leira) called CETA was contracted by the Orcasitas Residents Association to work on the plans for the housing and community facilities. Their design for the overall area is based on fairly standard residential blocks and was done with minimal resident input, mostly resulting in commercial space, children's playgrounds, park-type benches in each block, and a new community center.[7] The apartment design was done in a much more participatory fashion. A life-size model of the proposed apartment layout was built in concrete just outside the neighborhood association headquarters, with walls only 1 foot high so people could walk around inside the space and get a feel for the arrangement of rooms and the spatial distribution. The more detailed and less abstract the decisions,

[7] One note here is that the residents had decided to keep their original association headquarters as a symbol of their struggle and a reminder of their joint and clandestine self-help effort in building it. Somehow, over the two- or three-year period of design and new construction, that idea was lost and a new community center of modern design was substituted.

the more participation was involved, so that in terms of floor materials, bathroom and kitchen tiles, wall and ceiling texture, and interior color, the residents had total say.

While the new buildings and layout in Orcasitas look fairly standard, even unattractive, the precedent of a residents' group hiring its own technicians who are accountable to them and paid by the government was a significant victory. The challenge of meaningful citizen input into the overall urban design remains. Now Felix Lopez Rey, the long-term leader of the association, who makes and fixes watches and clocks, wants to go beyond that precedent and get the government to *pay* for the time of the community people who work with the technicians. He has established a system for keeping track of the hours of work he misses while attending to community projects and will try to get reimbursed accordingly.

Beyond the immediate physical success of the Orcasitas struggle, there are social, cultural, and political outcomes of equal if not greater significance. Socially and culturally the residents' association played a critical role in building a sense of community solidarity and in rekindling the tradition of local fiestas.

Figure 4. Orcasitas. One model of internal apartment layout with the original neighborhood association headquarters to the right and the new housing blocks in the background.

It also organized excursions and picnics, produced plays, and showed movies. It set up a consumers' cooperative and a sewing cooperative. And it encouraged all sorts of informal interactions that go to building strong links within a neighborhood.

Politically, the experience of fighting for their rights to a decent home in their own community transformed the residents of Orcasitas from a largely apolitical group to one with a quite mature political awareness and sensitivity. While the residents originally participated only when the particular demand in question affected them directly, over time they came to take collective interest in the entire community and gradually in other *chabolas* and low-income neighborhoods as well. As Felix Lopez Rey described it, many leaders from other neighborhoods now come to Orcasitas to learn how they organized, and various leaders and members from Orcasitas have gone to help start or strengthen residents' associations in other parts of the city.

In fact, the Orcasita group was instrumental in creating a Federation of Residents' Associations in Madrid, which now includes some 90 community groups. It is one of the 15 groups in the core committee. It has also lent solidarity to the movement, supported strikes in construction, and served as the basis of electoral mobilization in the April 1979 municipal elections. In the Orcasitas region, 50% of the votes were for the Communist party, which supported an autonomous neighborhood movement, and only 5% were for the formerly all-popular Union of the Democratic Center (UDC), which strongly opposed the movement.

A number of factors help to explain why the Orcasitas efforts were so successful and how they were able to create a neighborhood for working-class people that has better community services, a better student–teacher ratio, and better housing standards than most middle-class neighborhoods in Madrid today. Partially it was due to the extraordinary leadership of Felix Lopez Rey. He is an artisan by trade but, most of all, he is a natural, unschooled leader, so articulate that he can talk circles around the highest-level bureaucrats and so much in touch with his constituency that he can mobilize them without fail. He recently turned down a prestigious and well-paid job as city councilor (consejal) to continue his life and role in Orcasitas.

A related but distinct reason for success was the skill of the residents' association as a whole in negotiating, mobilizing, and waging a media campaign that captured favorable public sentiment across the city. They were able to build a widespread base of support, playing on the guilt of an administration that had been aiding the profiteering of speculators at the citizens' expense, was rife with corruption, had frustrated its own staff, and had little legitimacy within its own ranks and less outside.

Overall, the success of the popular participation in Orcasitas can be said to have helped change the basis of urban development from benefiting the spec-

Figure 5. Orcasitas. The mural declares: "We need decent but affordable housing!"

ulators to benefiting the residents. In this process, the residents began to regain their sense of dignity, to exercise their rights as citizens, and to affirm the values of their traditions, their community, and their social class. As Castells puts it:

> Orcasitas changed. Not only did it have new houses, gardens, and schools, but also it had new people, people who overcame a sense of dependency and powerlessness and developed pride at being the vanguard of a growing social movement.[8]

Felix Lopez Rey put it like this: "Aside from having a house, we are becoming citizens. . . . We have become persons, no longer ashamed but proud to say where we are from."

[8] Castells, *Ciudad,* p. 113.

Hortaleza: Case of a Peripheral Mixed-Class Residential Area Key Actor: Ignacio Quintana

Whether you take 45 minutes by car or almost two hours by bus, the approach to Hortaleza is not attractive. Once you have left the city of Madrid, you are surrounded by the unsightly effects of its population overflow and 20 years of rapacious, uncontrolled speculation and development. What was once fertile agricultural land is now a hodge-podge of treeless, ill-maintained, over-congested roads, power plants, warehouses, factories, and block upon block of ill-designed and cheaply constructed high-rise housing interspersed with squatter settlements, all baking relentlessly in the sun, devoid of greenery. In the dry season, everything is covered with dust; in the wet season, it all turns to mud. Mile after mile of the same unfolds. Many of the units lack running water; most lack adequate health care, schools, playgrounds, libraries, commerce and other urban services; and all are miserably served by the public transportation system.

For a long time, people were resigned to the inconveniences and put up with their lot. Then they began to fight back and joined in the creation of a powerful citizens' movement. Now they are working toward taking control of the planning process and investment decisions in their areas. Of all of the residents' associations that are engaged in this process, the Union of Hortaleza is the best known and, in some ways, the most successful. It was chosen by COPLACO to be the pilot experiment for the PAI, the participatory planning process described above.

The district of Hortaleza was created in 1948 when the city of Madrid incorporated three rural municipalities: Hortaleza, Cannillas, and Barajas. The population at present is 20,000; the majority are ill housed and ill served. There are 2,023 *chabolas,* 4,086 UVA's supposedly temporary emergency dwellings (built by the state in the 1960s) and 2,088 units without running water. There are few jobs; the vast majority of working men and women must commute into Madrid daily, using three overcrowded access roads and a totally inadequate bus system. The only remaining open space in the entire district is 5,000 acres of undeveloped land (of which 1,000 is a greenbelt); construction companies have already planned to put 60,000 units there in the near future.

These and other problems are the result of rapid and ruthless development facilitated by the General Plan of 1963, which designated the entire area as a dormitory suburb to "absorb Madrid's growth" and was subdivided into 40 uncoordinated "partial plans."

The PAI for the district of Hortaleza was formulated by the Union de Hortaleza plus 12 other neighborhood associations and one housewives' association.[9] These 14 groups compose the Coordination of Citizen's Associations,

[9] The neighborhood associations are Fuente de la Mora, Manoteras–Querol, Carcabas–San Antonio, Sa. Lorenzo, Minar del Rey, Canillas–San Fernando, Portugaleti, Villa Rosa, Barajas, Alameda de Osuna, and Barrio del Aeropuerto. The housewives' association is Nueve Experanza de Canillas.

Figure 6. Neighborhood Absorption Units (UVA) of the Union de Hortaleza.

representing a population of 20,000 in the district. Among them, the Union de Hortaleza (known simply as the "Union") is by far the best known. It, in turn, is composed of three distinct subgroups. The first is the Unidades Vecinales de Absorcion (UVA), built by the government as temporary emergency housing for displaced *chabolistas* in 1964. The 1,100 units are totally deteriorated due to their cheap construction and lack of maintenance and are totally dependent on the Ministry of Housing, insofar as any repair or modification must be granted approval.

The second part of the Union de Hortaleza is the shantytown, Casitas Bajas do El Carmen. Units here were self-built in the same way as Orcasitas and face the same problems, that is, the increasing threat of removal as land values in the area go up.

The third part is the housing blocks of Santa Maria, comprising 15,000 units all built, owned, and managed by a giant company called Cryssa. Whereas the residents of the UVA and El Carmen are mostly construction workers, those in Santa Maria are industrial workers, technicians, and professionals. Many of the women from the UVA work as maids in the houses of Santa Maria; some of the men do odd jobs for them as well.

Despite these not insignificant differences, the three areas are located very near one another and have a great many problems in common. In 1974, they joined together to form a joint association—the Union de Hortaleza. Their first demands were for simple services and amenities such as street cleaning, maintenance of the gardens and common spaces, and more shade trees. The organization grew so rapidly that within a period of $2\frac{1}{2}$ years it went from 40 to 1,200 members. It is composed of general commissions from each of the three areas plus special commissions on cross-cutting issues such as school, culture, and so on. It is led by an elected *junta directiva* (executive board).

Each area had its own demands. The people from UVA wanted to transform their provisional housing site into a well-equipped and well-serviced popular neighborhood, with the materials for upgrading donated by the Housing Ministry and the residents in charge of the renovation. They especially wanted a health clinic, a new school, and a cultural center. "Casitas Bajas" of El Carmen had basically the same demands as the *chabolistas* of Orcasitas—they fought to avoid displacement and wanted to be rehoused in subsidized public housing within their own neighborhood on financial terms within their means and conditions set forth in a local plan drawn up by technicians hired by and accountable to the residents' association. Santa Maria was initially united in anger against the Cryssa Company for unreasonable rent hikes and poor service. They demanded completion of the urban infrastructure, better transportation to the neighborhood, a

Figure 7. Santa Maria district of the Union de Hortaleza.

meeting hall for the residents' association (to be donated by Cryssa), a child-care center, and attention to their social and cultural needs.

Given the diversity of the three groups, not all of the issues found unanimity. For example, the UVA group wanted to tear down the dilapidated and ill-staffed grade school in its midst, send their children to the school in Santa Maria, and upgrade their building to a middle school. The parents' association in Santa Maria, however, opposed the overcrowding (and social mixing) this would create in their children's school and supported instead an improved but separate school for the UVA.

Luckily, such conflicts were fairly easily resolved and the union continued to grow. On the one hand, they fostered the revival of local traditions, community spirit, and informal friendship networks through a series of fiestas, picnics, and children's activities. On the other hand, through their political battles, they became increasingly politicized, aware of the dangers of localism and supportive of coordinated efforts with other citizens' groups at the district level. They eventually joined with the 13 other groups mentioned above to develop the PAI and began publishing an excellent monthly newspaper entitled *Union de Hortaleza,* which publicly documented their struggles, victories, meetings, debates and festivities as well as relevant electoral issues and results. A rough chronology of major events in the Union de Hortaleza is presented in the chapter appendix.

The evolution of the union followed a pattern similar to that noted in other chapters of this volume. Beginning as a demand-oriented organization almost exclusively concerned with short-range economic issues, it evolved into a sophisticated and powerful component of the institutionalized mechanism for the metropolitan area. It fostered a deep sense of community cohesion within its boundaries and created outside allies among professionals, the press, and the political parties of the left while maintaining absolute autonomy from any of them. As the leadership group explained, "We are delighted that our struggle for democracy has made it possible for left political parties to play the role they have now assumed, but we don't want to be used or controlled by them . . . the survival of the citizens' movement is premised on our independence."[10]

Through the PAI planning process that evolved, the residents specified what their problems were as well as what possible, necessary, and desirable solutions would look like. The Union members attained a high degree of awareness, self-respect, and planning capacity. They started a self-managed adult school, successfully pressuring the Minister of Culture to pay for the building and the Minister of Education to pay for the teachers. They use the Paulo Freire method of teaching[11] to ensure that political consciousness develops along with concrete skills. They continually link their specific victories—such as stopping a highway

[10] Interview with Soledad Quintana, Carlos Mulas and Juan Rey, July 12, 1979.

[11] Paulo Freire, *Pedagogy of the Oppressed* (New York: Herder & Herder, 1970), and *Education for Critical Consciousness* (New York: Seabury Press, 1973).

from bisecting their neighborhood (which they did through PAI) or getting pavement, street lighting, improved water supply, and drinking fountains (which they achieved through various separate struggles)—to the larger, ongoing issues of land use and political control. As Ignacio Quintana said:

> The citizen's movement has dynamized and totally reversed the planning process. . . . The municipal government now counts on the citizen groups as its major resource—they are getting knowledge about problems, diagnosing the issues, and developing the ideas for a solution. They cannot be ignored because of their high level of mobilization; when you liberate them from the level of necessity and quantity, they will be free to explore questions of quality. They are our greatest treasure.[12]

La Corrala of Lavapies–La Latina: Case of Historic Center Preservation Key Actor: Isabel Vilallonga

At 8:30 P.M. on a July evening, one of the most pleasant places to be in Madrid is at the park in Lavapies–La Latina. It is an oasis of cool green grass, tall trees, and well-situated benches and pathways set in the middle of one of the oldest neighborhoods of central Madrid. It overlooks classic Roman ruins on one side and a large anti-renewal wall poster on the other. One senses a general relief from the heat of the day as children play on the grass, young couples stroll hand in hand, and old folks sit smilingly, taking it all in. It seems as if all the narrow cobblestone streets of the neighborhood eventually wind their way past this little park and that those who are not there are sitting on their balconies looking down over it.

This particular July night some 40 of the people who might ordinarily be in the park are gathering in a storefront on a nearby street to attend a meeting of their neighborhood association. Isabel Vilallonga—a tall, striking young woman with flaming red hair and a broad smile—is rounding up the troops who linger over hot sausages and beer at the corner store. They come down the street jovially, reminding friends of the meeting. Once inside, the serious work begins. Some of the residents are about to be illegally expelled from their apartments and the association called La Corrala is organizing an action for the coming Sunday to protest. It is also busily developing its PAI.

La Corrala was founded in 1976 in response to the growing problems of deterioration, speculation, and displacement. La Latina and Lavapies, like other old central-city neighborhoods (the buildings are from 80–200 years old), were rediscovered by real estate speculators just about the time the economic crisis began, the boom in peripheral suburban development started to die down, and the citizens' movement began opposing new development. The central city, with its urban infrastructure already in place and just a few minutes from downtown,

[12] Interview with Ignacio Quintana, July 15, 1979.

seemed a promising site for new office buildings and expensive high-rise apartments—in short, private-sector urban renewal. The residents of the area, almost all renters with very low, controlled rents (protected against expulsion) are mostly elderly retired people, pensioners, and small artisans and shopkeepers, and were regarded as "in the way" of the area's revitalization.

Due to the rent freeze, landlords let their buildings deteriorate until they were declared "in ruins," which gave them the right to evict the tenants without relocation pay. City hall then housed these displaced people in "supposedly temporary pensiones" at a cost of 57 million pesetas per year.

Three kinds of actions displaced this population:[13] large-scale government-initiated renewal projects in entire areas, conversion of public spaces like railroad stations into new uses for a more upper-class clientele, and the systematic destruction of individual buildings by declaring them in ruins.[14] These lots were purchased not by individuals but by the largest real estate and development companies in Madrid. As the new population moved in, the small "mom and pop" store on the corner became a jewelry store; the shoe repair place was bought out by a chic boutique; the hardware store was converted by new owners into a houseplant and ceramic center; and bars, restaurants and nightclubs began opening up throughout the area.

Meanwhile, deterioration and lack of maintenance were taking their toll. In 1974, three unrepaired buildings collapsed on their tenants, killing a few and injuring many. The next year, fires broke out, also taking lives. Thus, by one means or another, by threat or deed, by speculation or deterioration, many lifelong residents were pressured out of their neighborhoods.

It was in response to this set of conditions that the idea arose of extending the citizens' movement from the periphery of the city to its core. Along with a group of 8 to 10 community residents, the Communist (and Socialist) party with Isabel Vilallonga in the lead founded La Corrala. It was named after the typical buildings of the area, which are four-story walk-ups arranged around an inner courtyard with communal balconies running around the inside walls.[15]

The goals of the group were (1) to relocate displaced residents within the neighborhood at prices they could afford and in adequate housing conditions (i.e., to get public housing), (2) to rehabilitate the existing stock rather than destroy it, (3) to preserve the social as well as physical infrastructure of the neighborhood, and (4) to enhance the cultural and social life of Madrid's center and preserve its artistic and historic monuments. They also wanted to raise

[13] For more detail, see Castells, *Ciudad,* pp. 170–180, and "La Interaccion," pp. 311–318.

[14] That is, when a building was ill maintained to the point that rehabilitation would cost half as much as the worth of the existing stock, it was to be torn down. The point is that many buildings declared in ruin were quite adequate and simply needed maintenance and rehabilitation.

[15] The building called La Corrala is an open one of these, with three sides only, bordering the park mentioned.

Figure 8. La Corrala. Like many other buildings in the neighborhood, this one was declared "in ruins" and boarded up.

questions about land ownership and about who pays for what. Ultimately, they wanted the revitalization of the city to be paid for by the state with laws forcing landlords to give their tenants rights of first refusal on sale, and a series of disincentives for speculation.

The catalyzing issue for the group was the relocation of 300 families who had been evicted when their homes were demolished and who were being kept (at city hall's expense) in run-down "pensiones," without adequate light, air, or sanitary facilities. Only 20 of the 300 families were willing to relocate on the periphery.

The association demanded inner-city public housing, rehabilitation while maintaining people in the neighborhood, and an official bargaining process between landlords, renters, and the state mediated by La Corrala. After much pressure and publicity, the Ministry of Housing agreed to build public housing for them if city hall provided the land.

La Corrala progressively expanded its demands and widened its base to gain support among conservationists, ecologists, and middle-class city residents interested in cultural and historic preservation. Though the leadership continued to consist predominantly of students and intellectuals, the neighborhood base grew rapidly among the residents. By 1979, the membership had grown from 400 to 1,800, each of whom paid 100 pesetas per month to belong.[16]

The organization's structure is quite typical of Madrid's neighborhood associations. It holds an annual general assembly which elects a *junta directiva* (executive board), which, in turn, appoints various working committees. The executive board and working committees meet weekly to carry on business and can call a general assembly whenever any important decision needs to be made. One does not need to be a dues-paying member to vote; simply attending the assembly is sufficient.

Over the three years of its existence, La Corrala has used a wide array of tactics, from mass demonstrations of 5,000 to 10,000 people to block parties and from legal and juridical measures to pamphleteering, wall murals, and city-wide media campaigns.

They succeeded not only in obtaining central-city housing for the 300 originally displaced families but also in getting a pilot program (funded at 500 million pesetas over a three-year period) for the rehabilitation of deteriorated buildings in the historic center. From 1977 on, they virtually put an end the practice of tearing down buildings declared in ruins by city hall. They did this by getting legal dispositions concerning the historic and artistic value on the one hand and by the use of massive direct action on the other. They also managed to impose fines on landlords who did not make necessary repairs on their buildings or let them run down below code.

They were so successful that they inspired the creation of a number of other neighborhood associations in the central-city area. Together, then, these groups joined into an association known as the Coordination of the City Center which, along with massive support from public opinion, succeeded in winning the most impressive victory to date. In 1978, they forced city hall to adopt a Special Plan for the Protection of the Historic–Artistic Complex of the City of Madrid. This Special Plan, as it was called, virtually suspended demolition activities and new building permits in the old city, requiring rehabilitation instead. Its mandate, based on the historic–cultural preservation argument, covers 12,000 buildings, about 70% of the total stock in the central city. It grew out of the limitations of an earlier proposal in 1977 to do a "precatalogue" of 2,000 central-city buildings of historic and architectural merit which were to be preserved. By halting all demolition and construction for an initial one-year period, it bought time to work

[16] Interview with Isabel Vilallonga and Eduardo Leira.

out more long-range solutions. In the words of a local planner, it "put Madrid in the refrigerator."

The special plan was by no means perfect and was criticized by some of the left parties as being too demagogic and not providing sufficient legal and financial tools for implementation. And, of course, it was bitterly opposed by the Madrid Chamber of Commerce and the Association of Madrid Landlords. However, the Coalition of Neighborhood Groups—working with renters, left and center political parties, conservationists, and ecologists—prevailed.

In November 1979, the one-year period of initial approval came to an end. Numerous lawsuits have been filed against it, declaring the plan illegal. Also, in the interim, the code enforcement procedure has been improved, so that if renters denounce their building, the city council can take the necessary steps to have the building brought up to code, but the tenants still have to put up the money for the repairs and try to get it back from the landlord. (Until now, however, code enforcement applied only to new construction.) Discussion is under way about how to hold landlords more responsible and how to lessen the appeal of speculation through prohibitive use and height changes so that new owners could not convert flats into hotels, luxury apartments, or office buildings. The basic question is how to reverse the incentive structure within the legal constraints of the 1963 Master Plan, which is geared towards "modernization."

Meanwhile, the Coordination of Central City Neighborhoods has begun to develop a PAI, with Bernard Yzenga (a graduate of Berkeley's City and Regional Planning Department) as their chosen technical assistant. They got 1.5 million pesetas to formulate a new housing and rehabilitation scheme and to revise the 1963 Master Plan in such a way that it would be legal to reduce densities and prohibit certain kinds of uses in specified residential areas. They are working toward the recognition of decent, affordable housing as a *right* and establishing some mechanism of guarantee.

Isabel Vilallonga is skeptical about the PAI but is willing to give it a try, because she sees that the entire conception of planning in Madrid is being renegotiated from the former top–down approach to a new bottom–up impetus based on the strength, sophistication, and independence of the neighborhood movement. During the planning stages, her main concerns are to keep serving the people of the neighborhood directly through La Corrala. She and a core of members spend Saturdays and Sundays visiting individuals in the neighborhood who have filed complaints about their apartments, advising them what steps they can take to get the necessary repairs done. They also spend considerable time and effort fostering the expression of popular traditions in the neighborhood, including children's games, folk dances, and religious celebrations. Her other concern is that the group maintain its independence from political parties. She wants the group to be free to develop its own demands and programs—to have the strength to keep the parties accountable to the group rather than being a part of any one of them.

She sums it up as follows:

> La Corrala as the vanguard of the citizens' movement in Madrid's center city has helped to change the entire notion of "historic and cultural preservation" from the preservation of monuments to the preservation of people, their activities, and their neighborhoods.[17]

The Message from Madrid: Conclusions, Update, and Reflections

Initial Conclusions

The saga of the citizens' movement in Madrid is an inspiring one for many reasons. It is a clear demonstration that grass-roots efforts can transform the logic of urban policy and planning. The combined effects of the various parts of the neighborhood struggle actually changed the norm *from squatter removal to on-site, high quality subsidized housing for the poor* (Orcasitas), *from reckless speculation to equitable participatory planning* (Hortaleza), and *from demolition and new construction to preservation and rehabilitation* (La Corrala). With the PAI, the people turned the city planning process on its head, initiating decisions from the bottom up rather than the top down. As many of the people interviewed said, it changed Madrid from a "city for profit into a city for people."

It is also an impressive example of the maturing of a movement. While the neighborhood associations typically started in response to an external threat and clearly defined adversary, they were able to evolve into vital, ongoing institutions that could nurture and sustain their members in the absence of an enemy and could move from conflict to cooperation when appropriate. They went beyond the immediate demands of direct action groups to build and reinforce social ties and cultural traditions within their neighborhoods in a systematic manner. Simultaneously, they built their self-confidence and capacity so that, as the left came into power, decentralized decision making could take on real meaning. In fact, one could say it was the citizens' movement that educated the left parties and not vice-versa, as theorists of the "vanguard" would lead us to believe. The form of direct democracy and autonomous participation that the neighborhoods enjoy today in Madrid—and the way this complements the representative democracy through elected political parties—is the result of the practice and persistence of the neighborhood groups, not the preconceived notions of the left.

During the summer of 1979, when this case study was being conducted, this issue of autonomy was the major concern on people's minds—from government officials to neighborhood residents. Upon this question hangs the survival of the citizens' movement as it has evolved to date. Whatever direction the movement takes in the 1980s, however, the impact of that movement on the

[17] Interview, July 14, 1979.

people who made it, on the communities it constructed, and on the city it transformed will not soon be forgotten.

Update Through 1980

The concerns expressed in the summer of 1979 continue to grow in importance. The PAI planning process has continued during the year and a half since completion of this case study. Thirteen technical assistance teams were contracted by the government to work with the neighborhood groups and help complete their study and implementation plans. However, the teams ran into severe difficulties due to lack of political support from central government. The PAI implies a fundamental change in the planning process. While it was proposed and developed by COPLACO, an organ of national government, its devolution of power to the neighborhood level posed such a contradiction and threat within central government that it was never fully supported. It was not institutionalized into law, for example, and could not provide any legal framework for the planning process. Local government in fact, supported it with a good deal more vigor than COPLACO, using the studies (as each one was completed) to ascertain needs and priorities and to help make daily operating decisions and resource allocations. To be truly effective however, PAI needs much stronger support from national government. The PAI could then replace piecemeal opposition to the 1963 Master Plan with a more thorough revision of its basic precepts, especially those regarding property rights.

Meanwhile, the neighborhood groups continue in their work. In Orcasitas, new housing construction has continued into Phase 2, but the final 700 units have been halted. What looked like a huge victory at the time of this study now has to be fought for and rewon every step of the way. Regarding Hortaleza, COPLACO published the Union's PAI and has since done nothing. People are justifiably frustrated and increasingly skeptical. Finally, in La Corrala, work has come to a (temporary?) standstill since the key person within the Housing Ministry—who was pushing for their pilot program—was fired. A new contract is now being negotiated between central and local government to deal with the issue of inner-city public housing, and a new pilot project for 400 low-income units in the town center is now being renegotiated from ground zero.

Further Reflections

The story of Madrid can provide lessons and ideas about the tradeoff between *conflict* and *consensus* and between *strategy and spontaneity*. It further helps us reflect on the *blessings and curses of professional help* and *roles for community-based organizations in urban development*.

Conflict and Consensus. While the Madrid message seemed to reinforce a general theme in this volume dealing with what we have called the progression from conflict to coproduction, on deeper reflection it dialectically shows the opposite; that is, that the success of coproduction rests fragilely upon the ongoing underlying threat of conflict and the ability—perceived or real—to produce it. The unfolding of PAI seems to demonstrate that at each stage the administration had to be pressured to take the grass-roots partner seriously and that each co-operative effort was the result of many months, even years of conflict. Furthermore, the stronger the entrenched interests and the greater the opportunity costs, the more likely cooperation is to be symbolic only. Finally, the more centralized the government is, the more strongly it will resist democratization efforts. Thus, while the supportive left-wing local government put PAI on its feet, the more conservative national government crippled it.

As Eduardo Leira concluded, "to really have a participatory process for planning implies transformation of a world view, including all established procedures, and probably implies a totally new administration."[18] He went on to point out that the local government was not only caught between the people and central government, but that, even with good will, they did not know how to incorporate a "bottom-up" perspective in their work.

Strategy and Spontaneity. While it is true that the neighborhood associations of Madrid are in every way authentic and autonomous expressions of the concerns of the local populace, it would be misleading to construe their origins and development as entirely spontaneous. This is true for the United States as well. Citizen action groups do not usually spring full blown from the impromptu anger or frustration of community residents but are catalyzed into action by professional community organizers who, as Saul Alinsky put it, begin by "rubbing raw the sores of discontent."[19] In the case of Madrid, much of this catalyzing was done by trained political cadres, well aware of the need for such organizing at the community level. Much of the neighborhood movement was led and sometimes even initiated by them. Isabel Vilallonga in La Corrala and Ignacio Quintana in Hortaleza were already Communist party cadre when their respective organizations were formed, and Felix Lopez Rey, while not originally in the party, joined later, after much thought and debate among friends in Orcasitas and the technicians who were working there. This does not imply control of the movement by the Communist party—far from it. The position of the associations remains fiercely autonomous; they do *not* want to become simply chapters of the party at the local level. A delicate line, then, is constantly being drawn and redrawn in the relationship between the party and the neighborhood associations. The leadership, resources, continuity, and clarity supplied by party support has been invaluable, but group members retain a healthy skepticism

[18] Comments by Eduardo Leira at the German Marshall Fund Conference, Washington, D.C., April 28 and 29, 1980.

[19] See Saul Alinsky, *Rules for Radicals* (New York: Random House, 1971) and *Reveille for Radicals* (New York: Vintage Books, Random House, 1946).

about the dangers of control, the need for flexibility and innovative creativity, and—above all—a system of accountability downward to membership rather than upward to a party hierarchy.

One alliance the groups *have* tried to forge horizontally is with labor unions, though this has not yet been very successful. The links between quality-of-life issues in the community and quality-of-work issues in the factory have not yet been fully made, but many individuals on both sides are beginning to discuss the possible mutual advantages of working together (with or without party ties).

Roles for Professionals: The Helping Hand Strikes Again? It appears that even when sympathetic, politically astute technicians (such as those of CETA) work with neighborhood groups, the tendency is to relegate major decisions to the "experts" while leaving only the initial abstractions and final details for community input. This is not simply a question of intentions but of a pervasive mental set concerning "professionals" and years of reinforced deference to them.[20] Even where planning decisions are in fact open for debate, it is often difficult to obtain widespread participation by residents, especially if they consider the matter a "technical" one. What is needed is a self-educating incrementally designed process by which residents can make meaningful decisions throughout the stages of planning and implementation and gain confidence in the process. Two types of processes come to mind. One, the Paulo Freire technique, was first used in northeast Brazil as an adult literacy training method.[21] It is based on people's recognition of the worth of their own culture, on the perception of choice, and on the dignity and rights of human beings as actors in their environment. Freire calls it *conscientização,* which, loosely translated, might mean "consciousness-raising"; but it is much more subtle than the type of "political education" often attempted by the left. It is a process of critical thinking and analysis emerging from people's own experience which would be immensely interesting if incorporated into the planning process.

The second source of ideas comes from an experimental process called "Ecologue" carried out in Cambridgeport, Massachusetts in the mid 1960s.[22] To begin with, the process identified some 40 categories of people within the community according to common interests along age, sex, class, and occupational lines; one person from each category designated nine friends who fit the same description—for example, teenagers, students, working mothers with young children, retired men, and so on. The issue at hand was what to do with a vacated storehouse area, the only open space in the community, which MIT wanted to buy for high-rise faculty housing. Each group of 10 people met once a week,

[20] See John McKnight, "Enabling and Disabling Help in the Service Professions," Northwestern University, Chicago.

[21] See Paulo Freire, *Pedagogy of the Oppressed* (New York: Herder & Herder, 1970), and *Education for Critical Consciousness* (New York: Seabury Press, 1973).

[22] The project was directed by Stephen Carr, who is now working on a book about it.

first discussing what they liked and disliked about their neighborhood, then taking polaroid photos of places they most enjoyed and felt comfortable using in their neighborhood and those they most disliked or felt uncomfortable in. Finally, they did short videotapes along similar lines in each stage, developing a common vocabulary regarding their preferences and a frame of reference for what they would like to see in their community. Those groups most similar in values and priorities were then paired up, with five members of each dropping out to create 20 new groups of 10 each. The process of discussion and pairing went on until three very distinct groupings remained. Among them, they were then able to negotiate an acceptable plan for the site and, with the massive support plus technical skill and confidence they had developed, they defeated MIT's plan in favor of their own.

While sketchy, these two examples provide an idea of a participatory planning process in which the professionals are actual *facilitators* rather than *directors* in the process. Clearly, not everyone in a community or even in a group will have the time or interest to invest in such a process, but for those who do, the experience will have real meaning as opposed to the token majority vote between two preconceived alternatives presented by the experts (the most common form of input where any at all is elicited).[23]

The challenge to productive roles for technicians, professionals, or even party cadre in self-help organizations is to find some workable middle ground between the elitism and paternalism of the all-knowing expert, and the "co-pout"—under the guise of "what do I know? I'm simply here to learn from you"—of witholding understanding and skills that could be useful.

Roles for Community Organizations. Finally, the Madrid case raises the entire gamut of questions involving the potentials and limitations of neighborhood associations. Clearly in Madrid they served as (1) "free spaces" that provided training grounds for democracy even before it was publicly possible; (2) reinforcers of popular culture and folk traditions that had been undermined in the push towards modern homogeneity; (3) advocacy agents pressuring for legislation and demands of people and local communities in urban policy and regulation; (4) service providers, dealing directly with sanitation, child care, recreation, and so on; and ultimately (5) local planners engaged in needs assessments, priority setting, planning, implementation, management, and evaluation.

Four lessons became clear from observing the organizing process: (1) it is always easier to stop, block, or veto a proposed measure than to initiate and

[23] For example, in the Tondo upgrading project in Manila, Philippines, the World Bank Urban Projects staff presented a choice for site planning between (1) total demolition and a grid street pattern or (2) service installation around existing dwelling units to be upgraded without destruction. Despite a much greater financial burden and the fact that they would be displaced and homeless during the upgrading process, people voted for the former for reasons of greater symbolic prestige.

implement an alternative; (2) it may be easier to organize (as Alinsky did) around a clear common enemy, but eventually the tie must become not the hatred of "they" but the solidarity of "we" (as Chavez showed among the farm workers by reinforcing the ties, links, and bonds among them);[24] (3) as separate local groups, community organizations are too weak to effect any major changes in the distribution of resources or the logic of urban policies until they become a movement and forge a coalition; and finally (4) once they were all organized at the local level, they needed some mechanism of social justice or the "common good" to mediate among their own parochial demands. That is, the other side of neighborhood self-determination is some renegotiated form of the social contract. If this is not attended to, all decentralization means is that those units with the most power and wealth to begin with (as in the case of New York City's community boards) can best take advantage of the new channels opened up by devolution of power. Eventually, this widens rather than closes the gap between the privileged neighborhoods and the rest of the city. This future development of the neighborhood movement in Madrid will depend not only upon its internal strength and autonomy and on the evolution of government responsiveness but also upon the discovery of mechanisms that deal with competing and conflicting demands within the movement itself.

ACKNOWLEDGMENTS

Madrid is an extraordinarily powerful city, and the people who are struggling to make it a "city for people, not for profits" have undertaken an admirable and exceedingly difficult task. The more time I spent with them, the more humbled and impressed I became. All I can do here is to acknowledge my gratitude for the time they spent with me and for their willingness to share with me their stories, their homes, and the neighborhoods they love. I want particularly to thank Felix Lopez Rey from Orcasitas, Ignacio and Soledad Quintana from Hortaleza, and Isabel Vilallonga from La Corrala. I am also grateful to Fernando Teran of COPLACO for helping me to understand the PAI; Mayor Tierno Galvan and Vice Mayor Iamanes; Eduardo Mangaba, the head of the Urban Planning Department; Javier Angulo from CIDOR; and Angel Hernandez Craqui, the councilman responsible for neighborhood action services in Madrid.

Finally, the entire case study would have been impossible without the insights and access provided by the two study associates—Manuel Castells and Eduardo Leira. What is written here can only embody a small part of their constant committment to and deep understanding of the neighborhood movement and battle for social justice in Madrid.

[24] This point was first mentioned to me in 1973 by Professor Friedland of the Community Studies Board, University of California, Santa Cruz.

APPENDIX

Chronology of Union de Hortaleza[25]

October 1974: Founding of the Union de Hortaleza with Ignacio Quintana as president.

July 1975: The first contacts between the various associations within the district yield a preliminary joint report entitled *Program revindicativa de zona*.

December 1975: Representatives of all the associations demand a solution to their problems through an "open letter to the president of the municipal council."

Spring 1976: Published *Manifesto for a Democratic City Hall.*

September 1976: Presented the municipal council (Concajalia del Distrito) a detailed list of "urgent demands."

November 1976: Asked city hall to define a plan of municipal action for the Hortaleza district.

April 1977: Coordinating body begins its study of the neighborhoods, including a complete population census, housing survey, and detailed listing of all public spaces and equipment deficits. One year later, this is published as the now-famous "blue book" entitled *The District of Hortaleza: Today and Tomorrow*.

July 1977: The civil governor of Madrid visits the Union de Hortaleza and holds a working meeting with community representatives to listen to the findings of their study and their recommendations. This in itself is a victory. Also, around this time, Cryssa agrees to install the demanded civic center and to provide a fund for "fiestas," recognizing that the union's activities have raised the quality of the neighborhood and therefore, the value of the residents' land.

February 1978: COPLACO publishes its book, *Problems and Perspectives for Metropolitan Madrid*. This contains an analysis of the history of planning in Madrid and a critique of its present state. This leads to the proposal for a new planning process, the PAI.

April 1978: Coordination convokes a meeting on its own turf (in the residents' association of Portugalete) to make a public presentation of the blue book to the legislators and government officials. The executive committee of the neighborhood association meets with the Madrid

[25] For more information, see *Cambiar el Presente, Ganar el Futuro,* (Madrid: Coordinadora de Asociaciones Ciudadanes del Distrito Municipal de Hortaleza, November 1978), plus monthly newsletters, fliers, and documents. One of the ways they showed research results, plans, and priorities was in two complex matrixes. All the citizens' groups and neighborhoods involved were listed down the sides, and across the top were, on the first one, general questions (legal zoning and code, infractions, urban services and infrastructure, access and roadways, transportation, etc.), and, on the second one, specific services (education, health, open space, social services, administrative services, community services). Needs in specific neighborhoods are indicated by circles in the appropriate box.

deputies (deputados) from each party and the key city officials from COPLACO, the civil government, the district municipal council (Concejal de Distrito) and the municipal delegate (Delegado Municipal). A representative from the Developers' Association also attends. All approve the work of the coordinating group and agree to set up, within COPLACO, a working group composed of the *administration, political parties,* and *citizen associations* to follow up on the blue book study and recommendations.

April 13, 1978: The first meeting of the joint working group and COPLACO approval officially to begin the PAI in Hortaleza as a pilot project for Madrid. (Since then, the citizen associations have been actively elaborating the PAI and meeting regularly with COPLACO.)

May 1978: The coordinating group's public works committee presents a series of detailed maps and detailed programs.

July 1978: Group prepares "urgent actions" and proposes an eight-point program.

November 1978: Publish special newsletter on PAI entitled, *Change the Present within the Future.*

December 1978: Ignacio Quintana is named technical director of Madrid's Agency for Urbanism and leaves the presidency of the union to assume this position.

January 1979: The urbanism subgroup resents a proposal for an historic park and theater for Hortalcza, making public the beautiful grounds of a formerly private mansion.

February 1979: "The Day of the Tree" is declared by the ecology subgroup. Massive field trips and picnics are held throughout the district.

March 1979: In preparation for the upcoming elections, the coordinating group publishes and circulates *Ten Points to a District,* which raises a series of key questions for candidates to answer. The points deal with housing, the urban plan, the illegalities and abuses of real estate companies and developers, the lack of urban services, the lack of jobs, the scarcity of open space, the PAI demands for a democratic and accessible municipal government, and the role of citizen associations in control over their neighborhoods.

May 1979: As a follow-up to the ten-point general program, the Union de Hortaleza circulates its own sixteen-point flier entitled, *Principal Solutions That the Neighborhoods of the "Union" Expect.*

June 1979: Special supplement to the Union of Hortaleza newspaper on election results breaks down voting patterns in each of the fourteen neighborhoods and shows a considerable swing to the left.

July 1979: The PAI is finished and in print at COPLACO, which refuses to circulate a draft copy for this researcher to see.

8

NORTH SEA
THE NETHERLANDS
The Hague
FEDERAL
REPUBLIC OF
GERMANY
Brussels
BELGIUM
Bonn
FRANCE
LUX.

8

FOSTERING EFFECTIVE CITIZEN PARTICIPATION

Lessons from Three Urban Renewal Neighborhoods in The Hague

MARC DRAISEN

Dutch Society and Urban Renewal

There is no doubt that Dutch and American cultures and political institutions are very different. However, in the field of urban renewal, there have been many important similarities—both procedural and historical. Planning in both countries is conducted essentially at the local level by bureaucrats and members of popularly elected city councils. Physical plans are broadly influenced by policies adopted at the state (provincial) and national levels; regional and national authorities oversee (and in many cases must approve) planning decisions made by municipalities. Finally, most of the money that finances urban renewal stems from the central government and private developers (both profit and nonprofit).

Even more important, the history of urban renewal in both countries is very similar. In both the United States and Holland, early renewal plans called for the construction of high-rise office buildings, luxury apartment houses, and major highways. Significant displacement of residents was proposed and actually occurred in many cities. Postrenewal rents usually proved too high for the original residents of the neighborhoods involved. Citizens' groups organized to protest renewal plans and eventually to participate in the revision of those plans. Most municipal governments have generally accepted citizen participation as an unavoidable political reality, but effective participation is still elusive. Difficulties have arisen in getting "average" citizens to become seriously involved and to sustain their involvement over time. Finally, many if not most major plans are still made without significant citizen participation.

However, despite these similarities, Dutch and American societies have responded quite differently to calls for citizen participation in terms of both philosophy and methodology. These differing approaches to similar problems should prove most instructive. To focus attention on these approaches, this researcher spent sixteen weeks analyzing four participation experiences in The Hague, the Dutch seat of government.

The Hague has had successes as well as failures in citizen participation. The events in the Schilderswijk neighborhood were among the most critical occurrences in the history of modern Dutch urban renewal; the capacity of citizen groups in this neighborhood to substantively alter government plans on two separate occasions had reverberations in every major city in Holland. The Hague (like most other large Dutch cities) has recently adopted a formal procedure for citizen participation, but it was not as sweeping a reform as in Amsterdam, Rotterdam, or Groningen. On the whole, The Hague is a conservative city with a generally docile population, a bit less spectacular than some Dutch cities but perhaps more typical, and that is precisely why we can best learn from it.

Understanding the cases to be presented requires some background knowledge of Dutch society. Naturally, it is impossible to analyze a very complex social system thoroughly in one brief chapter. We can only highlight basic political and cultural facts that have bearing on the case studies and on the implications to be drawn from them. The discussion will emphasize the rudiments of Dutch politics and social traditions and two fundamental aspects of Dutch culture: the accomodation of differences between social blocs and deference to authority. It will also explain the role played by planning in the political process, recent stresses in Dutch social fabric, and the rising demand for political participation and democratization.

Political and Social Organizations

The Netherlands is a constitutional monarchy governed by a parliamentary system. The royal House of Orange is quite popular, and a new Queen ascended the throne in April 1980. Dutch society is traditionally organized into four major pillars, or *verzuiling*. These pillars represent the three major religious groups of the Netherlands: the Catholics (approximately 40%), the Dutch Reformed (a Calvinist denomination comprising about 28% of the population); and the Re-reformed (a stricter Calvinist group of about 9%); plus the secular (or "humanist") Dutch citizens who have no formal religious affiliation (18%).[1] Five percent of the population belong to other religious denominations not included in the three major religious groups.

The division of society into these four separate pillars pervades all aspects

[1] Arend Lijphart, *The Politics of Accommodation: Pluralism and Democracy in The Netherlands*, ed. 2 (Berkeley, Calif.: University of California Press, 1975), p. 16.

of Dutch social life. The most obvious aspect, perhaps, is the existence of political parties associated with each group. Support for these political parties from their respective constituencies has been remarkably constant since the beginning of the century. The division of Dutch society according to bloc spreads beyond political party, however. Individual blocs have their own newspapers, television stations, schools, hospitals, social welfare agencies, and labor organizations; even retail stores are often informally labeled by the bloc membership of their owners.

Despite deep divisions among the pillars of Dutch society, the nation is stable and passionately democratic. Equity among the groups is manifested in many ways throughout society. The existence of parallel institutions (social, economic, and political) for each bloc indicates that no group has been prevented from developing its own "internal society" to whatever degree it sees fit. Air time is provided to bloc television stations in proportion to their memberships; financial aid to church schools is appropriated according to enrollment.

Accommodation and Deference

The roots of this anomalous situation of separate but equal roles can be found in a complex system of accommodation that developed during the early years of this century. At that time, three critically divisive issues dominated Dutch politics: state aid to religious schools, extension of the franchise, and collective bargaining and labor rights. Ideological and religious debates threatened to destroy the Dutch state. (Details of the conflicts need not be presented here; a fuller explanation can be found in *The Politics of Accommodation* by Arend Lijphart.[2]) In 1913, Prime Minister P. W. A. Cort van der Linden attempted a solution by establishing commissions to seek compromise solutions. After months of deliberation, a compromise was reached. Lijphart cites three factors as key to the achievement of a solution:

> (1) the pre-eminent role of the top leaders in recognizing the problems and in realistically finding solutions in spite of ideological disagreements; (2) the participation of the leaders of *all* blocs in the settlement; and (3) the importance of the principle of proportionality in the substance of the settlement.[3]

This "peaceful settlement" (known as the *Pacificatie* in Dutch) represented a critical political turning point in the Netherlands and ushered in an era where the accommodation of differing interests became the watchword of the Dutch state.

Lijphart identifies seven "rules of the game" that allowed the Dutch to

[2] Ibid., p. 53.
[3] Ibid., p. 111.

continue the process of accommodation for fifty years following 1917. These rules are as follows:[4]

1. Government is very much like business. It is a serious means toward a serious end and not a game in which individuals play differing interests off against one another regardless of risks to the system as a whole.
2. The major blocs agree to disagree about major ideological and religious questions without dragging them into daily politics.
3. Summit diplomacy among the elites of the blocs serves to maintain communication and achieve compromises on important issues.
4. Proportionality governs key aspects of state life. Jobs in the civil services, air time for television and radio broadcasting companies, financial assistance from the central government, and other scarce resources are all allocated with bloc proportions in mind.
5. When an issue cannot be neutralized by proportionalizing an allocation decision, it is often "depoliticized" by resorting to legal or constitutional principles or even modifying facts and figures deliberately to preserve the peace.
6. Secrecy governs negotiations among the elites of the four blocs to prevent the need for face-saving showdowns.
7. The government has the right to govern. This attitude reflects the high level of Dutch deference to authority. The right of the government, once elected, to lead the nation as it sees fit is manifested in the general quiescence of Parliament regarding ministerial actions and in the docile way in which most citizens have accepted the dominant planning role of the Dutch municipality.

This seventh rule of the game leads us to a discussion of Dutch deference to authority, a key element in the Dutch code of civility. Dutch civil behavior precludes violent displays of emotion and rejects the indulgent pursuit of personal goals. Civility and tolerance go hand in hand in the Netherlands provided that no group violates basic codes of conduct. This "code of orderliness" was preserved by middle-class Indonesians who moved to Holland in the 1950s. It has been similarly maintained by other ethnic groups who have found havens in the country, such as Spanish Jews and English Pilgrims. This norm is currently being violated by South Moluccans (Ambonesians), Surinamers (immigrants from this former Dutch colony in Latin America), and Mediterranean guest workers who live in styles very different from that of the Dutch majority. There is significant discrimination against these groups, and integration appears unlikely—apparently proving that the Dutch are willing to accept people who think differently but not those who act differently.

Perhaps the most critical aspect of Dutch civility for our study is deference

[4] Ibid., pp. 122–38.

to authority. Dutch citizens are not likely to act singly to redress social grievances; in fact, like many Europeans, the individual citizen is most likely to do nothing at all, contrary to the more activist outlook prevalent in the United States. The idea of organizing an informal group to seek redress is far more acceptable to the Dutch. They are also more likely than other nationals to work through existing formal organizations like political parties or labor unions.[5] These attitudes indicate a high level of deference toward leaders, skepticism about the correctness of one's own opinions (or about one's power to induce change), and a willingness to approach leaders through formal or informal organizations *if* an issue is serious enough to demand attention.

The system of accommodation may seem an unlikely setting for the development of a movement for political participation. The acceptance of summit diplomacy, the conspiracy of silence, and the intensity of deference to government leaders suggest that the Dutch system of accomodation is not likely to spawn demands for citizen involvement. In the past, this has been true. As long as accommodation worked to the satisfaction of most citizens, they were willing to allow bloc elites to control the country's destiny—and the future of individual cities and neighborhoods as well. In recent years, however, the process of accommodation has decayed, and with it popular deference toward government decision making.

Planning in the Dutch Political System

Planning in the Netherlands is deeply rooted in the age-old need for neighborhoods and communities to join forces to protect the country from the ravages of sea and storm. As one observer has commented, this traditional need to plan the physical environment has made the Dutch people significantly more amenable than Americans to planning as a national (and local) public policy tool. Today, Dutch land-use planning revolves around a series of plans that are completed in succession. A typical series of plans for the renewal of an urban neighborhood would include the following: a *structuurschets* ("structure sketch") describing long-term goals for the neighborhood and outlining ways in which the municipality might achieve these goals; a *bestemmingsplan* ("destination plan," more loosely translated as "allocation plan") explaining the land uses that will be permitted in various parts of the neighborhood; the *verkavelingstudie* (or "parceling-out study") which gives the plan spatial detail; and a *bouwplan* ("building plan") that provides architectural details for individual buildings, parks, and roads.

In addition to the public notice and review requirements of the 1965 Physical Planning Act, national government policy supports citizen participation through the urban renewal appeals process and by providing subsidies for participatory

[5] Ibid., pp. 145–54.

efforts. A central part of the Dutch accommodation system is a procedure allowing any Dutch citizen to appeal virtually any government decision to municipal, provincial, and Crown authorities. After publication of a destination plan, for example, anyone can lodge a complaint with the town council, seeking some amendment to the plan. There are no restrictions regarding who can lodge a complaint or as to the nature of the complaint. Legal counsel is not required for the complaint procedure. If the plan is adopted (*vastgesteld,* more literally "established") by the town council, complaints about the plan can be lodged with the Provincial Executive. If the plan is out of accordance with the regional plans of the province or if the Executive believes that one of the complaints against the plan is justified, it can send part or all of the plan back to the municipality for amendment. If the plan is approved (*goedgekeurd*) by the Executive, it is still possible to lodge a complaint with the Crown. The Crown appeal procedure is not token in impact; about 50% of the Crown appeals are granted annually, and many of these assist common citizens who would have been adversely affected by municipal plans.

In addition to municipal urban renewal subsidies that may be used to finance participation efforts, the central government directly finances citizen participation through two programs. First, it funds experiments in citizen participation. Communities must apply for these funds through the Ministry of Housing and Physical Planning, which judges applications not only on the objective potential of the procedures suggested but also on their newness or experimental quality. More importantly, the Ministry of Culture, Recreation, and Social Work pays the salaries of about 500 "social/cultural workers" (*opbouwwerkers*), who are active in renewal projects throughout the country. The Ministry also subsidizes the operating expenses of many organizations run by these *opbouwwerkers,* providing money for supplies, news bulletins, and even political demonstrations (generally against the local government). In many cases, subsidies are provided to organizations run by neighborhood residents who are not *opbouwwerkers,* as is the case in the Schilderswijk neighborhood which we will examine later.

The payment of salaries to social workers who organize demonstrations against city administrations may seem to be an unlikely political situation. However, it stems from the special place social/cultural workers hold in Dutch society and the high value which the Dutch attach to "welfare work" (*welzijnwerk*). As urban residents have formed action groups to deal specifically with complaints against the municipality, they have demanded (and usually received) subsidies to hire *opbouwwerkers* and other staff, many of whom espouse radical ideas and spend much of their time organizing political dissent. Both central and municipal authorities have found it politically disadvantageous to move against these workers, fearing reprisals from the action groups and their voting members.

In 1979, however, the entire system of financing social work underwent a major change. Instead of dispensing subsidies from the level of the central government directly to the organizations that hire social/cultural workers, the Ministry is now disbursing its subsidies to municipal authorities according to a mathematical formula. The decentralization scheme was intentionally designed to give municipal governments maximum flexibility in determining how the money will be spent. The new scheme bears striking resemblance to our own Community Development Block Grant Program, which seems to enjoy quite a good reputation in the Netherlands—at least among government officials.

Many *opbouwwerkers* and the organizations they serve are openly hostile to the decentralization scheme. Although workers for the more established and "respectable" social service agencies claim their finances are safe, the employees of smaller and more dissident action groups fear municipal authorities will cut off funding from any groups that "make too much noise." They would much rather receive their money from distant bureaucrats in the Ministry than from the aldermen and planners they are regularly fighting in their own cities. An even greater worry is that neighborhood groups will begin to fight among themselves as welfare funds grow more scarce.

Secularization, Protest, and Urban Renewal: The Growing Demand for Citizen Participation

Now that we have explored some of the underlying attributes of Dutch culture and politics, it is possible to say a few words about the growth in demand for public participation. It was not the purpose of my research to determine the roots of this demand, so I am hesitant to claim the discovery of causal links. However, my research in Holland led me to believe that three trends in recent Dutch history may help explain why citizens are demanding greater involvement in government at this time: (1) the increasing secularization of Dutch society, (2) the legitimation of protest as a political tool in Dutch politics, and (3) insensitive urban renewal planning. We will discuss each factor separately.

Increasing Secularization. The powerful position of the churches as pillars of society depended heavily on popular devotion to religion and deference to religious leaders. This preserved not only bloc segregation but also public acceptance of the role of religious elites in shaping national policies. During the 1960s, however, as part of the general retreat from religion, the Dutch willingness to be led by religious elites declined. Support for the five major political parties waned in favor of numerous new splinter parties, with the three religious parties suffering the greatest losses.

These forces indicate a decay in the strength of the *verzuiling* system in the Netherlands and the gradual shift of Dutch public support to secular media, political parties, and other institutions such as schools and unions. Not only are the people less willing to have religious leaders negotiate key issues in Dutch politics but these leaders are also more reluctant to accept a system of accommodation. As religious issues became less important in national life, secular issues have caused deep rifts within the blocs themselves—especially with the weakening of religion as a binding factor.

At the same time, tensions have sharpened between the centrist forces of the religious parties and the socialist Labourites. By the early 1970s, the Labour party had ceased to support the system of accommodation. Under pressure from the left wing of its own membership, the party began reemphasizing traditional socialist demands. In 1977, the three major religious parties formed the Christian Democratic Association (CDA), a new coalition whose primary purpose was to prevent further losses by the religious parties to Labour. This increasing polarization has been evident in recent elections, which present voters with the kind of "clear alternatives" that rarely occurred in the bland elections of the accommodation era.

Acceptance of Protest as a Legitimate Political Activity. Dutch deference to authority has waned during the past 15 years. Action groups to deal with political problems (particularly in urban areas) have proliferated, frequently led by social/cultural workers (and often by Marxists in the nation's larger cities). These groups have become increasingly willing to engage in demonstrations, acts of civil disobedience, and violence as they find the *verzuiling* and electoral systems notably impregnable to their demands.

The movement sprang from real dissatisfaction with a process governed by religious and secular elites. While urban and highly educated young people are those most likely to participate in direct action, the tendency to be involved in such actions has "spread to the older and less educated residents of cities."[6] This trend can be seen as evidence of the increasing legitimacy of protest as a form of political activity in the Netherlands.

Insensitive Urban Renewal Planning. A direct link can be drawn between the insensitivity of government plans for urban renewal and the demand for citizen participation. Urban renewal began later in the Netherlands than in the United States. After World War II, the Dutch government concentrated energy and money on the construction of new housing to solve the critical housing shortage, which remained as a legacy of the occupation. Much of this new construction occurred in urban peripheries, helping to spur the decline of inner-city neighborhoods. It was not until the early 1960s that the Dutch government began to pay attention to its "blighted" central cities. The chief mechanism for renewal, however, was demolition.

In early cases, Dutch deference to authority (and the unwillingness of the major parties to object to the destruction of inner-city neighborhoods) muted significant public response. Massive displacement occurred in several neigh-

[6] Social and Cultural Planning Office, *Social and Cultural Report 1978* (The Hague: State Publisher, 1978), p. 205. The surveying process involved only people who regarded themselves as members of a church.

Figure 1. Postwar housing created during the building boom of the 1950s.

borhoods of The Hague with relatively little opposition. Gradually, however, students, left-wing politicians, *opbouwwerkers,* and neighborhood leaders organized and demonstrated to fight the demolitions.

Convinced that the government would never take their demands seriously unless residents participated directly in the planning process, the leaders of these confrontations generally demanded far greater citizen involvement in urban renewal planning. Municipalities have not responded uniformly to these demands, but many have recognized that citizen participation is indispensable to the modern renewal process.

I would postulate that these three factors (secularization, the legitimation of protest, and insensitive urban renewal) were catalysts for the growing demand for citizen participation. In the past, institutions associated with the pillars of Dutch society could mediate an individual's complaints about the system. One would seek the redress of grievances through political parties, trade unions, and other institutions associated with a church or the secular bloc.

In recent years, however, these associations became less helpful to Dutch citizens, partly because the religious elites failed to provide assistance where it was desperately needed (as in their failure to prevent massive displacement in the inner city) and partly because people's expectations had increased (which helps to explain the call for major expansions of public services in the 1960s). As a result, people drifted toward the secular bloc or toward splinter groups associated with no bloc at all. Since they may well have felt abandoned by leaders they had previously trusted, deference toward authority declined and protest became an acceptable political instrument.

However, ingrained cultural traditions die hard. Although they are no longer afraid to express their disagreements strongly, the people of the Netherlands are still deeply committed to preserving the system around them for the sake of the "common good." Accommodation may no longer rule the political life of the country, but it remains a respected mode of social conduct.

This condition gives rise to two factors that are both critical to the growth of citizen participation as a national movement: (1) recent developments provide an impetus for citizens to recognize and defend their own interests vigorously and (2) Dutch traditions, on the other hand, provide a willingness to sit down and reach compromises in an atmosphere of mutual respect and trust.

This does not mean, of course, that conditions for citizen participation are universally excellent throughout the Netherlands; our cases in The Hague will certainly indicate that participation does not always work flawlessly. Nonetheless, the country does appear to provide a healthy climate for the growth of effective citizen involvement in government decision making.

Evaluating the Effectiveness of Citizen Participation

Researchers and practitioners who have written about citizen participation usually explore techniques that can increase or improve citizen involvement in government decision making. The literature on how one might evaluate the *effectiveness* of citizen participation efforts (and the specific techniques in use) is much more limited.

In a recent paper, Judy Rosener lamented:

> Too few evaluations generate data on the effectiveness of techniques in the context of some set of goals and objectives. For the most part, existing evaluations are "after-the-fact," philosophical, and lack support for a determination that there is a relationship between a technique and some desired

> outcome. Even in those cases where the term effectiveness is used (implying that a specific technique produces some intended effect) the criteria for measuring effectiveness are not spelled out.[7]

Before we left for Europe, members of our project team were specifically asked to address the question: What makes citizen participation effective? In attempting to answer this question, I am guilty of at least two of the charges leveled by Rosener. My evaluation is after the fact and largely philosophical. It is after the fact because it is difficult for Americans to induce European governments to launch controlled experiments in social science research. The fact that some provocative cases can be studied only ex post facto should not, in my opinion, stop us from studying them. Nonetheless, I carefully avoided drawing causal links between given techniques and outcomes when I believe that only correlation can be safely supported by the evidence I have gathered. In addition, the study is also somewhat philosophical in searching for implications in the data. This seems easily justifiable, since the struggle for greater citizen participation in government has involved serious conflicts between differing value systems over such questions as these: Who will determine the future of residential neighborhoods? What are the limits of representative democracy? How will the interests of the inarticulate and disadvantaged be represented in society?

Later in this section, I will attempt to overcome the third charge that Rosener has leveled against the evaluators of citizen participation; I will present a scheme of criteria through which we can evaluate the effectiveness of citizen participation in The Hague.

Defining Citizen Action and Citizen Participation

Before moving to the evaluation criteria, I would like to explore possible definitions of citizen participation itself—and contrast it with citizen action.

In the cases I studied, and in many of those examined by other members of our project team throughout western Europe, citizen action and participation seemed to be ends of a continuum that described the historical evolution of participatory processes. Stuart Langton argues that citizen action "is initiated and controlled by citizens for purposes that they determine. This category involves such activities as lobbying, public advocacy, and protest."[8] So defined, citizen action can include neighborhood organizing, coalition building, and the development of alternative plans.

Citizen participation, on the other hand, is a process wherein citizens and

[7] Judy Rosener, "Evaluating the Effectiveness of Participation Techniques: Whose Criteria? Who Cares?" Presented at the Symposium on Citizen Participation: Models and Methods of Evaluation, February 4, 1980; Washington, D.C., p. 2.

[8] Stuart Langton, "What is Citizen Participation?" in *Citizen Participation in America,* ed. Stuart Langton (Lexington, Mass.: Heath, 1978), p. 21.

government officials jointly plan or implement public policies. This process contains at least the following components:

1. Both citizens and government tacitly or explicitly agree to cooperate with each other in the formulation of policy; although cooperation need not be complete at all times, an atmosphere of mutual trust and respect must dominate the process.
2. Both citizens and government have real power to affect public-policy decision making; although one side may have more power than the other, the power of neither side may be trivial.
3. As a process rather than a stable state, citizen participation may evolve or regress. An example of evolution would be the expansion of subject areas covered by the process or a deepening of cooperation between government and citizens; an example of regression would be a lessening of cooperation and a return to conflict.

One possible criticism of this definition is that it is too broad, because it overlaps with the definition of citizen action discussed above. Certain forms of lobbying, coalition building, the development of alternative plans, and many other aspects of citizen action are not ruled out by the existence of a participatory process, nor do I believe they should be. The notion of a citizen action–citizen participation continuum necessitates some definitional overlap.

A second criticism (paradoxically) could be that the definition is too specific, because it rules out forms of alleged participation (such as information sharing and ex post facto opportunities for public reaction) that characterize many government efforts to appease citizen activists. Such efforts can be labeled "pseudo-participation" because they only *appear* to be participation although they actually are not.[9] They are reactive and often cosmetic. Such attempts can be *part* of a broader participatory process, but they cannot stand on their own. Since this study aims at isolating examples of effective citizen participation, we should not hesitate to define the phenomenon strictly.

Evaluation Criteria

My choice of criteria has been influenced by existing literature in the field of citizen participation and by my own judgment of what the goals of participation ought to be. In addition, I have given careful consideration to the normative values of the people I studied—that is, Haagenaars themselves. City workers, community organizers, neighborhood residents, independent planners, politicians, and academics have each explained their views on what constitutes effective citizen involvement and have commented on whether they thought the participatory experiences I studied were examples of this.

[9] The dictionary defines *pseudo* to mean: "being apparently rather than actually as stated." *Webster's New Collegiate Dictionary* (Springfield, Mass.: Merriam, 1973), p. 922.

Five major criteria for evaluating the effectiveness of citizen participation can be isolated.

The Diffusion of Conflict within the Community and/or between the Community and the Municipal Government. For purposes of this report, we define conflict between residents and the municipality rather broadly to include active efforts on the part of residents to discredit or embarrass the government in the public media, to legally maneuver to delay the implementation of government plans, to demonstrate publicly (legally or illegally), and to commit acts of violence, including not only violent demonstrations but also acts designed to stall the implementation of certain plans (such as tearing up cobblestones or squatting in vacant buildings). It is important to note that conflict is not only resident-initiated—municipal officials also "wage conflict" against residents through efforts to deceive the public about the stage of renewal planning or the actual content of renewal plans, through plans that aim to displace large numbers of residents (directly through demolition or indirectly through severe rent hikes) without adequate compensation or relocation assistance, and through attempts to discredit or embarrass neighborhood leaders by attacking their legitimacy in the public media, and by attempting to divide resident groups or coalitions through under-the-table deals with individual groups or persons.

Conflict is costly both in terms of time and money. Generally speaking, little renewal was actually accomplished in most Dutch cities during periods of active conflict. In addition to this practical consideration, the Dutch place a very high value on the maintenance of stability in society. Although the past 20 years have witnessed a marked increase in the willingness of Dutch citizens to oppose government authority openly, the overriding Dutch sentiment remains that conflict is an unnatural state which ought to last only until a reasonable accommodation among parties can be reached.

The diffusion of conflict applies not only to conflict between the community and the municipality but also to conflict within the community itself. Renewal issues are sufficiently complex and Dutch neighborhoods are sufficiently diverse that conflict can develop between differing factions within individual neighborhoods over both substantive and procedural issues.

Municipal Recognition of the Current Residents' Goals for Renewal of the Neighborhood. Recognition is a two-step process of legitimizing the residents' goals and implementing plans that flow therefrom.

In order to legitimize the goals of the current residents, the municipality must agree to a series of process-oriented reforms, assuming that residents were heretofore essentially excluded from the planning process. These reforms should provide residents with increased access to official information, involve residents in the plan-making process as well as the process of implementation, and enable citizens to have a *real* impact on the decision making process. In short, the consensus of the community should substantially shape the future of the neighborhood.

Reforms such as these comprise the whole range of participatory techniques that have been the focal point of most studies of citizen involvement in government decision making. Frequently, residents in conflict with the municipality will not only demand substantive changes in plans for their neighborhood but also insist upon participation in the plan-making process itself. In fact, the search for these procedural changes may become the central focus of the group's struggle.

However, the creation of a participatory process considered satisfactory by the neighborhood does not necessarily mean that the process will be effective. An effective participation process must also attain concrete results that improve the physical and/or social qualities of the neighborhood, such as housing improvement, greater traffic safety, more park space, reduced crowding, greater equality of housing opportunity, expanded educational opportunities, and/or reduced social tensions. Without such accomplishments (i.e., actual accomplishments, not merely planning documents generated by the participatory process), effective participation will remain merely a promise to those who were involved.

If participation is to be successful, government must recognize that residents' desires for their own community must play a major role (perhaps *the* major role) in shaping the future of that neighborhood. Nonetheless, a complete acceptance of the majority view (or even community consensus) can damage not only the general public welfare but also the interests of minority groups within the community itself whose views (even if vocally expressed) may not be "loud enough" to have significant influence on the outcome of the participatory process. Since government is charged with protecting the interests of such people, the need to balance general neighborhood desires with the interests of underrepresented groups must be carefully considered in evaluating the effectiveness of participation.

Strengthening the Organizational Structure of the Community. Although a "strong organizational structure" is difficult to define, it is possible to list certain developments that will improve a community's ability to articulate and defend its interests during current and future planning. A representative, cohesive resident organization(s) with a dedicated community leadership will evolve. Useful contacts between residents and municipal officials will be established. Resident concern for and involvement in planning the neighborhood's future will also expand.

The analysis of a community for evidence of a strengthened organizational structure is a task fraught with pitfalls. Use of the following three questions can be of help:

Has the Entire Community Been Strengthened or Only One Organization? In the Schilderswijk neighborhood of The Hague, there are legitimate reasons to believe that most of the power flowed to one particular group which did not necessarily represent the opinions of certain segments of the population.

Who Has Been Strengthened—The Community as a Whole or a Leadership Cadre Alone? In Schilderswijk, much of the power accrued to a select group of organizers.

Has Strengthening Occurred in a Way That Will Benefit the Community in the Future? Much of the organizational infrastructure recently developed in Molenwijk will probably decay during the next decade as residents move out of the neighborhood.

Democratization of Community Institutions. Paradoxically, citizen participation—touted as a great example of modern-day democracy—frequently imposes authoritarian structures on the neighborhoods involved. We have seen the dominance of charismatic leaders and militant elites in many U.S. cities, and Dutch cities follow a similar pattern.

Ideally, participation offers a unique opportunity to expand the democratic institutions existing at the grass-roots level. The participation process can include groups that had not previously been active in neighborhood politics. In The Hague, this would mean particularly blacks, guest workers, women, the elderly, and children. Powers and responsibilities within citizen groups for group administration, policymaking, public relations, and negotiations with the city can be widely delegated. In addition, direct lines of authority from group members *to* group leaders, enabling the membership to influence thc participation process routinely, can be developed. This use of participation to increase neighborhood democracy is near and dear to the hearts of many planners and activists.

While action groups are frequently charged with undemocratic activities in order to discredit them and derail the participation process, many activists do, in fact, resist pressure to democratize their institutions. Activists may oppose democratization (although rarely in public) for differing reasons. Some are merely jealous of their own power. Others, however, sincerely believe that increasing openness will put them at a disadvantage against their natural enemy: the municipality. After all, if the city can function as a bureaucracy (with bosses making decisions and subordinates carrying them out), how can an action group be expected to keep pace when its leaders are restricted by the need to send recommendations through public, democratic channels before taking action?

These complaints, although understandable, may be exaggerated. Numerous experiences in the United States and in the Netherlands seem to indicate that significantly more democracy is both possible and practical in urban communities. To the degree that citizen participation in government decision making can serve as a catalyst for democratization, such trends should be encouraged. I believe, therefore, that the effectiveness of a participatory process should be judged partly on its record as a democratizing influence on the community involved.

Community-wide Learning. Community-wide learning is likely to be the most elusive of all outcomes of participatory efforts, as observers rarely agree on what it is. The proponents of participation cite it as one of participation's chief benefits; opponents claim it is very overrated. Learning is defined here to include a series of skills that would help people defend their interests more effectively in the next round of conflict or participation.

Skills that action groups and residents could benefit from learning and which participation can, in fact, teach are numerous. An understanding of political relations, the planning process, and the economic realities of urban life is an essential basis for learning to cope with them. Skills in bargaining and mediation, in setting priorities, in group process, and in identifying both personal and community-wide goals are also needed.

In addition to examining the skills learned, however, it is equally important to ask *who* learned them. The importance of particular skills varies among different types of people, and certain groups clearly manifest greater or lesser ability to learn in the course of the participatory process. Communities appear to break down into three distinct groups according to the type and amount of learning accomplished: the leadership cadre (such as action-group leaders and social/cultural workers); the average residents of the community (the Dutch working class); and the most disadvantaged members of the community (such as black immigrants from Dutch colonies, guest workers, and the elderly). Although these groups clearly learned at different rates, I do not presume to compare their actual learning abilities. Instead, we should recognize that the degree of learning depends not only upon the talent and dedication of the group involved but also on external conditions such as past educational opportunities, fluency in the Dutch language, previous exposure to democratic institutions, and the degree of involvement afforded by the participatory process to different classes of people.

Renbaankwartier/Scheveningen: Conflict under the Guise of Participation

Scheveningen is one of the most famous seaside resorts in western Europe. It was partially reconstructed to suit the needs of the tourist industry between 1918 and 1938, but the period after 1960 saw a steady decline of the resort. Decay along the waterfront and in the surrounding residential areas resembles that of many ocean resorts in the cooler parts of Europe and North America.

The process in this neighborhood was not really a participation process at all, in light of our definition of participation. Conflict was never replaced by cooperation; an atmosphere of respect and trust did not develop between the city and the activists. The procedure adopted for involving citizens was unsophisticated and very brief, providing little opportunity for long-term citizen impact, community-wide learning, or the democratization of neighborhood institutions.

Figure 2. The Hague, showing the location of the three neighborhoods discussed in this chapter relative to the city center.

Although some immediate gains were achieved, the process ended in acrimony: it was a clear case of a participatory effort in regression back toward open conflict. The beach resort in Scheveningen is surrounded by a ring of residential neighborhoods, which are, in turn, separated from the center of The Hague by parks. The residential neighborhoods include multi-unit rental structures, attractive single-family homes, boardinghouses, small hotels, and several modest business districts. One such neighborhood is Renbaankwartier, the focus of this case.

Unlike the other districts I studied in The Hague, Scheveningen is not poor, even by Dutch standards. Most of the people could be called middle-class, with a good number of wealthy residents as well. There are, however, a few pockets of immigrants and lower-class Dutch workers who perform the low-skilled tasks required by every resort. Although some of the housing stock could use rehabilitation, it is for the most part in good shape.

When Scheveningen's largest landowner began to lose money and put his property up for sale, the municipal government and a development company generated a sweeping renewal plan for the area, including the construction of a major highway to accommodate the increased tourist traffic that the renewal was expected to generate. Neighborhood residents were outraged by the plan. Although they were eager to see the beachfront rejuvenated, they were equally concerned about the integrity of their residential neighborhoods. The proposed plan would have completely changed the residential character of Scheveningen. Houses would have been torn down and not replaced. Luxury apartment complexes and office towers would have destroyed the physical integrity of the neighborhood and brought a wealthier class of residents to Scheveningen. Traffic would have increased and safety for children would have declined.

The government claimed that the revitalization of Scheveningen was essential to the city's economic well-being. If the neighborhood was to be renewed, certain sacrifices were required. In addition, the government noted that it had set up a special citizen advisory panel to assist in formulating the beachfront plan; this group was called Inspraakgroep Scheveningen. The residents, however, charged that most of the group's members came from outside the neighborhood and could not therefore presume to speak for the residents.

A public hearing held in October 1976 revealed the depth of citizen hostility toward the beachfront plans and the proposed highway. The residents at the hearing demanded changes in the beachfront proposal, a reversal of the plan for highway construction, and a new, detailed plan to protect the residential parts of the neighborhood. The city, however, refused to budge on the beachfront and highway plan, and it refused even to discuss the possibility of a new plan for the rest of Scheveningen.

The four months after the October meeting saw intense citizen mobilization. Among the most militant activists were the representatives from Renbaankwartier, an old district particularly near the beachfront. The city had steadfastly refused to discuss the future of Renbaankwartier with the citizens, who were certain the government planned eventually to seize most of the prime land in Renbaankwartier, tear down the housing, and extend the beachfront plans to include this area. In fact, the municipality refused to include Renbaankwartier in the planning process for the other residential neighborhoods of Scheveningen, heightening citizen mistrust and apprehension.

In February 1977, a major public meeting was held in the Circus Theater in Scheveningen. Over 1,000 people showed up, and some were refused admittance due to the size of the crowd. Residents came from all parts of Scheveningen and formulated three demands to present to the municipality: (1) the beachfront plan must not be extended further, and the road must not be built;

(2) demolition of houses must cease; and (3) the physical harmony of the neighborhood must be maintained (i.e., no more high-rise office or apartment towers, and replacement housing at affordable rents for those displaced).

Meanwhile, the people of Renbaankwartier formed their own organization, the Residents' Council for Renbaankwartier (Bewoners Beraad Renbaankwartier, or BBR), to deal specifically with their own area. This organization began action in earnest to press not only for the demands of the meeting but also for a separate city plan that would specifically protect the residential integrity of Renbaankwartier. The organization held sit-ins on the resort's major pier. It drafted a letter, signed by respected members of the community, asserting that Mayor Schol of The Hague was not welcome to set foot in Renbaankwartier. BBR sought out the help of the press and organized strategy meetings.

American observers may be surprised by the mild form of protest. In the context of Dutch society, however, such actions were major departures from normal deference to government authority. Surprising as we may find it, Dutch officials were shocked at such brazen protests (not only in Scheveningen but in other neighborhoods throughout the country), and they were eventually forced to respond.

BBR divided into five separate working groups. The first publicized the group's demands outside the neighborhood; the second published a newspaper for the neighborhood itself. A third group coordinated protest actions. The final groups documented changes which were taking place in the neighborhood and explored the history of redevelopment in Scheveningen.

After arguing about the powers and composition of a governing board, BBR agreed to have no single board in control. As might have been expected, however, a cadre of active members rose to positions of control, calling frequent neighborhood meetings to discuss strategy and positions. The leader of the group was L. Pronk, a doctoral candidate from the Royal University at Leiden. Significantly, Pronk's dissertation dealt with the complex Dutch procedure for appealing municipal decisions to the province and the national government. In the course of the next several years, BBR would use the appeal process masterfully to stymie municipal plans.

After several months of action, BBR convinced the city council to pass a resolution calling on the municipal government to include Renbaankwartier in the plan area. The mayor and aldermen at first refused, but they finally agreed to prepare a separate plan for Renbaankwartier. The council then insisted that the executive accept the three demands of the February 1977 meeting, and complete the plan for Renbaankwartier within one year. The mayor and aldermen capitulated, and a deal was struck. However, they would not soon forget that the deal was forced on them. Although the municipal officials involved in the Renbaankwartier process had agreed to participation schemes in other parts of the city, they began the participation process in Renbaankwartier on a uniquely sour note.

Reluctantly, the city officials contacted BBR to prepare for a process of consultation. Immediately, a conflict arose concerning the boundaries of the area to be included in the plan for Renbaankwartier, with the city demanding that certain streets be excluded. BBR agreed, but explained that if the city did anything to those streets that damaged Renbaankwartier, there would be an immediate return to protest.

The "consultation group" was to consist of one resident from Belgische Park (another Scheveningen neighborhood, part of which was included in the Renbaankwartier planning district), two residents from Renbaankwartier itself, one storekeeper, one hotel owner, one representative from the building company involved in the planning process, and one representative from the City Development Agency. Although initial contacts began in July 1977, the actual meetings did not begin until November. The major concession extracted by BBR regarding the process itself was that all meetings would be open to the public.

Six meetings were held, all during the month of November. There is disagreement about attendance. Pronk claims about 100 people attended each meeting; city officials believe the figure was closer to an average of 30. Once, the procedure appeared on the verge of breaking down completely, as members of the group charged that the city's representative lied about municipal plans for the adjoining neighborhood of Seinpostduin. The city countered with the charge that it could not possibly reveal everything BBR wanted to know if the meetings were kept public. At one point, Michel Hardon, Alderman for Physical Planning and Urban Renewal, invited Pronk to his office for a private meeting; but Pronk arrived with 30 citizens, boldly demonstrating his insistence on the public character of all contacts with the city.

Finally, after a month of agreement on certain points and persistent bickering on others, BBR told the city to finalize its plan. The plan set a maximum height for future construction in Renbaankwartier: no building may be more than one floor higher than the current housing stock. There will be no more hotels, pensiones, or office buildings constructed in the quarter. There is no ban on demolition, but the building restrictions mentioned above greatly reduce the financial incentives that could lead to massive demolition. Although limits have been placed on rent levels, there are still quarrels about the details of this stipulation.

On other issues, however, BBR was less successful. The city has adopted no official policy to preserve the low-income housing that currently exists in Renbaankwartier or to ensure replacement housing for people who may be forced to move. There are currently several factories and schools in the area; the residents do not mind if these remain. But if they close, BBR wants the city to promise to construct low-income housing in their place. The city claims it does not have the funds to make such a commitment. The group will take these remaining issues to the city council or the provincial and national authorities.

In many ways, the Renbaankwartier participation procedure fell short of an ideal process. It failed to involve a large number of citizens in a long-term planning effort for the neighborhood. After the one-month participation process, the City Development Agency worked by itself for another six months before producing a draft plan. Although the residents accomplished some of their objectives, critical guarantees about housing, rents, and demolition were not obtained.

Conflict between the residents and their city government has not been replaced by trust—not even by tolerance. Conflicts continued throughout the process, and BBR has promised to carry the fight even further now that the plan is finished. Although Renbaankwartier itself is safeguarded by aspects of the new plan, the surrounding neighborhoods of Scheveningen are gentrifying rapidly, with luxury apartment houses and office buildings sprouting along most major streets. The roadway has been stopped, but it was demonstrations and protests that stopped it, not the participation effort.

In fact, most of the successes of BBR and the other residential organizations of Scheveningen appear to stem from conflicts with the city rather than from participation. This does not mean that participation was useless: on the contrary, it served to formalize and tie down victories that the residents had already won in the streets and meeting halls of the neighborhood. Furthermore, BBR's insistence on public meetings confirmed the residents' determination not to be coopted by attending carefully orchestrated meetings in the conference rooms of municipal agencies. However, the insistence on public disclosure may well have reduced the degree to which participation could actually influence critical planning decisions. In Schilderswijk, for example, many important proneighborhood compromises have been struck behind closed doors where both city and citizens were less concerned about "saving face."

Participation in Renbaankwartier did not replace conflict as the dominant mode of action in the community. A very different situation occurred in Schilderswijk, where participation led to real accomplishments for the neighborhood and the city as a whole.

Schilderswijk I: Coproduction in Action

Schilderswijk is generally considered to be the largest continuous urban problem area in the Netherlands.[10] Housing density is very high, and most of the structures were built in the second half of the nineteenth century, before strict building codes were in force. Schilderswijk is inhabited largely by low-income groups, including immigrants from Surinam and guest workers from Turkey, Morocco, and other Mediterranean countries. The city claims that im-

[10] H. van Dijck *et al.*, "Reconstructing the Schilderswijk: A Study of an Attempt to Introduce Innovations into the Making of an Urban Reconstruction Plan" (Paris: Organization for Economic Cooperation and Development, September 1975), p. 6.

migrants account for 40% of the population of Schilderswijk, although some native Dutch residents claim the figure is closer to 50 or 60%. Young people form an unusually large percentage of the population of the neighborhood.

The neighborhood is bordered on one side by the central business district of The Hague and on the other side by a railway complex. It is both physically and socially isolated from the rest of the city. Since 1960, the area has suffered a considerable reduction in population and housing stock, although the rapid influx of immigrants during the last five years may be pushing the population above its low mark of 40,000 in the early 1970s.

The Schilderswijk case can be divided into two separate participatory experiences. The first involved city officials and the organization called Payable Rents in the development of a plan for two low-income housing complexes in the Oranjeplein area of Schilderswijk. The second concerned the development of a "structure sketch" for all of Schilderswijk; the principal (but not the only) citizen group involved in this participatory effort was the Residents' Organization Oranjeplein–Schilderswijk (Bewoners Organizatie Oranjeplein-Schilderswijk, or BOS), an outgrowth of Payable Rents.

Concerted protest by the residents of Schilderswijk preceded each participatory effort. In each case, municipal willingness to set a participatory process in motion resulted directly from the conflict waged by the citizens. First, we will consider the Oranjeplein participation process, which developed housing complexes along the Jacob Catsstraat and the Gortmolen, replacing city plans that would have displaced many people and radically changed the character of the neighborhood.

There are several important themes to the Oranjeplein experience. Both the city and the action group Payable Rents agreed in writing to a series of goals for the neighborhood—and a process for achieving these goals—*before* participation began. Once the participatory process commenced, conflict between the city and the neighborhood declined significantly, although a certain level of conflict among various groups in the neighborhood continued.

The procedure resulted in significant changes in plans for the neighborhood, and these plans have already been implemented. Payable Rents and, more generally, the citizens of the neighborhood, became legitimate parties to renewal planning and implementation, although this status proved temporary in nature. Significant learning has taken place within the community and among city workers as well. The organizational base of Schilderswijk has been strengthened. Unfortunately, there were only minor signs that the process democratized community institutions.

The Oranjeplein participation process did not spring from a vacuum. For several years prior to the unveiling of the city's plan for the area, relations between the city and Schilderswijk were tense. Efforts by municipal planners to widen streets and route more traffic through Schilderswijk had already sparked substantial citizen protest, which forced the city to back down on at least one critical occasion. In addition, the people of Schilderswijk were intimately familiar with renewal plans already implemented by the city in the neighboring section of Stieltjestraat and Kortenbos, where as many as 12,000 people (according to one estimate) were displaced to widen roads and to build a hospital, a technical school, and apartments that were too expensive for the previous residents.

Thus, when the city presented its dismemberment plans for Schilderswijk, the community already possessed an indigenous cadre of protest leaders and two nearby examples of what would happen if they remained silent. Stieltjestraat and Kortenbos were vivid proof that the low-income working-class population of the central Hague was being uprooted. If the process was allowed to begin in Schilderswijk, the social base for protest would swiftly be reduced. Not only buildings and streets would be lost but also the people needed to protest further evictions. The time for action had arrived, and it could not be delayed.

The plan put forward by the city in 1971 for the renewal of Oranjeplein was the latest in a series of ill-fated plans dating back to 1953. This newest plan, the brainchild of a powerful private development company called MAVOB, called for the demolition of the entire neighborhood, replacing it with high-rise towers filled with high-rent units. MAVOB agreed to put up all the money to prepare a more detailed plan plus a substantial amount of capital for the project itself. Through its close ties to the provincial planning agency, MAVOB could guarantee the city that the plan would receive provincial approval—a particularly important bargaining point, since the province had recently been quite displeased with planning efforts in The Hague.

Protest in Schilderswijk took several forms. Of course, there were demonstrations and skillful use of the media. Perhaps most important, however, was the decision of the action group Payable Rents to produce a viable alternative plan for Oranjeplein. At the time of the protests, the city asserted that it was impossible to create a plan for new housing in Schilderswijk at rent levels that the current tenants could afford. Payable Rents might have been content merely to protest the city's plan. This was, after all, the tactic that had been adopted by many citizen groups in The Hague. However, this particular organization decided to challenge the assertions of the politicians and their planning experts by drafting a plan of its own. In a sense, it decided to play ball in the opponents' park.

Despite the distrust residents displayed for city officials, they accepted the informal (and sometimes secret) assistance of several city workers and employees of the architectural firm that had made the original plan. These professional helpers joined the residents in the evenings to expose flaws in the plan and to recommend alternative solutions involving lower rents. This cooperation indicated that the leaders of Payable Rents were willing to work with outside allies. On the other hand, the group retained all decision making authority and would not allow the outside professionals to become anything more than strictly advisory. This balanced policy toward outside assistance would serve the group well in the future, encouraging contacts with the outside world (and the city government specifically) while avoiding domination of the group by outside experts.

The alternative plan was not a comprehensive planning document. It was a political strategy—one in a series of tools for conflict with the city. It outlined three key principles, conditions which the residents considered indispensable for a satisfactory renewal of their neighborhood: rents must be affordable by the current inhabitants, current residents must have priority for new units, and any new architecture must blend in with the old (that is, no high-rise apartments would be built in Oranjeplein).

It also presented some ways in which these goals might be achieved, enough to be persuasive but not enough to fulfill the rigorous demands of a real plan. The plan was valuable not because it provided a definitive explanation of how to renew Oranjeplein but because it drew a baseline along which future planning might proceed.

For months, the city refused to bargain with the residents. It insisted that any plan featuring lower rents was impractical and that the MAVOB plan was necessary for the economic welfare of the city as a whole. In an effort to calm the troubled waters of Schilderswijk (and save the MAVOB plan), B. J. Udink, the new Minister of Housing and Physical Planning, decided to visit the neighborhood for a personal tour. Wherever he went in Schilderswijk, people greeted him with black flags flying from their windows—a sign of the residents' united disapproval of the MAVOB plan. The incident attracted nationwide media attention. Shortly after his visit, Udink agreed to increase the subsidies available to the development project, thereby reducing the projected postrenewal rents. The Hague could now argue that the current residents could afford the new units.

The additional subsidies however, failed to defuse the situation. The leaders of Payable Rents would not budge from their original position. They insisted that if the subsidies were applied to their alternative plan, rents would be lower. Further, they argued that the buildings in MAVOB's plan were still too tall and entirely out of character with the surrounding neighborhood. Finally, the residents remained committed to more rehabilitation and less demolition.

The impasse did not break until the spring of 1972, when W. Nuy, the alderman in charge of urban renewal, finally agreed to a series of four working sessions at which the city and Payable Rents would work out a set of mutually acceptable renewal goals. Nuy and city planners would represent The Hague and representatives from Payable Rents would speak for the residents.

Undoubtedly, Nuy was tiring of continued protest over the MAVOB plan. Perhaps he feared political damage to his own reputation if he allowed the impasse to continue. These factors seem inadequate, however, to explain Nuy's sudden turnaround. An additional factor may have been the action group's success in stymieing Nuy's attempt to hold a public hearing to discuss the plan. Nuy felt he could prove that the action group was not representative of the people of the neighborhood; but Payable Rents mounted a boycott of the public hearings. When only journalists and city workers showed up, Nuy began to understand the depth of resident feeling against the MAVOB plan.

Whatever additional convincing Nuy may have needed was provided by the same group of architects who had been secretly helping the action group. They drafted a letter from their firm (Buro van Tijen) to the local government. In it they warned The Hague that the plan was unreasonable for the current residents and that, even with Udink's new subsidies, the current specifications could never produce reasonable rents. They brought it to one of their directors and threatened to resign and go to the newspapers if he did not sign it; he signed, and the letter was sent.[11] Nuy was left with little choice but to enter into negotiations with Payable Rents.

The outcome of the four working sessions was a *doelstellingennota,* or a "declaration of intents" agreed to in writing by both sides. The declaration committed the parties to set up goals (affordable rents, priority for the current residents, improved housing conditions, a harmonious physical environment, adequate park space, and no increase in through traffic) *and* a series of steps for jointly reaching these objectives. The first step was to make a global plan that would explain, in approximate terms, what the area would look like after renewal and how that endpoint would be achieved. Step two involved adding detail to the global plan. The third step was to set up a mechanism for distributing the new housing units. Finally, the buildings would be constructed, along with various amenities, and the units would be distributed.

The process for consultation which the parties agreed on was far from spectacular. It was a two-tiered process, with representatives of the action group, the municipality, and the architects working together on a building/design committee which had to review all aspects of the plan. After proposals were worked out in the building/design committee, they went for review to a steering committee, on which residents were initially not allowed to sit.

Payable Rents soon became very dissatisfied with their lack of members on the steering committee. They therefore decided to stall every proposal which came before the building/design committee until Nuy granted them positions on the steering committee as well. After that, the planning process proceeded smoothly. Eventually, plans were prepared for two low-rent housing developments in the Oranjeplein. The town council approved the plans, and the first pile was driven on August 3, 1973. Both developments are now complete and inhabited.

Despite the simplicity of the participatory structure, several characteristics of the process indicate that it truly represented bona fide coproduction. First,

[11] Interview with Henk van Schagen, Department of Architecture and Town Planning, Delft Technical University; Delft, December 14, 1979.

the declaration of intents committed both sides to rights and obligations not only about goals but also about the process itself. Accepting this agreement required an enormous leap of faith for both sides. The city was in essence admitting that it had made a grave error when it stated that low-rent development was impossible in Oranjeplein. The residents, on the other hand, were agreeing to sit down and bargain in trust with an opponent whom they had fought for more than a year.

Second, as far as Payable Rents was concerned, conflict with the city came to an end. Protests and other efforts to embarrass city officials ceased. The organization was responsible for attracting the largest possible number of people to the meetings with the city—but for the purpose of constructive dialogue, not for a brute show of force.

(Cessation of conflict was limited in two important ways. First, Payable Rents was at all times ready to return to protest if the city reneged on its commitments. Second, other groups in the area, especially a large rival organization known as Renters' Association of Schilderswijk [Huurdersvereniging Schilderswijk, or HVS], continued their protests and refused to accept the notion of coproduction. In an important respect, this continued conflict helped Payable Rents' position vis-à-vis the city, because the municipality was forced to recognize the pressure being applied to Payable Rents on its left flank. Unless the city kept its promises and compromised on substantive planning issues, the moderate leadership of Payable Rents could have been overthrown and the whole process might have fallen apart.)

Third, although many public meetings were held, important decisions were made in behind-the-scenes negotiations between the city and the leaders of Payable Rents. The organization never sent less than two representatives to these meetings, and they were required to report back to frequent public meetings which were held afterwards. While we may argue about the advisability of joining the city in "smoke-filled rooms," there is little doubt that city workers would try to circumvent a process that committed them to make all decisions in public (as was precisely the case in Renbaankwartier). The interests of the residents of Oranjeplein seemed to be well served by a combination of private negotiations and frequent public meetings.

Finally, the residents and the city did not merely coplan but actually *coproduced*. The buildings were constructed according to specifications determined by the city and the citizens, and units were distributed according to agreements reached by all parties. Although the degree of renewal may not have pleased all members of Payable Rents (in fact, there are still many run-down buildings and vacant lots in Oranjeplein), there is general satisfaction with the physical and social outcome of the new buildings.[12]

[12] For a detailed account of the Oranjeplein participation process and related events, see van Dijck (note 10).

Schilderswijk II: Ossification and Recovery

In 1974, Payable Rents decided to expand into a new organization that the Dutch call a *stichting*, a foundation with the legal right to receive government subsidies. Members of the board of directors pick their own successors. The members named their new organization Residents' Organization for Oranjeplein–Schilderswijk (Bewoners Organisatie Oranjeplein–Schilderswijk, or BOS).

With the arrival of government subsidies, BOS opened a permanent office and hired several staff members. The staffers were nonprofessional, and most were from Schilderswijk. However, they performed functions that made close consultation with the residents less vital. Aad Kuypers, the leader of the group, was particularly skilled as an organizer and as a housing advocate. He began, eventually, to go to negotiating sessions with the city alone. Meanwhile, the residents' council *(bewoners raad)* or primary governing body of the organization dwindled in number of attendants; eventually, the council stopped meeting—a moratorium that lasted for about two years.

BOS workers, especially Kuypers, developed closer associations with a number of city departments. People began to say that Kuypers liked the new-

Figure 3. New moderate-income housing built in Oranjeplein after a process of citizen participation.

Figure 4. Rehabilitated housing in Schilderswijk.

found respect he had acquired in city hall; he enjoyed being able to call up an alderman and say, "I need to see you in an hour," and have the appointment granted.

In 1976, the organization became involved in a process to develop a "structure sketch" (a fairly general preliminary planning document) for the entire Schilderswijk neighborhood. This process also arose out of conflict, this time concerning a structure sketch that municipal planners formulated entirely without the participation of the citizenry, sparking bitter fears that the city was trying to reverse the gains made by Payable Rents in Oranjeplein.

The city's structure sketch was similar to the MAVOB proposal, only on a much grander scale. It called for massive demolition throughout Schilderswijk and the construction of new units whose rents would be far beyond the means of the current residents. A grandiose linear park was to weave its way through the district, and roads were to be widened to accommodate increased traffic. The plan was geared to serve people who lived "individuated" rather than "communal" lives, according to one observer.[13] In other words, the neighborhood

[13] Interview with Paulinus Kuipers, Jacob Maris Group, Schilderswijk; The Hague, December 3, 1979.

visualized by the structure plan would be excellent for young families with few children, a car, and most of their friends living outside of Schilderswijk. The plan was uniquely unsuited to the needs of larger, poorer families with close ties to other people on the block or down the street. For a year, the city tried to get the residents to discuss the plan on its merits, but they refused. Finally, the city agreed to drop the plan and involve citizen groups from Schilderswijk in a new process to formulate a second structure plan.

We can isolate several major themes in the participation experience that followed this decision. First, the citizens themselves were involved in shaping the process by which they would participate. The process that evolved was more sophisticated than the Oranjeplein procedure, involving more people and broader issues. At the same time, however, BOS was growing further away from its constituency; it was becoming increasingly dependent on the talents of one charismatic leader. The organization began to ossify, failing to keep in touch with citizen needs and frustrations about the time it took to translate plans into action. By the end of my study, however, BOS had begun taking steps to restore its legitimacy in the community.

The participation process for the structure plan differed from the process that was used for the Oranjeplein developments in three ways. First, a detailed survey was conducted throughout the neighborhood by workers for the City Development Agency, who solicited resident opinions about current conditions in the neighborhood, people's perceptions of Schilderswijk vis-à-vis other neighborhoods where they might choose to move, and how people wanted their housing circumstances to be improved. Issues such as rehabilitation versus new construction, parking availability, height of buildings, and tradeoffs between amenities and postrenewal rent levels were explored. The research indicated that the leaders of BOS did, for the most part, accurately represent the views and desires of their members, with two important exceptions indicating that the information channels within the organization were not completely clear. First, the leaders wanted a higher rate of demolition and new construction than did their constituents, who frequently liked their old houses and preferred rehabilitation. Second, the leaders were less concerned about parking places for cars than the residents; in fact, residents were willing to pay higher rents in order to have carports placed underground.

Second, the residents of Schilkerswijk were directly involved in formulating the participation process itself. In fact, the process of negotiation frequently generated as much controversy as substantive planning issues. The leaders of BOS and other neighborhood organizations sat down with workers from the City Development Agency to set up details. They agreed on a step-by-step participation process, for which they divided the neighborhood into ten districts. The first step was to disseminate information about the neighborhood and the structure-sketch planning process throughout the quarter. District consultation groups were then organized in all 10 parts of Schilderswijk, and five neighborhood-wide working groups were created to deal with the substantive concerns of housing and parks, traffic, recreation, shops, and schools. Membership on these committees was open to anyone who wanted to attend, but most of the attendants represented individual neighborhood organizations. The neighborhood-wide working groups conducted research and came up with basic proposals, which were then reviewed by the 10 district consultation groups and fleshed-out by the City Development Agency. Finally, the proposals were reviewed one final time by the working groups and district consultation groups, after which they were routed to the mayor, aldermen, and town council for final approval. The town council approved the new structure sketch in January 1979.

This process was obviously much more complicated than the earlier process involving Payable Rents. Plans went through several iterations, with different groups being involved on different occasions. In addition, several neighborhood organizations were involved, with different opinions about what ought to happen in Schilderswijk. Disagreements within the coalition of groups opposing the original structural plan were largely resolved, and agreement with the city was reached.

Finally, and perhaps most importantly, the agenda of issues for this participation effort had been extended far beyond the housing concerns of the earlier conflict. Residents were now asked to comment on green space—not for two housing complexes but for an entire neighborhood. They were concerned not with the volume of cars on a few streets but with traffic loads through and around all of Schilderswijk. Recreational facilities and schools, subjects which were hardly discussed in the first participation process, now had working groups of their own. In the first participation effort, there had been only one question to settle regarding the neighborhood economy: Will stores be concentrated within the housing complexes, or will they be moved down the street? Now, however, a whole range of economic and business-location decisions had to be considered to spur the economic revitalization of Schilderswijk. Toward the end of the first effort, observers had noticed that certain residents (although not an overwhelming number) were becoming more interested in issues "larger" than just housing. The new participation process confirmed that interest in these issues had grown to the point where citizen opinions could no longer be ignored.

While the three factors indicated above suggest that the participatory process was maturing, still other events threatened to derail the structure-sketch effort or at least to make similar participatory attempts in the future unlikely. Perhaps the most significant development was the growing distance between the leaders of BOS and the residents of Schilderswijk. The residents' council had been an integral part of Payable Rents and BOS during its early period. Gradually, however, people lost interest in the regular meetings of the *bewoners raad.* Meetings became less frequent and eventually disappeared altogether. When officers of the organization tried to resurrect the *bewoners raad,* Kuypers and other BOS workers stiffly opposed the effort and delayed reconvening of the *raad* for several months. The organization's board of directors saw its power diminish as Kuypers became more and more powerful. He controlled the hearts of many if not most of the organization's members. Even more importantly, he had amassed tremendous knowledge about urban renewal and city planning. He was BOS's foremost spokesman and negotiator with The Hague, and his contacts in the city bureaucracy were invaluable. The organization could not, or thought it could not, get along without him. His abrasive character and willingness to negotiate agreements on his own increased complaints about him, while his successes simultaneously increased his control over the organization.

Attacks on the representatives of BOS and criticism of Kuypers's strategies from left-wing forces in the neighborhood became increasingly vicious. HVS, disenchanted members of Payable Rents, and certain social workers in Schilderswijk were particularly outspoken. In certain districts, opponents of BOS were able to disrupt the planning process entirely for weeks at a time, influencing people not to attend meetings and to demand that the city do something about the immediate problems of the neighborhood (such as crime, traffic, and unsanitary housing conditions), rather than all this "useless long-term planning."

The demand for immediate remedial action as a substitute for endless months of planning struck a sympathetic chord throughout the community, highlighting a key fact which BOS's leaders had begun to forget. In their excitement about the structure sketch, they forgot that it was only a plan. Plans, after all, are only as good as the city's word and the money allocated to back it up. The people of Schilderswijk were tiring of plans. They remembered old promises that the city had made at the time of the original process for the Oranjeplein, promises of concrete action to fix leaky roofs, erect new buildings in vacant lots where old residences had been demolished, and make streets safer for children. Then, without realizing these promises, the city had come right back with a secretly produced structure sketch that required a major return to the conflict of the past. Many residents, even long-term supporters of Kuypers, began to ask why they should return so readily to the bargaining table before the city made good on some of its old promises.

In some cases, BOS could no longer stem the course of conflict, even during the participation process. When people got tired of vacant buildings standing next to their own homes, they would burn the buildings down; arson as a form of protest is Schilderswijk's worst-kept secret. When people tired of their own unsatisfactory quarters, they squatted in more desirable, but vacant buildings—some of which were scheduled for demolition as part of plans BOS had been instrumental in formulating. When the city failed to respond to complaints about street safety, the social workers sometimes led people out into the street to tear out the cobblestones, making the roads impassable. Actions such as these were particularly serious in light of the mild forms of protest characteristic of Schilderswijk up to that time. The serious confrontations now occurring were quite out of tune with Dutch traditions and indicated the depth of people's anger. Each action of protest embarrassed BOS and its process of participation.

The troubles plaguing the structure-sketch process were manifestations of ossification in BOS. I have chosen the word *ossification* deliberately. Its meaning is: "a tendency toward a state of being molded into a rigid, conventional, sterile, or unimaginative condition."[14] This word accurately describes what was happening to BOS as it embarked on a process which, on the surface at least, was far more sophisticated than the process for Oranjeplein which had preceded it. An unwillingness to concentrate on the immediate needs of the residents, the loss of legitimacy flowing from the disbanding of the residents' council, increasing domination by one man (albeit a talented and dedicated man)—these factors created an organization that was increasingly rigid and authoritarian, unable to meet attacks on its legitimacy from the left and, in many important respects, increasingly deaf to complaints from the neighborhood it was trying to serve.

Fortunately, BOS and its leaders (including Kuypers) had invested too much time and effort and learned too much about the danger signals of organizational decay to allow ossification to continue indefinitely. They recognized that the structure-sketch planning process was essentially sound; they were firmly committed to the future of Schilderswijk, and the people remembered that their successes in negotiation with the city were not inconsiderable.

At the time I left the Netherlands in January 1980, there were distinct signs that BOS was beginning to rebuild its legitimacy and strengthen its bargaining position with the city. Kuypers had agreed to the reinstitution of the *bewoners raad,* which has met several times since the summer of 1979. A new chairman of the *raad* had been appointed.

In addition, the organization has begun to reemphasize small, street-level meetings to discuss neighborhood problems—one of its initial organizing methods. One such effort, on the Naaldwijkstraat and the Falckstraat, involves intensive conversations with about 80 families to determine their current problems and desires for the future. A number of people who were skeptical of BOS's recent activities cited this as a very positive step.

Finally, the city has made certain efforts that have helped defuse a potentially explosive situation in Schilderswijk. By appointing a special committee to deal with immediate problems of the residents rather than just planning for the long term, it may solve some of the inhabitants' most pressing problems. The structure sketch itself, which could easily have been a very general plan with few time limitations, states rather explicitly the steps to be taken in each district and by what dates, presenting short-term actions as well as long-term plans.

[14] *Webster's New Collegiate Dictionary,* (Springfield, Mass.: Merriam, 1973), p. 805.

The sketch differs from the original one in several important respects. Displacement is not encouraged. Massive demolition has been replaced by selective demolition; for example, the old plan recommended that all houses in the central district of Schilderswijk be torn down, while the new plan calls for the destruction of only two-thirds of these units. City traffic will be routed around (instead of through) the neighborhood. Finally, the linear park has been replaced by a series of smaller parks scattered throughout the neighborhood and more suited to the recreational needs of children.

The planning process has now shifted from the general structure sketch for the neighborhood to specific building plans for blocks and individual housing complexes. If the momentum toward real renewal can be maintained, the process of participation in Schilderswijk may well be rejuvenated. If it cannot, however, many of the gains accomplished at great expense during the past decade could be lost.

Molenwijk: Participation with an Ulterior Motive

Molenwijk, "the windmill neighborhood," is an area just south of the city center, filled with low-rise apartment buildings. The housing stock is solid, but the units are rather small, frequently lacking showers and gardens (in Holland the latter may be more important to most people than the former).

The participatory process in Molenwijk is highly elaborate and thoughtfully constructed. There is a good potential for community learning, and special efforts have been made to include women, children, and foreign immigrants in the planning process—with varying degrees of success. The actual outcome of the procedure, however, will result in significant displacement as well as hardship for foreign residents of the community.

The presence of foreign immigrants is a particularly thorny issue in Molenwijk. Both guest workers from southern Europe and north Africa and immigrants from the former Dutch colony of Surinam began moving into Molenwijk at a rapid pace during the late 1960s, to the distress of longtime Dutch residents. The foreigners have many habits and customs that the Dutch find offensive. They live together in very large families, speak different languages, and compete with Dutch natives for jobs. On the whole, the Dutch do not like the outsiders, considering them dirty and uncivilized, and they are willing to say so.

Molenwijk is a relatively poor neighborhood. About 60% of the residents are below Dutch minimum income standards. Unlike that of the other neighborhoods in this study, the initial action of Molenwijk's citizens did not occur in response to an existing government plan. Actiegroep Laakkwartier (AL; the primary action group in Molenwijk) was apparently started five or six years ago simply because people wanted the city to fix up their deteriorating and inadequate housing, especially in light of the fact that a large percentage of the units were publicly owned.

A consultation group consisting of AL members and city development representatives conducted the participation process. Together, the members and the city chose an architect to work on the plan. Eventually, the residents hired a consulting group called Planwinkel (or the "plan store") from Delft to provide them with technical information, to establish a better participation process for new construction in certain parts of Molenwijk, and to run a project seeking to involve schoolchildren in planning the neighborhood's future.

AL and Planwinkel have attempted to broaden the base of the participatory process. They have established subgroups to do initial planning in specific subject areas (such as welfare services and schools) or to perform certain functions (like providing information about the planning process to residents). Each month, the consultation group meets in two public sessions: the first is a general meeting to work with the city on neighborhood plans and the second provides an in-depth discussion of a particular planning issue. Many of the meetings are held during the day, giving housewives in the area maximum opportunity to participate (very few Dutch wives work outside the home).

Special efforts have been made to involve foreigners. All of Planwinkel's documents are printed in Dutch, Turkish, and Arabic. Simultaneous translation in meetings has been attempted. One social worker keeps in regular touch with foreign residents of the community. On the whole, however, few foreigners come to meetings, and none is a regular member of the action group.

At first, the apparent passivity of the people in dealing with the municipality was surprising; major conflicts have not occurred. Although the techniques of participation were highly elaborated, the meetings routinely accepted the recommendations of city workers. The new houses which are being constructed will be fancier than those in Schilderswijk, and the rents will be much higher as well. In fact, Planwinkel estimates that about half the people currently living in Molenwijk will not be able to afford the new housing. Why would a well-organized group of residents demand that the city renew their entire neighborhood and yet not complain when the rents are forced to exorbitant levels? In other parts of the city (and throughout Holland), citizen groups have demanded and often obtained extra subsidies from the national government to cover increased postrenewal rents, or they have demanded less fancy renewal in order to keep rents low.

Discussions with two city workers and one planner from Planwinkel led to the same astounding answer. The people did not really care, because they did not plan to remain here after the renewal process is completed. Under Dutch planning law, the government must provide a family with a comparable housing unit at similar cost if it cannot provide housing in the renewal zone. Had there been no renewal, the people would not have been able to leave Molenwijk—and leaving was the real objective for most of them. In Scheveningen and Schilderswijk, by contrast, most of the people (especially those who were active in the participatory process) were eager to stay in their neighborhood.

Molenwijkers do not like living in their present neighborhood. After all, the housing is crowded, the infrastructure is substandard, and they find the foreigners annoying. If the neighborhood is renewed, the government will have to find them a living situation elsewhere. The new unit should be comparable, and it might be better. It is not surprising, therefore, that few people are overly concerned about the future of Molenwijk. Some will remain, of course—those who enjoy living near the city center and have the money to pay the higher rents. In one part of northern Molenwijk, an area of 500 families is being demolished; only 40 of these families are expected to return to the new units.

The strategy of the Molenwijkers to escape from their neighborhood is not without risk. Despite the letter of the law, it is often impossible for municipalities to find new housing of comparable quality and cost for displaced residents; there may be years of waiting. For the white Dutch residents who are eager to leave Molenwijk, the risk is worth taking. Although they might not be fully satisfied with the quality of their new unit, they can always apply for subsidies to reduce the cost, and at least they will be living in an area with fewer foreigners.

The picture is less rosy for the foreigners themselves. In Holland as in the United States, government policy is frequently bent to the will of the strongest, and the foreigners are last on the totem pole. There are many neighborhoods that actively and successfully resist "immigrant incursions," and most housing associations (nonprofit construction/management organizations that control a large percentage of Dutch housing units) impose low quotas on the numbers of foreigners they will admit. There are also native Dutch residents who will suffer: those who have low incomes and want to remain in Molenwijk. The elderly, who are rarely involved in Dutch participation schemes, are one probable example. Planwinkel claims to have conducted research into the satisfaction of residents who have recently left Molenwijk, finding out that 80 to 90% of them are satisfied with their new surroundings.[15] Foreigners, however, generally do not respond to such surveys, and the destination of many Molenwijkers is hard to trace.

The Molenwijk case left me with mixed feelings. A sensitive and elaborate participation process had been created, one that held great promise for improving the neighborhood in accordance with community wishes and fostering a great deal of citizen learning about the planning process and the urban environment. The neighborhood is undergoing a significant physical renewal, through carefully balanced rehabilitation and new construction. The future of Molenwijk as a physical space looks bright.

On the other hand, the purpose of participation is being corrupted by a quirk in legal regulations. Many people who are doing the planning are not even planning to stay! Many of those who may be forced to leave are not involved

[15] Interview with Patrick Boel, Planwinkel; The Hague, December 21, 1979.

in the process or cannot understand how the process will affect them. The contradictions of this situation made one of Planwinkel's major efforts in the community seem all the more ironic.

Planwinkel, in cooperation with AL and the local schools, is conducting an experiment to involve children in the planning process for Molenwijk. When buildings are torn down, the children paint murals on blank walls. The teachers divide their classes into small groups, which are asked to come up with solutions to simple renewal problems based on information provided by Planwinkel and the instructors. The children produce pictures, reports, and exhibitions used to explain the renewal process to members of the community. Perhaps most importantly, the children bring information about the renewal process to their parents, trying to get the entire family involved. This is especially useful in the case of foreigners, since their children are able to present information to the parents in their own language.

The idea is to get the children involved in the life of their community, to make them feel a part of the change which is occurring in Molenwijk. Since youth vandalism and arson are problems in the neighborhood, Planwinkel hopes that, if young people learn to take greater pride in their neighborhood, these crimes can be reduced in years to come.

This hope may be valid, but only if the children and their parents remain in Molenwijk. If they leave, the lessons (if they are at all transferable) will leave with them.

Assessing the Effectiveness of Participation in The Hague

Reviewing the Evaluation Criteria

A set of five criteria that can be used to assess the effectiveness of citizen participation were presented in the early part of this chapter. The cases will now be analyzed for each of our five criteria: diffusion of conflict, recognition of residents' goals for the neighborhood, strengthened organizational structures of the community, democratization of community institutions, and community-wide learning. For each criterion, I will examine the possibility of causal links between the participatory effort and the end result; I will also discuss the role that various antecedent conditions may have played in determining the outcomes.

Diffusion of Conflict. For the most part, participatory efforts were accompanied by reductions in the level of conflict within the community and between the community and the municipality. Both forms of conflict declined as divergent community elements jointly focused their attention on coproduction efforts with the municipality. There were two major exceptions: in Renbaankwartier, open forms of conflict (such as sit-ins and demonstrations) ceased, but subtle confrontation politicking continued in full force. During the second effort in Schilderswijk, conflict within the community reasserted itself as BOS suffered from increasingly severe legitimacy problems.

In most cases, the reduction in or elimination of conflict was directly tied to the participatory process. In several cases, in fact, conflict between the people and the municipality centered more heavily on the demand for participation than it did on the susbtantive renewal issues under consideration. Therefore, it was only logical that the beginning of a process should see some reduction in the level of conflict, especially if representatives of the community had been involved in designing the process.

In all cases, however, community organizations left open the possibility of a return to conflict. The "open door" policy with regard to conflict was critically important to the success of the participation process, at least from the residents' point of view. City politicians who gave orders to the negotiating agencies were very nervous about the potential impact of further conflict on the next elections. This fear of the adverse political results of conflict applied particularly to Schilderswijk, where the action groups were powerful and very well-organized.

Within the city bureaucracy, the view toward participation was by no means uniform. Although younger planners tended to see the need for participation (some actually have made their reputations by skillfully dealing with participation efforts), the older planners and city bureaucrats were vehemently opposed to the notion of citizens (mere planning amateurs) questioning their professional judgment. When the professionals working for the city were at odds on this matter, the potential threat of resurgent conflict was one of the strongest arguments younger planners could use to bring other bureaucrats into line and continue support of the participatory efforts during times of stress.

City workers involved in the participation efforts in The Hague faced no greater challenge than breaking through the endless red tape and bureaucratic delay that characterize Dutch municipal governments. Nothing is more likely to prompt a return to conflict than bureaucratic delay; it is prime fuel for citizens' insecurity about entering into a participation process with "the enemy." Delay generally is interpreted as deception.

In many cases, the City Developement Agency and other departments of the municipality had to develop special techniques to deal with this problem. These techniques included the assignment of specific contact persons in the City Development Agency and various resident groups to act as troubleshooters if the process hit a snag. In the case of the Schilderswijk structure-sketch planning process, the agreed-upon procedures provided for a special "signal team" to be called into action if negotiations broke down on any particular issue; the signal team was never used.

An important if informal agreement between the City Development Agency and various citizen groups involved in participatory efforts was the strict adherence to deadlines (for both sides) in order to engender trust and give the proceedings an air of professionalism. In addition, the city developed a strict policy of never promising more than it could deliver, hoping to avoid raising expectations that might later be disappointed. Through mechanisms such as these, the city and the participating communities sought to ensure smooth bargaining and to minimize the chance of a return to conflict. Perhaps one of the most important precedents was set by Alderman Nuy during the Schilderswijk Oranjeplein process, when he gave his on-the-spot negotiators wide discretion to make promises and act on their commitments rather than having to report back to city hall for permission on every small point.

Although techniques such as those discussed above can certainly help diffuse conflict (and keep it from returning), conditions antecedent to the beginning of participation are also critical. The attitudes of city workers and neighborhood leaders seemed particularly important in the cases I studied. In Schilderswijk, city workers and citizen activists initially approached each other with grave misgivings. A working relationship quickly developed, however, largely due to Kuypers's reliability as a negotiator and his willingness to bargain behind closed doors. In Renbaankwartier, on the other hand, an atmosphere of mutual respect never developed.

Recognition of the Residents' Goals for the Neighborhood. Ideally, a participatory process should legitimize residents' goals for their neighborhood by providing them with a meaningful role in the planning process. In most of these cases from The Hague, a meaningful role was achieved, but the credit seemed to go to the protests that preceded participation rather than to the adopted processes themselves. In some instances, however, particular techniques deserve special note for enlarging the roles of residents and allowing for a more precise articulation of desires.

In Schilderswijk, three such techniques stand out. The first is the in-depth survey of resident attitudes which the City Development Agency conducted prior to the structure-sketch process. An enormously comprehensive survey, this research provided planners with detailed information about the feelings of Schilderswijkers—most importantly those who were not vocal enough actually to participate in the more public aspects of the process. The second technique involved the division of the neighborhood into 10 small districts, each with its own consultation group, to handle the unique problems and concerns of that district. Third, the process expanded participation beyond the realm of housing, the subject area which was the sole focus of most participatory procedures in cities throughout Holland. By involving working groups on schools, recreation, traffic, and businesses, the process ensured that resident views would be heard on a broad range of issues. Such techniques made the planning role of residents more meaningful and enabled them to articulate their desires more precisely.

Above and beyond the processes themselves, however, what have the residents actually accomplished? The key political issue at stake in the struggles over participation revolved around the displacement of current residents, which the city was recommending (explicitly or implicitly) in most of its plans. In cases where new housing was proposed, there would be far fewer units and far higher rents after renewal. The most disadvantaged and politically weak residents of other quarters feared there would not be enough housing for them in other parts of the city. They worried that they might be paying far higher rents in their new homes. They did not want to lose the familiar faces and streets they had lived with for so long, and they feared being moved into some of the sterile, cardboard-construction neighborhoods which had recently been built as repositories for "undesirables" being moved away from the city's core, which was now to be put to "more economic use."

The single greatest achievement of the residents' struggles, therefore, has been a series of commitments from the municipality to abandon their relocation schemes and plan renewal of the old quarters for the current residents, not for newcomers. (The single major exception to this rule is Molenwijk, where the new neighborhood will probably house a substantially wealthier population than currently resides there.)

Was the participation responsible for this commitment, however? The answer seems to be partly, but not entirely. The initial commitment flowed from the conflict that in all cases was a necessary precedent to participation. Without conflict, the city would certainly not have changed its plans. However, participation gave the commitment detail and ensured its duration. Backsliding from the commitment on the part of the city became significantly more difficult when city workers had to attend meetings with citizens several times a week. In addition, the participatory process allowed citizens to influence the precise ways in which the commitment would be carried out.

Strengthening the Organizational Structure of the Community. Generally, these cases from The Hague involved a strengthening of the organizational structure of the neighborhood. New action groups were formed, coalitions (however tenuous) were created, and old groups grew stronger (or were replaced by stronger successors). A powerful leadership cadre regularly emerged to represent the interests of communities that had previously been underrepresented on the municipal level. The politicization of the population increased the attention city politicians paid to these neighborhoods, and in some cases indigenous organizers emerged to run for public office on the city level (although this created loyalty conflicts that occasionally alienated these leaders from their natural constituencies).

However, most of these results can be traced to the stage of conflict rather than the period of participation. Conflict, if it is to be successful, demands strong organization. Participation, on the other hand, can sometimes induce a softening in community power institutions (as witnessed most noticeably in BOS during the second phase of participation in Schilderswijk). Furthermore, much of the strengthening which did occur, whether during conflict or cooperation, was centered in particular groups, with questionable spinoff effects through the rest of the communities.

Democratization of Community Institutions. Only limited democratization occurred in The Hague cases. More frequently, militant elites or charis-

Figure 5. The old makes way for the new in The Hague.

matic leaders came to dominate the process of participation. A lack of democratic operating principles is not without cost. The single greatest cost is the risk that group leaders will not accurately represent the views of residents, leading to renewal that the inhabitants may not find suitable—an event which could lead to conflict in the future. An excessive dependence on individual leaders can also adversely affect the learning process. In addition, the failure to democratize a participating organization can have negative repercussions within the leadership cadre itself. In Schilderswijk, the conflict over the *bewoners raad* split the leadership, forcing one active member to remove herself from the organization; an outside advisor to the group was also planning to leave soon, partly because of this controversy. In short, a hierarchical, authoritarian style of operation can deprive an action group of allies and members who might otherwise contribute to the effort.

Our cases in The Hague give us a bleak outlook on the role of participation in democratizing neighborhood institutions. There are some bright spots, however. Although BBR in Renbaankwartier is still under the control of a militant elite, citizens play important roles on the organization's working committees, and they are frequently called on to attend meetings with the city (which are always public), where they can influence proceedings by their very numbers. In

Molenwijk, some attempts have been made to include in the process groups that have largely been ignored in the past, specifically women, immigrants, and children. In Schilderswijk, Payable Rents tried very hard during its early stages to develop a democratic institution. It created a residents' council and a "daily board of directors" *(dag-bestuur)* as part of this effort; regular meetings with the community were held, and BOS organizers worked on the street and block levels.

In all three neighborhoods discussed above, democratization was correlated with the arrival of participation. In each case, the efforts made to increase the democratic nature of the action group related to the group's need to prove its own representativeness and its need to delegate the complex and time-consuming tasks of participation. Little democratization was observable during the period of open conflict. Participation can lead to organizational rigidity, however, and a trend away from democratic values, as proven in the Schilderswijk structure-sketch process, where the leaders of the organization became so wrapped up in the process itself that democratization took a back seat (although efforts are now underway to reverse this situation).

Community-wide Learning. On the whole, skills possessed by the leaders increased markedly in the communities studied. People with natural organizing and planning talents emerged rapidly and set up positions of dominance in the community. For example, Schilderswijk, which has now gone through two phases of conflict and participation, has developed an especially sophisticated leadership cadre. Laborers with no more than a primary-school education have risen to positions of great power in their communities, based not merely on the strength of their personalities (although this has played a role) but also on real knowledge and skills.

It is equally important, however, to analyze learning among those members of the Dutch working class who did not rise to positions of leadership. Generally, the average citizen who attended a number of meetings and offered opinions about the future of his or her block and street has learned a good deal about setting priorities, the economics of planning, and planning procedures. Increased organizational abilities and deeper understandings of political realities are not so strongly evidenced, however. If they were willing to put in some time and buy into the process, they would come away with greater knowledge, which might help them defend their own personal interests better in the future.

This phenomenon was generally observable in most of the neighborhoods studied, although different processes used different techniques to induce greater learning among average citizens. In Renbaankwartier, all meetings with the city had to be public, a fact which may have had adverse political implications but which gave many residents an opportunity to learn about the planning process. BOS in Schilderswijk hired part-time (sometimes volunteer) experts from inside and outside the community to provide technical information and help with drawing up responses to city proposals; BOS members and workers learned much from these outside experts. The working groups instituted in the structure-sketch process also enabled citizens to learn about subject areas other than housing (such as schools, economic development, and traffic).

On the whole, however, these techniques were not very useful in involving the disadvantaged members of the community in the participation process. The elderly, handicapped, and immigrant members of the community were not, on the whole, deeply involved. These groups are prevented from participating in at least three ways. First, they are less able to involve themselves due to infirmity, a lack of skills, or an inability to articulate their desires in Dutch. Second, (especially in the case of minorities), the dominant participants are not eager to see them involved and make few efforts (if any) to encourage them to overcome their participatory disadvantages. Finally, these groups lack natural leaders and spokespersons who can bring them into the process or at least represent their interests.

As with the other criteria we have examined, the role of antecedent conditions cannot be discounted. Although certain techniques seemed linked with a high level of learning, an impetus was required to put these techniques into effect in the first place. This makes the role of community leadership particularly critical. In the case of Renbaankwartier, for example, the leadership was not apparently interested in encouraging learning within the community and has installed few techniques to do so.

The Primacy of Attitudes, Conflict, and Leadership

Practicing planners and academic researchers who try to improve planning practice seem ever to be involved in a search for the correct mix of "participation techniques" that will ensure "effective participation" (or at least promote it with a reasonable frequency of success).

My research in The Hague—and briefer visits to Amsterdam, Groningen, Nijmegen, Rotterdam, and Utrecht—has convinced me that conditions antecedent to the beginning of a participation process have much more to do with its likelihood for success than the actual format of the process.

Specifically, I would identify three key antecedent conditions: the attitudes of city workers and citizen participants toward each other; the level and form of conflict preceding the participation (and chances for the reemergence of conflict during or after the participation process); and the presence, strength, and opinions of an indigenous neighborhood leadership. In the next few pages, I will explore these conditions, citing The Hague examples as well as others I examined more briefly in the Netherlands.

Attitudes. The history of planning practice in Holland, as in the United States, makes participation an unnatural form of behavior. Planning professionals and politicians usually prefer conflict to be mediated through the electoral process or through standardized, although informal, mechanisms of political reward and punishment. The idea of having all parties to a dispute sit down at a table to plan the future is a new and frightening phenomenon to many associated with municipal government. Similarly, citizens are not used to the process or to the very idea of bargaining for neighborhood renewal. Naturally, they enter the process with trepidation and mistrust.

As they begin participation, both sides will make mistakes which hurt their negotiating counterparts. The level of respect and trust the parties bear for each other will directly affect their ability to overlook problems early on and proceed with the process. The Schilderswijk and Renbaankwartier cases in The Hague are opposite examples of how the attitudes of citizens and city workers can affect a participatory process.

The city of Groningen, in the northern part of the country, was in many ways Holland's laboratory for participation, largely because of the efforts of Max van den Berg, the alderman in charge of urban renewal. A member of the socialist Labour party (and currently national party head), van den Berg believed that participation was the best way to ensure that neighborhoods were renewed without displacing the present residents. He and his associates spent several years encouraging action groups to form throughout Groningen, followed by efforts to actually plan renewal efforts in accordance with citizen desires. Although van den Berg did not have a completely smooth relationship with the groups he helped create, his own personal commitment to participation laid a firm groundwork for an atmosphere of trust and cooperation.

Nijmegen, a staunchly Catholic and conservative city on the German border, provides a counterpoint to the Groningen experience. City officials long sought to undermine organizing efforts and repeatedly supported plans which displaced residents (or threatened to do so). For years, they refused to recognize action groups as legitimately representative. A recently adopted code to govern future participation in Nijmegen is viewed with distrust by many residents of the neighborhoods, who believe it will do nothing more than coopt their interests. A number of action groups fear that the city will try to cut their budgets, since the bulk of social service funding was recently decentralized from the national government to local municipalities.

Conflict. The typical American typology of citizen participation techniques (most notably Sherry Arnstein's "ladder of citizen participation") describes participation according to a series of possible power relationships between the city and the residents. The greater the real power of the residents, the higher the form of citizen participation. These systems, although they provide accurate snapshots of various stages of participation, have always been rather unsatisfying to researchers who recognized that the evolution of citizen participation does not flow evenly up the ladder. The stages do not necessarily follow each other in a logical order, and the various ladders that have been proposed tell us little about the evolution of participation.

Professor Hugo Priemus of the Delft Technical University has concluded substantial research on citizen participation in the Netherlands, with particular attention paid to Crosswijk in Rotterdam, one of the neighborhoods discussed in Chapter 9 by John Zeisel and David Godschalk. Priemus has advanced a historical explanation of the participation process. In the first three stages, the absence of a role for citizens in the renewal process gives way to protest and conflict and eventually to the establishment of an informal role for citizens (or certain citizens). In the two final states, a formal structure is created and usually considered a victory by citizens; eventually, however, the inadequacy of formal structures for participation is recognized. Priemus explains that Dutch groups have not yet developed a coordinated strategy for dealing with this last stage, although some are beginning to recognize that formal structures often work against their interests.

Like any model, this one is far from perfect, and several Dutch scholars have criticized it. In my opinion, however, it makes two significant contributions to our understanding of the dynamics of citizen participation. First, it recognizes the integral role of conflict as an antecedent condition for meaningful participation. Participatory schemes may be set up in the absence of conflict, but they are rarely meaningful and genuine processes in which citizens receive and exercise real power. Before the municipality is willing to share power, it must be made to see the price of retaining all power to itself.

Priemus's second major contribution is his recognition of the fact that the participation process can go sour, necessitating the reintroduction of conflict strategies. We saw several cases of this in The Hague. In Renbaankwartier, the participation process ended on schedule, but without complete agreement. Instead of proceeding to a higher level of bargaining, the confrontation process was reintroduced almost willingly by the action group. Since conflict is both costly and time-consuming, its return in Renbaankwartier can hardly be applauded. However, knowledge on the part of the government that BBR could swiftly reintroduce open conflict probably contributed to the city's willingness to compromise on issues in the participation process. A municipality which recognizes that conflict can easily be resurrected is more likely to make the concessions necessary to move a process along. If a group is divided, tired of protest, or financially unable to mount more confrontation, government will have little incentive to bargain in good faith—a process that requires concessions from both sides.

Presence of a Neighborhood Leadership. In all the neighborhoods of The Hague that I studied, an effective leadership was present for purposes of organization and, eventually, participation. Since the integration of average residents into the participatory process is difficult and time-consuming, an existing leadership is vital to get a new participatory process off the ground.

The leadership formulates citizen opinion into rational positions which can be negotiated with city workers and politicians. The leaders can facilitate integration of other citizens into the process (although there is evidence in many neighborhoods that the leaders prefer to handle things themselves). Perhaps most importantly, an effective leadership makes the threat of renewed conflict credible to the municipal authorities. Community leaders shape participatory procedures in innumerable ways, both in formal discussions with municipal officials and in the way they handle the day-to-day operation of their organizations.

Learning from the Experiences in The Hague

This final section will briefly explore what activists and government officials (particularly in the United States) can learn from the experiences recounted in this report.

Developing Techniques for Effective Participation

As United States citizen activists and government officials consider the creation and implementation of participatory procedures, they will undoubtedly review the techniques they should employ, even though antecedent conditions may be more important. Do the experiences of The Hague indicate any particular techniques (or, more generally, kinds of techniques) that seem highly correlated with effectiveness, or (since it is often difficult to draw correlations from only a few case studies) techniques that the participants themselves found useful? The answer is yes. Although it would not be possible to describe each technique definitively, I will present a few that I believe are particularly likely to be transferable to the U.S. scene.

1. *Use written agreements* between the municipality and residents' groups to clarify a set of goals and determine a mutually acceptable process for attaining those goals in a fixed space of time. The written agreement (based on the *doelstellingennota* from the Schilderswijk case) is undoubtedly difficult to achieve, especially when parties have been bickering for a long time. However, these problems are generally easier to solve than the substantive issues, and a few months of working out a procedural agreement can go a long way to cooling tensions and building trust before the hard bargaining on substantive renewal questions begins.

This period of time can also be used to allow both sides to decide who will serve as key persons in the participatory process and what the division of responsibility will be within city agencies and within the neighborhood group(s) as well as to locate various "third parties" who should be represented in the process but who have not yet been involved.

2. *Citizens must have time by themselves* to think, meet, iron out their own differences, and talk freely without municipal presence. Overbearing city admin-

istrations in the United States frequently demand that municipal representatives be present at all meetings; they will often send people to neighborhood meetings uninvited. This can be extremely counterproductive because it reduces the level of trust between both sides. Just as municipal officials can meet alone in their offices, citizens should also have a chance to take care of their own business *in private* and prepare positions for negotiating sessions without municipal interference.

3. Hold special neighborhood meetings (with the attendance of municipal planners where appropriate) *to discuss specific planning issues, provide information, and compare the relative merits of differing opinions*—without the need to make a decision on that day regarding any particular stand. Such meetings organized by Planwinkel in Molenwijk were very useful in explaining key issues to residents, expanding community-wide learning, and making the decision and negotiating sessions more efficient.

4. Employ a host of mechanisms to *cut down on red tape and bureaucratic delays*. Special troubleshooters, adherence to strict deadlines, intra-agency coordination teams, and other such methods can be useful. Since delay breeds mistrust, it must be minimized.

5. Use innovative techniques to *involve children, the elderly, and the handicapped* in the process. Special outreach workers are probably needed in this regard; but even more can be done, as evidenced by Planwinkel's efforts to involve children via the schools and to arrange meetings at times convenient for housewives. In Schilderswijk, part of the Oranjeplein development was targeted to the elderly, and Payable Rents tried to get their opinions about how their units should be designed.

6. *Make certain that all interested parties* to a renewal project *are represented* in the process, and encourage democratic operating procedures within the citizen organizations that are involved. Both of these steps strive to ensure that the participation process takes into account the true feelings of as many residents as possible in order to locate a true neighborhood consensus (if that can be achieved) and to prevent future conflict by limiting the number of parties who will feel that the process left them out.

7. Urban neighborhoods are large and diverse. It is frequently useful to *divide them up into small districts,* with a planning team to consider the problems and needs of each. This will help isolate localized issues, encourage people to become involved even though they are only concerned with their immediate area, and take advantage of small working groups of people who probably know each other.

At the same time, the interests of the entire neighborhood should not be ignored. The neighborhood-wide working groups organized in Schilderswijk to deal with particular subject areas represent one way of bridging the gap between localized, district issues and concerns of the entire neighborhood.

Affecting Antecedent Conditions

Making the all-important antecedent conditions more favorable to citizen participation is even more difficult than developing effective participatory techniques. Such steps are hard to locate and, once located, they are hard to accomplish. However, I believe there are three particular classes of actions that United States activists and officials could benefit from knowing about:

1. *Leadership is not preordained.* Group members must choose their leaders carefully, not based merely on who knows the most, who speaks the best, or who has the most forceful personality. Other qualities to consider (and to examine carefully if they are not obvious at first glance) include respect for the opinions of others, willingness to work *with* groups, concern for the knowledge and understanding of colleagues, and capacity to change personal theories and assumptions about the world. At one point or another, groups usually have the chance to decide who will be in charge, and they must choose with care.

2. Municipal governments should *encourage neighborhood self-organization*. Dutch traditions of government-supported social work plus recent political developments like the organizing efforts of Max van den Berg in Groningen create a social expectation in Holland that the municipal government will help (or, at very least, will not hinder) the efforts of neighborhoods to develop coherent resident organizations to lobby on behalf of local interests. This expectation does not yet exist in the United States, where municipalities are allowed (and even expected) to undermine neighborhood organizing. Despite being repeatedly attacked by government in an attempt to destroy or weaken them, neighborhood groups usually organize themselves anyway and respond with their own attack. This process makes the growth of trust and cooperation virtually impossible.

In the Netherlands, the right of citizen groups to organize and protect their interests is recognized. The result is a set of resident groups that are generally more docile than American groups but also more accepted as political entities speaking for recognized constituencies and having guaranteed rights as organizations. Dutch municipalities (and the central government) help these organizations to grow by providing subsidies, meeting spaces, social workers and other staff persons, and regular contacts with municipal personnel. Naturally, some Dutch municipalities are more generous and cooperative than others; Rotterdam is very helpful, while Nijmegen is obstructive. On the whole, however, municipal officials adhere to a policy of "self-organization" which states that government will help neighborhood groups to organize themselves while reserving the right to disagree with and oppose these neighborhood organizations in relation to specific policy issues. A greater effort on the part of American municipal officials to aid in the development of neighborhood organizations (or at least not interfere with such efforts) would probably improve the climate in which participatory efforts can develop.

3. Neighborhood groups must *leave the door to conflict open* even during a participatory process. If they do not, they are likely to suffer attacks from more militant organizations in their own community. In addition, the municipality is always more likely to grant concessions to an organization that it knows is capable of reasserting protest at any time. Without an open door to conflict, a participatory process is likely to stagnate when the municipality realizes that it no longer has anything to fear from the residents.

A municipality has many weapons to use against communities, but neighborhoods have only protest. Once they give that weapon up, municipal officials will have little reason to listen to them. A good participatory process demands a balanced power relationship. Without an open door to conflict, this balance disintegrates, as the leaders of BOS painfully learned during the Schilderswijk structure-sketch effort.

ACKNOWLEDGMENTS

During the sixteen weeks I spent in Holland, I interviewed approximately 60 persons and visited urban renewal sites throughout the country. It would be impossible to thank all the people who enhanced my knowledge and understanding of the Dutch scene during that time.

Particular appreciation is due to my friends at the International Union of Local Authorities in The Hague, who provided me with office space, library resources, and much sage advice during my stay. Much thanks is due to The Hague City Development Agency (Gemeentelijke Dienst voor de Stadsontwikkeling) for the patient assistance of staff members through four months of badgering. Finally, my appreciation is extended to citizen organizers throughout The Hague, who were always too busy to meet with academic intruders, but who found the time to do so anyway.

The following people deserve special thanks for their contributions to my research: Wolfgang Beck (University of Amsterdam); Patrick Boel (Planwinkel); Adri Duivesteijn (The Hague City Council); Philippine Heering and Aad Kuypers (Residents' Organization for Oranjeplein–Schilderswijk); Paulinus Kuipers (Jacob Maris Group); Hans Leeflang (National Physical Planning Agency); J. G. Meester, Nico op de Laak, and T. M. van der Caaij (The Hague City Development Agency); Hugo Priemus and Henk van Schagen (Delft Technical University); and L. Pronk (Residents' Organization for Renbaankwartier).

A few personal acknowledgments are also in order—to Paul, Gerda, Sander, and Barbara Citroen as well as Maurits and Maria Kapper for the joy that was mine in coming to know a family from far away; and to Eileen and Goty Harloff for their help with the Dutch language and with many other problems that researchers face in foreign lands. Naturally, none of my research could have been accomplished without the generous financial assistance of the German Marshall Fund of the United States.

Bibliography

Bagley, Christopher. *The Dutch Plural Society*. London: Oxford University Press, 1973.

Bailey, Anthony. "The Little Room." *The New Yorker*, 8 and 15 August, 1970.

Beck, Wolfgang *et al*. *From Acton Group Payable Rents to Residents' Organization for Oranjeplein–Schilderswijk*. Institute for Science and Andragology, University of Amsterdam, 1977 (Dutch language document).

Crijnen, Ton. "Living in Schilderswijk." *De Tijd*, 21 December 1979 (Dutch language document).

Haberer, P.; de Kleijn, G.; Nicolas, F.; and de Wit, C. *The Neighborhood Approach: Making Old Neighborhoods Better, Through and For the Residents*. Research project completed for the Ministry of Housing and Physical Planning. The Hague: State Publisher, 1978 (Dutch language document).

Hague, The, City Development Agency. "Structure Sketch for Schilkerswijk: Proposal." The Hague: State Publisher, 1978 (Dutch language document).

Hague, The, City Development Agency. "Structure Sketch for Schilderswijk: Reactions." The Hague: State Publisher, 1979 (Dutch language document).

Hague, The, Office of Information. "Rules for Input." The Hague: State Publisher, 1979 (Dutch language document).

Hale, Robert. "In The Netherlands, Rehabilitation Is a National Effort with New Ideas for the United States." *Journal of Housing*, November 1978.

International Federation for Housing and Planning. *Urban Renewal in The Netherlands*. Papers and proceedings from a seminar organized by the IFHP, held in The Hague, September 1975.

International Union of Local Authorities. "Symposium on Citizen Participation and Local Government." *Studies in Comparative Local Government* 5, no. 2 (winter 1971).

Langton, Stuart, ed. *Citizen Participation in America*. Lexington, Mass.: Heath, 1978.

Nelkin, Dorothy. *Technological Decisions and Democracy: Sweden, the Netherlands, and Austria*. New York: Sage, 1977.

Rosenbaum, Walter. Evaluating the Policy Impacts of Public Participation. Presented at the Symposium on Citizen Participation: Models and Methods of Evaluation, February 1980, Washington, D.C.

Rosener, Judith. Evaluating the Effectiveness of Participation Techniques: Whose Criteria? Who Cares? Presented at the Symposium on Citizen Participation: Models and Methods of Evaluation, February 1980, Washington, D.C.

Shingles, Robert. The Effects of Citizen Participation: A Program for Evaluation. Presented at the Symposium on Citizen Participation: Models and Methods of Evaluation, February 1980, Washington, D.C.

Van Dijck, H. *et al*. *Reconstructing the Schilderswijk: A Study of an Attempt to Introduce Innovations into the Making of an Urban Reconstruction Plan*. Paris: Organization for Economic Cooperation and Development, 1975.

9

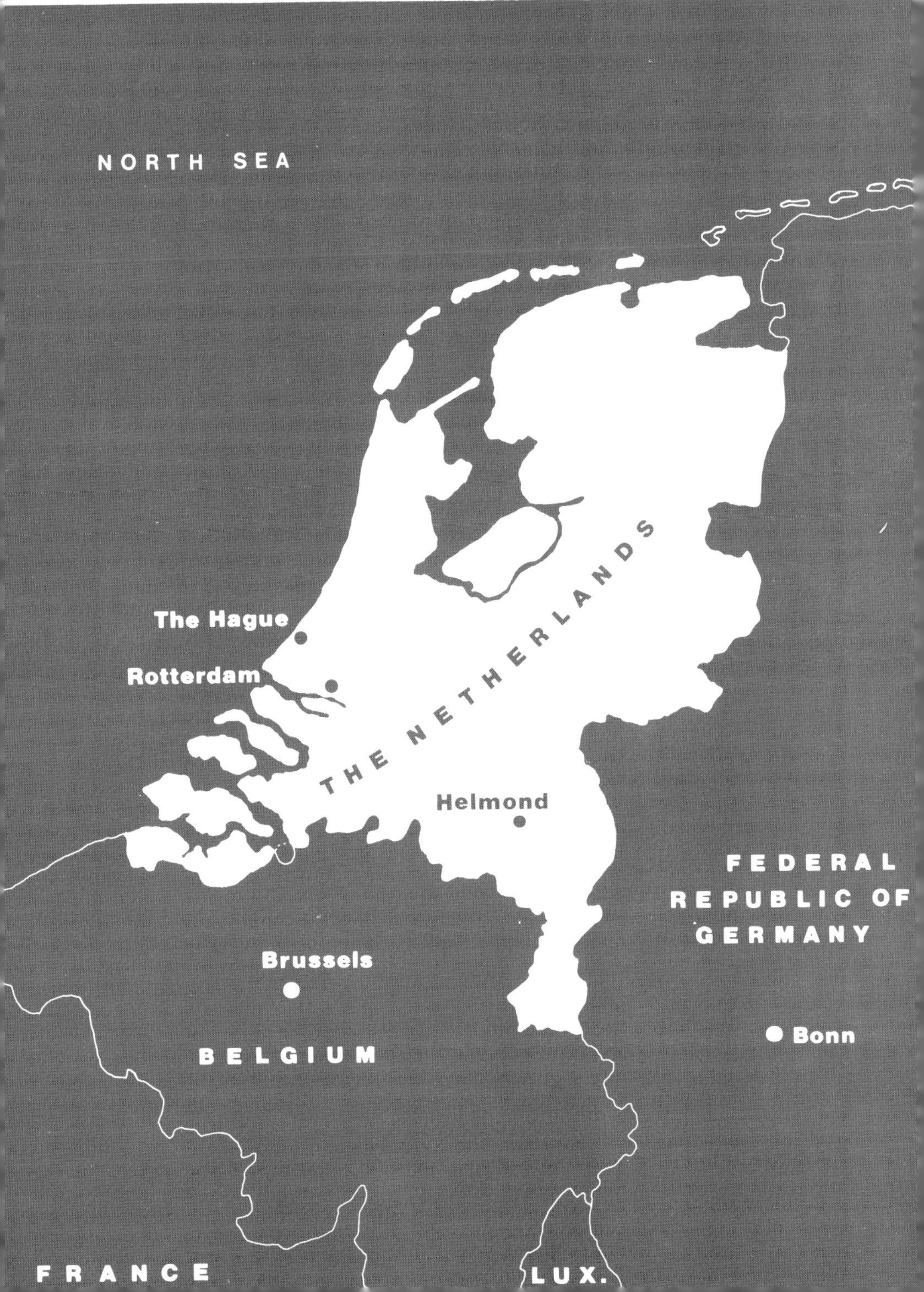
NORTH SEA
The Hague
Rotterdam
THE NETHERLANDS
Helmond
FEDERAL REPUBLIC OF GERMANY
Brussels
BELGIUM
Bonn
FRANCE
LUX.

9

COPRODUCING URBAN RENEWAL IN THE NETHERLANDS

DAVID GODSCHALK AND JOHN ZEISEL

Involving the Burghers

To an outsider, the participants did not seem to belong together. The bearded young man talking at the lectern wore old jeans and had arrived on a motorcycle. The clean-shaven older men in their coats and ties, listening attentively, and their wives in nice print dresses had arrived at the auditorium by bus. A couple of professional-looking men were passing out information sheets. What were these disparate people doing together on a summer evening at the Technical University at Delft?

The short answer is that they were working together to plan rehabilitated housing for a part of the Vreewijk neighborhood, one of Rotterdam's urban renewal areas. However, behind that apparently simple answer lies a series of intriguing paradoxes that tell a story about one of many techniques the Dutch have invented to bring diverse and often conflicting interests together to plan the renewal of their older urban areas.

Before the evening was over, a number of paradoxes had become evident. The young man was an "external expert," whose salary was paid by the Dutch national government but whose orders came from the neighborhood group *(wijk-orgaan),* an arm of the local citizens of Vreewijk. The listeners were a group of a dozen residents—retired schoolteachers and other middle-income people whose houses were to be completely rebuilt but who would continue to live in their neighborhood. The professionals included an architect, hired by the local housing corporation but directed by the residents in the design of their houses, and a research sociologist from the technical university who helped people find out what housing designs they preferred not by abstract questionnaires but by putting them into life-size model houses. The young external expert espoused a radical political philosophy. He was opposed to government planning. He lived on a minimum salary in an unheated windmill without electricity or running water. Yet his salary was paid by one of the most pervasive and highly organized governments in western Europe, and his clients and sometime bosses on that occasion were 12 solid Dutch burghers.

The reason for the meeting was citizen participation in public planning, often the genesis of dull and tedious speeches by government officials justifying already-made decisions with ambiguous information to residents who were either bored and apathetic or fighting mad and protesting. Yet this meeting had the stimulating flavor of a well-organized seminar on a topic of keen interest. People were there to learn, to debate alternatives, and to influence plans.

In the course of the meeting, the citizens were provided with written and graphic information about alternative housing plans, including costs, advantages and disadvantages, and evaluations by housing and planning experts. They had opportunities to question their architect and their external expert and to form their own judgments about the alternatives. Finally, they were able to walk through full-scale models of two alternative house designs and to rearrange walls, doors, and other elements of these models in order to try out and debate variations of the original plans.

Only a few years ago, Dutch people like these either had been fleeing their old central neighborhoods to new suburban housing projects or had been banding together to fight against government renewal schemes. They had seen public planning either as something so long-range and abstract that they could not be bothered with it or as something carried out by insensitive bureaucrats in which there was no role for ordinary people. What had happened to change the minds of these Dutch citizens about participation in urban renewal?

This study explores techniques and strategies the Dutch have invented to make planning real and immediate to ordinary people and at the same time to convert into the urban renewal production process resident energies that were formerly poured into conflict. Our aim is to draw insights for U.S. urban revitalization from the Netherlands experience.

From the start, our goal has been to discover and describe Dutch citizen participation approaches that have effectively made urban renewal planning and implementation more productive. This is not a comprehensive evaluation of all participation in Dutch urban renewal; it concentrates on selected cases where participation has worked. For every successful instance of participatory planning, there are unsuccessful ones. The examples here are exceptional. In most cases, urban renewal is marked by tension and confrontation between residents and local government. However, out of these struggles in some places has come a new understanding of the potential for productive collaboration between people and public officials interested in rebuilding old neighborhoods.

Our approach has been to concentrate on common principles and ideas present in instances of productive collaboration. We have not attempted to reconstruct historical events, roles, and motivations, as in a typical political science case study. Rather, our intent is to synthesize from a carefully screened set of Dutch experiences those central features that may be useful to renewal planners and activists in other countries involved in designing participatory strategies and processes.

Comparison of U.S. and Dutch Urban Renewal

Urban renewal in the Netherlands has evolved through three stages during the past decades (Haberer & Vonk, 1978). In the early 1960s, renewal focused on center-city redevelopment and slum clearance. Renewal goals were largely economic. Reaction against the effects of this approach led to a focus on housing rehabilitation, neighborhood preservation, and more socially oriented goals in the late 1960s and early 1970s (Hale, 1978). In the late 1970s, a pragmatic mixing emerged, resulting in urban revitalization projects that combine rehabilitation and clearance and that seek both social and economic development. Although the National Urban Renewal bill introduced in 1976 to formalize these changes was not passed,[1] lessons learned from earlier experience have been incorporated into the practice of many Dutch local authorities.

Each urban renewal stage had its dominant style of planning and decision making. During clearance and redevelopment, planning was dominated by paternalistic government control. In this stage, government structured and managed "citizen participation." Efficiency was the watchword as entire neighborhoods were bulldozed and their inhabitants relocated, some to other old neighborhoods and others to new high-rise suburbs. During the following stage, political protest against the results of widespread housing demolition and neighborhood dislocation led to citizen counterplans. Renewal projects were stalled in conflict as neighborhood groups initiated "citizen action." More recently, a diverse range of citizen/government "partnerships" has been created to plan and implement renewal projects. The overall nature of citizen participation in Dutch urban renewal evolved from early control by government through a second stage of aggressive partisan conflict and resistance by citizen action groups to the recent appearance of citizen–government collaboration in some cities (Turpijn & Veenema, 1978).

Different Dutch cities have moved through this evolution at differing speeds. Rotterdam is the clearest example of a city that has designed collaborative planning institutions. Amsterdam, on the other hand, has had far greater difficulties in bringing about cooperation between resident action groups and mu-

[1] Currently, the urban renewal process in the Netherlands falls under the land use procedures of the Physical Planning Act of 1965. A revision of the 1976 Urban Renewal bill was under consideration in 1980.

nicipal planners. Different cities have used different strategies, depending on local political and socioeconomic situations.

Those familiar with U.S. urban renewal history will recognize many parallels with the Dutch experience. Early U.S. programs also stressed total, large-scale redevelopment of central cities in order to expand business districts, improve automobile access to core areas, and clear residential slums. These early plans, drawn by local government officials and legitimized by blue-ribbon citizen commissions, resulted in the bulldozing of many low- and moderate-income neighborhoods that might have been saved. In reaction to the insensitivity of these plans, the Model Cities approach stressed housing rehabilitation, neighborhood preservation, and resident influence over public plans. Conflict increased and many projects were blocked (for an account of the U.S. experience, see Frieden & Kaplan, 1975). Current efforts in the United States represented by the Community Development Block Grant Program (which allows for flexibility by individual local governments), somewhat resemble the mixed rehab–renewal, social–economic, citizen–government planning strategies of many Dutch cities, although with a more tentative commitment by the national government.

In both countries, the long-term evolution of urban renewal can be viewed as an oscillation between opposing poles of centralization versus decentralization, standardization versus diversity, and government versus citizen control. The movement has been from action to reaction to interaction. The resulting long-term trend line takes the form of a gradually changing curve, as shown in Figure 1.

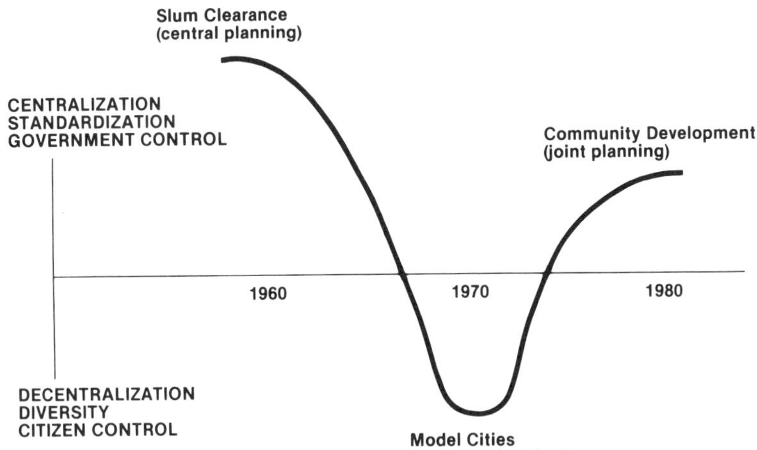

Figure 1. General participation trends in the urban renewal efforts of the United States and the Netherlands.

Despite similarities, there remain important economic, political, and social differences between the U.S. and Dutch experience in providing housing and community development. The level of government subsidies for housing and neighborhood development in the Netherlands is many times greater than in the United States. The Dutch central government subsidizes 76% of all new dwellings under a national goal that a proper dwelling must be available to all persons over 18 at a price compatible with their income (Ministry of Housing and Physical Planning, n.d., p. 5).[2] The political philosophies of the two countries echo this difference between a commitment to the welfare state in the Netherlands versus the market in the United States. In Rotterdam, the most extreme example, the city government is the largest landlord, having acquired 50% of all the dwellings in renewal areas (City of Rotterdam, 1979, p. 10). Elsewhere in Holland, most of the subsidized housing is built and operated by nonprofit housing associations that act as landlords for their members.

The Dutch planning and participation systems also are different than those in the United States, tending toward the British model. The Physical Planning Act of 1965 set up a three-tier system of local allocation plans, local authority structure plans, and regional plans (Haberer & Vonk, 1978). Allocation plans are the most specific and legally binding. They indicate physical development intentions and budgets for renewal districts or urban extension areas. Structure plans specify strategies for linking sectoral plans for housing, education, transportation, and the like. Regional plans, the most general and weakest type, are prepared by the provincial governments. Citizens can comment on these plans either by means of complaint procedures directed at the municipal, provincial, and finally ministerial levels or by working through local committees and action groups. The central government operates the appeal system, provides information on planning problems, and provides subsidies for community organizers and social workers *(opbouwwerkers)*, social action groups, and community planning services. Local citizen action groups can receive subsidies for community organizers and social workers, many of whom espouse radical ideals, like the young professional working with the residents at the Technical University at Delft.

There are enormous social and cultural differences between Holland and the United States, including the Dutch practice of "columnization," or forming special organizations and providing public facilities for every religious group. As Beck (1976) notes:

> By having separate political parties, unions, schools, radio stations, farmers' cooperatives, and everything else imaginable, one each for Protestants, Catholics, non-believers and often several other kinds of interest groups as well,

[2] However, due to a housing shortage, many Dutch people seeking public housing have not been able to obtain it.

> they have contrived to avoid religious clashes in what is still a deeply religious society. (p. 4)

Obviously columnization complicates the institutional arrangements for participation in Dutch renewal.[3]

Because of these and other differences, it would be foolish to attempt a wholesale transfer of Dutch urban renewal practices to the United States or any other country. Instead, we can look at the Dutch experience in terms of its ideas and principles and the results of their application. Interestingly, we often find similar ideas and principles reflected in American experience, and the Dutch parallels offer a chance to ask: What if we had tried the Dutch way of applying that principle? Or what if we had taken our approach as far as the Dutch have? The value of the Dutch approach is not as a product to be copied literally but as a process that may enlarge our vision of our own possibilities for increasing the productivity of renewal programs.

The Coproduction Concept

A recurring lesson from experience in both Dutch and U.S. cities is that over the long term, high productivity in urban renewal tends to be linked with high levels of cooperation between residents and government. Productivity depends on bringing together the necessary resources to do a job and on learning through the participation of those affected what job needs to be done. Because urban renewal is a complex and interdependent process and because its outcomes are determined by public and private organizations and social groups acting together and separately, its success requires an institutional framework that provides continuous opportunities for communication and collaboration in planning and decision making. The essence of such a framework is a commitment to coproduction.

Coproduction of urban renewal occurs when government agents and neighborhood residents work together to define problems and needs, to devise and evaluate alternative plans and strategies, and to carry out renewal actions. In this way, citizens are not simply the clients, the targets, or the consumers of community change. Instead, they are active participants and partners in the production of neighborhood renewal. By contrast, when government takes on the role of sole producer of urban renewal, as in the early 1960s, citizens are typically cast in advisory or target roles. When citizens opposed government proposals, as in the late 1960s, they cast themselves as adversaries. In the former case, the goals were unacceptable to citizens; in the latter case, the resources available to citizens were inadequate to plan and carry out renewal projects. Coproduction approaches offer the opportunity to combine acceptable goals with

[3] As secularization has increased in the Netherlands, however, the importance of columnization has declined.

adequate resources. On Sherry Arnstein's "ladder" of citizen participation, coproduction as defined here would fall on one of the higher rungs of shared power over plans and decisions (Arnstein, 1969).

The concept of coproduction has recently been applied in evaluations of the provision of human services, particularly community safety and law enforcement (Percy, 1978; Whitaker, 1978, 1980). One community safety analyst points out that "citizens, according to this concept, help provide services that professionals are formally responsible for, by taking cooperative self-help action" (Sharp, 1978). She compares two forms of collective activity: (1) coproduction in which neighborhoods cooperate with the police by helping to prevent crime through block watching, property marking campaigns, and mobile citizens's patrols, and (2) advocacy in which neighborhoods seek to obtain better police service by petitions, demands, and other pressure-group tactics. She finds that neighborhood organizations often pursue both advocacy and coproduction strategies effectively.

Our use of the coproduction concept in urban renewal is similar to that of the community safety analysts in several respects. Both emphasize actions that take place within the limits of the law, that are territorially based in urban neighborhoods, that stress citizen–government cooperation and shared responsibility to improve public service productivity, and that recognize that both government resources (such as funds and technical expertise) and citizen resources (such as local knowledge and the ability to mobilize) are necessary and legitimate.

However, our approach expands the concept in emphasizing that the stage preceding coproduction in urban renewal is likely to be neighborhood–government conflict and that conflict may reappear periodically during urban renewal planning and implementation. Another difference is the nature and size of the stakes at issue; urban renewal outcomes involve large-scale, relatively permanent, and indivisible capital investments that change the physical environment—not relatively low-cost extensions of existing human services designed to transform behavior. Deleting a budget line will stop a citizen–police project; a Pruitt-Igoe must be dynamited. Citizen entry into planning is more likely to result from a challenge to the professionals' plan and from a new claim to be a partner in decision making than is the case in law enforcement, with its different expertise, risks, institutions, and authority relationships. And once allowed into the renewal game, citizens must deal with the same complex investment, design, and allocation issues as government planners; it is more difficult to separate out "citizen" responsibilities, such as block watching or property marking, from the overall planning and community redevelopment process.

The coproduction that we found in Dutch urban renewal rests on power shared between government and citizens. The sharing is not simple; it is dialectic and dynamic. Coproducing parties differ in their ability to exert top-down control. In the Netherlands, as in most societies, government authorities can exert control

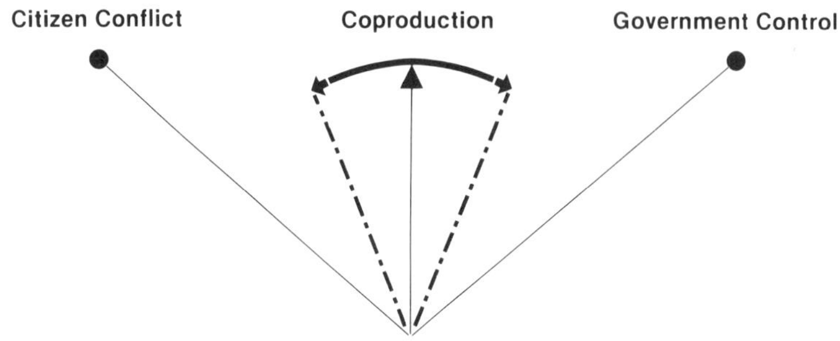

Figure 2. The coproduction dialectic.

if coproduction efforts break down. On the other hand, citizens' groups retain the threat to resort to open conflict in order to keep a balance with government. For example, when the government mandates participation as a requirement for accepting a plan, citizens may gain some control or force new negotiations by refusing to participate. (See Figure 2.)

Dutch coproduction, then, contains both the threat of conflict and of top-down control. Throughout a coproduction effort, all groups retain the right and the ability to destroy the process. One result is that coproduction appears in various guises—sometimes verging on open confrontation, sometimes verging on traditionally imposed hierarchical control. Coproduction involves concentrated efforts and active mechanisms to take advantage of these tensions rather than merely to manage them. It does not sweep under the rug citizen–government differences over ends and means.

Productivity in neighborhood redevelopment and rehabilitation cannot be measured simply in terms of numbers of new buildings or public facilities; rather, it must be reckoned in terms of the value over time of the resulting social, physical, and economic outcomes. Preservation and development of social networks may be as valuable as adding new schools or houses. One practical way of judging the value of the output from urban renewal is in terms of the long-term satisfaction of the community and urban area with the quality of the total environment produced by these efforts. This quality must be evaluated longitudinally in both objective and subjective terms in order to capture its full significance.[4] Bulldozing an old center-city neighborhood and moving residents to suburban high-rise apartments may produce dramatic short-term change but will not necessarily generate long-term community satisfaction. On the other hand, low-cost rehabilitation of deteriorated housing may result in immediate

[4] Models of public service production include both objective outcomes, such as impacts on community conditions, and subjective outcomes, such as changes in citizen perceptions and attitudes. See Ostrom *et al.* (1978).

satisfaction but may not stand the test of time.[5] Coproduction aims at balancing these two types of goals rather than sacrificing one for the other.

Dutch Coproduction Strategies

In the Netherlands, citizens participate in urban renewal in diverse ways. Even within those selected instances in which coproduction is involved, there is considerable diversity. At first glance, citywide participation efforts in the large North Sea port city of Rotterdam seem to have little in common with local organizing going on in Helmond, a small working-class town in the south. Closer inspection of these and other Dutch coproduction examples, however, reveals interesting patterns of action. The diagram in Figure 3 summarizes these patterns.

Two Goals

The coproduction approach to citizen participation in Dutch neighborhood renewal projects achieves two significant ends. First, the process frankly recognizes differences between people and organizations and does not gloss them over; it makes their resolution a key goal of the planning process. In this way, differences may even become productive planning resources; they are seen as sources of information about needs and preferences rather than simply sources of conflict and stalemate. For example, in the community of Schilderswijk in The Hague, community organizers hired by the city to work with citizen groups used the renewal planning process to express citizens' attitudes on a range of

Figure 3. Dutch coproduction strategies are based on two goals (recognizing differences and making renewal efforts real), out of which five strategies for action were developed.

---

[5] The Dutch subsidize two levels of rehabilitation: a minor level with a ten- to thirty-year remaining life span and a higher level with a thirty- to fifty-year remaining life span (Hale, 1979).

issues and to secure changes in the goals of the municipality's official planners. By articulating neighborhood differences at the start, citizens put planners on notice that the original objectives of the municipality were unacceptable.

A second end that Dutch coproduction achieves is that the process makes planning real and immediate for local participants. Organizers do this by centering participation around issues affecting people's lives immediately and visibly: their rents, their streets, the timing of improvement to their block. Starting with issues like these, which lie on their own doorsteps, citizens in the future may be able to extend their participation to neighborhood, citywide, and national issues. In the meantime, they take an active interest in the renewal at their doorstep.

Five Mechanisms

Citizens and planners turn differences into resources and make renewal real by employing a few central yet often unarticulated mechanisms:

1. *Contextual diversity* involves an expectation that planning processes and organizations will vary to fit different local conditions. It means that although ends may remain the same, the means and resources employed are adapted in each case to the skills of people involved and to the demands and opportunities the particular situation presents.
2. *Self-organization* means that groups are encouraged to organize themselves to achieve what they want rather than being organized by others. When "outside organizers" are involved, they try to foster self-organization.
3. *Extended doorstep issues* are used by planners and organizers to mobilize participants. These problems are anchored in the daily lives of the citizen groups that tackle them, no matter if the resolution lies in larger—even national—actions.
4. *Coproduction contracts* are agreements of principle held in common by all participants involved in coproducing a final product, even if one is the officially elected city council and the other a self-organized group of citizens.
5. *Give-and-take planning* refers to a process whereby several groups of people develop plans by passing documents back and forth—at each pass commenting, changing, adding alternatives, and enriching the product.

Each of these five mechanisms has an important role in coproduction.[6]

[6] Those familiar with U.S. community organization literature and practice, such as that of Saul Alinsky and his followers, will recognize similarities with the Dutch mechanisms, especially the use of self-organization and doorstep issues.

Contextual Diversity. If you want to create misfits, employ strictly standardized procedures. Standardize glove sizes at 8 1/2 and you create a host of misfits—everyone with either larger or smaller hands. Dutch citizen participation efforts avoid this by exploiting the diversity among people, neighborhoods, and cities to help determine how citizen participation is implemented around the country. Contrary to what some would expect, there is no standardized urban renewal process or organizational structure in the Netherlands.

In the economically sound, large city of Rotterdam, the Labor party has initiated a structured and well-financed citywide citizen participation effort. In 11 neighborhoods, the city funds local action groups to hire their own planning experts. The experts, often design students or architects who live in the area, work to organize revitalization activities as employees of the action groups.

In the smaller and less well-to-do town of Helmond, there are hardly any action groups. Consequently revitalization efforts there have centered around a citywide community development board funded largely by the central government. The board has hired local community development students and an independent citizen participation organization, Workgroup 2000, to provide organizing impetus for participation. Eventually Workgroup 2000 removed itself from the scene. The board hired the former students—now graduated—to work in local storefront offices, and the town hired a project leader to coordinate all its revitalization planning.

In all such activities, the involved students, planners, organizers, experts, and citizens adapt the way they carry out their tasks to the context they are in: the awareness of the citizenry, the skills of the planners, and organizers, and the resources of the city.

Self-Organization (with Encouragement). Neither national nor local government organizes citizens. The Netherlands national government does not have a citizen participation office. Nevertheless, citizen participation is a national public policy that is vigorously pursued. The government at all levels does this by encouraging groups to organize themselves and supporting them when they do. In Rotterdam, in neighborhoods like Crooswijk, Oude Westen, and Feijenoord, the government funded existing groups that had organized themselves around actions such as fighting rent increases or improving local buildings.

The government also supports self-organized citizen participation assistance organizations. Throughout the country, a network of community development centers exists at the provincial, local, and neighborhood levels. Several of these centers are actively engaged in supporting citizens. In addition, the central government subsidizes organizations like Landelijk Ombudsteam Stadsvernieuwing (LOS), founded in 1973 by town planners, architects, citizen activists, sociologists, politicians, and others to provide technical and, eventually, citizen participation advice to towns with revitalization projects. The national government wholly supports LOS, whose six staff members in turn provide advice at no cost to tenant and other action groups. The government controls how money is spent but not LOS's philosophy of action. By supporting such self-organized groups, the government provides decentralized, diverse, and motivated urban revitalization services to those who want them.

Support for self-organization does not preclude attempts by the government to learn from and build on others' self-organizing experience. For example, the Ministry for Housing and Physical Planning together with the Ministry for Culture, Recreation, and Social Work funded a group of researchers to document how a sample of neighborhoods organized themselves. The resulting document, *The Neighborhood Approach* (Haberer, deKleijn, Nicolas, & deWit, 1978) attempts to make visible some underlying similarities in the planning and community development processes. While the booklet begins to standardize certain methods and chooses some value orientations over others, it maintains the goal of neighborhood self-organization.

Self-organization has pitfalls. Neighborhood action groups tend to be more visible and successful the more people they represent. In ethnically and economically homogenous areas, majority groups may represent nearly a consensus. Large Dutch city neighborhoods like Rotterdam's Oude Westen, however, are not homogeneous; they include minority populations of Indonesians, Chinese, Turks, Moroccans, and Surinamese. (See Berg-Elderling [1979] and Schoonenboom [1979] for a review of this problem and its policy implications.) While not transient, these minorities typically are not well organized and do not necessarily see their interests reflected in official government actions. For groups like these, the principle of self-organization in citizen participation may be counterproductive. It may exclude them.

Extended Doorstep Issues. When Workgroup 2000 began to work with citizens, they asked people to develop their own housing policies. How many privately financed units ought there to be and how many public? How much urban housing and how much rural housing? The citizens objected. They wanted to be asked questions they could use their own expertise in answering. They could deal with policy questions more comfortably when these were rooted in issues with which they were personally familiar: their homes, their rents, their streets. As a result of experiences like these, groups like Workgroup 2000 and LOS as well as experts hired by neighborhood action groups tend to start with and expand on immediate citizen problems: doorstep issues.

Using a *Woonerf* or "residential yard" is just such a doorstep mechanism (Heeger, 1979). Developed by planners at Delft University, the *woonerf* idea is first to raise a local street with bricks so it is at the same level as the sidewalk. This flat surface, now one wide sidewalk or street, is then regulated with signs to indicate that children, pedestrians, bicyclists, and cars all have equal right of way. The car is not dominant and must obey special rules. In *woonerven* like these, cars are supposed to be parked only in assigned spaces; also, benches and tree planters may be installed. See Figure 4.

Figure 4. A *woonerf,* or "residential yard," has a special sign to announce that cars, pedestrians, children playing, and bicycles all have equal rights.

Workgroup 2000 organizers in Helmond used the planning of the *woonerf* as a doorstep topic of discussion around which to organize groups of local citizens. They discussed what ought to be done, made a videotape of the discussion, and discussed the videotape as a means of promoting group discussions.

Coproduction Contracts. A diverse set of actors is involved in most Dutch urban revitalization efforts. Among these are citizen groups, their paid experts, developers, city planners working for the local municipality, and building corporations—development organizations formed decades ago by labor unions and other groups. These groups begin working together by agreeing on what their mutual responsibilities are. Agreements may be formal; for example, a municipality that is going to finance personnel to work in a neighborhood storefront would develop a contract with the neighborhood group. When both renters and house owners in a neighborhood are going to be involved in a revitalization project, they work out formal procedures for working together. Agreements like these represent a point of view toward urban revitalization that includes shared responsibility for coproducing a product. Where citizen participation is part of the planning and implementation process for urban revitalization, diverse groups can jointly produce the outcome. Contracts made between groups can formalize the coproduction planning approach. The city council maintains its powers during coproduction, as do developers, citizens, and experts. Coproduction contracts are not so much exchanges of power as they are mechanisms to make more visible the responsibilities each group expects the other to meet during coproduction. Coproduction contracts are explicit declarations of mutually agreed upon principles which form the basis of coproduction. (See chapter appendix for examples of contracts.)

The participation consulting group, Workgroup 2000, uses negotiation of coproduction contracts as a tool to help decide where its involvement will be most useful. When a municipality or citizens group asks Workgroup 2000 to consult on a project, its members begin by seeing whether all parties in the process will agree in writing to such principles as the following:

1. Planners of the municipality will perform services for the community as well as the government.
2. The city council will avoid limiting the planning process by not stipulating in a policy document what they want to do; an open discussion is a condition.
3. Alternative plans will be developed; not just one plan.
4. Information—including third-party relationships—will be open to all involved parties.

Principles such as these serve as the basis for coproduction contracts between a municipality, citizen action groups, and process consultants. Because of differences over such principles, it is not surprising that only one of ten initial Workgroup 2000 proposals ends up in a signed coproduction contract.

Coproduction contracts may also be less formal, more like accords. In the Schilderswijk neighborhood of The Hague, a coproduction steering committee was formally constituted. It included an alderman, municipal bureaucrats, building corporation representatives, and citizen action group members. After the participation process was well under way, the group identified an issue potentially debilitating to further group discussion: proposed physical improvements were likely to push rents up so high that people living in Schilderswijk (and coproducing the plan) might well be excluded from living there. Civil servant participants on the steering committee could not assure citizen-participants of the city council's intentions on this score. Planning stopped. To get the process started again, the city council sent a letter to all residents of the Schilderswijk stating that the council did in fact want to keep rents down and help the present residents to stay on. No promises were made, but a statement of principles was presented. Residents accepted this letter as an informal coproduction accord. The council eventually helped the group prevent gentrification of the Schilderswijk neighborhood—stopping well-to-do people from buying up houses and kicking out low-income renters.

Give-and-Take Planning (with Deadlines). In the small town of Helmond, as in the Hague and Rotterdam, citizen and government coproduction of plans is carried out in stages. Citizen groups articulate problems and wishes, planners draw up alternative plans, citizens discuss plans, civil servants comment on the citizens' discussion, plans are reformulated, and so on. During each stage, one or another group "has the ball in its court." At the end of each stage, a written document is produced and shared with the other participants. The product of this give-and-take sequence is a set of documents that reflects the diversity of opinion in the groups, as well as plans or actions meant to respond satisfactorily to the totality of opinion. This is not consensus building, not negotiation, and not multiple input to a single decision making source. The give-and-take process is geared to developing alternative plans and actions over time by having participating groups formally and sequentially make explicit both their own preferences and their responses to other people's ideas.

Organizers who use this approach report increased impacts on both people and plans. The Helmond participation workers report that the more plans are recycled, the more citizens know about what is actually going on. They are not merely reacting to the confrontation rhetoric of large group meetings. In turn, planners tend to be more willing to modify their own ideas—since in this process ideas are initially presented as tentative proposals, not completed plans. As a result of give-and-take planning in Helmond, a factory that planners originally felt ought to remain was relocated, a townwide shopping street was extended into a new town center previously expected to serve only local interests, and additional studies of residents' shopping needs have been carried out.

Give-and-take participatory planning, together with the four other mechanisms discussed above, makes citizen participation in the Netherlands real to citizens and can turn differences between groups into a productive resource. So far, the coproduction process has worked only in some cities, and there only with predominantly Dutch residents. As will be seen in the following cases, one of the major problems confronting advocates of participation in the Netherlands is the lack of involvement by non-Dutch residents in older neighborhoods.

Participation Is Not a Free Good in Rotterdam

Rotterdam, with a population of 700,000, is the Netherlands' second largest city. At the start of World War II, the German Luftwaffe bombed Rotterdam unmercifully, destroying over half the city's buildings and most of its housing. Today, 40 years later, Rotterdam's seaport is the largest in the world. Income from seaport operations makes Rotterdam the Netherlands' richest city. The seaport also brings sailors and other workers to Rotterdam from many parts of the world, making it a culturally diverse city.

Rotterdam's history, together with its wealth and cultural diversity, have presented the city's residents with unique urban problems and opportunities. Citizen participation is central to Rotterdam's urban renewal program.

Why Is Rotterdam Interesting?

Rotterdam's citizen participation program is a successful example of government and local residents coproducing neighborhood renewal.

In Rotterdam's Crooswijk neighborhood, there are at least half a dozen new architect-designed housing complexes, all planned by boards that included resident representatives. In Feijenoord–Noordereiland, lower income residents have helped to develop a housing project on a riverside dock site, providing them with water views and access previously reserved for more wealthy Rotterdam residents. In the Oude Westen neighborhood, community project leaders and citizens were involved in establishing an innovative type of "congregate" housing for older residents (to be described below).

The success of participation in Rotterdam is also manifest in the way people involved relate to each other and to urban revitalization projects. Government officials, project leaders, "external" experts paid by the municipality, and some residents expect to confer with each other when decisions are made. Participation is not necessarily a part of every decision and does not necessarily lead to consensus, but it is an accepted way of doing business in Rotterdam.

Citizens and government representatives take for granted that regular community meetings take place to discuss ongoing development projects and to resolve social problems such as drug-related crime as they come up. But the fact that meetings are held is not in and of itself a measure of success. Rather, as Wim van Es, the coordinator for project group leaders in Rotterdam stated, "citizens began to trust in the project groups and this enabled project leaders to

Figure 5. Rotterdam. Central districts and part of the port facility.

be on speaking terms with people of the district." Mutual trust and open communication with those who want to be involved is perhaps the most important success of Rotterdam's citizen participation efforts. This success is not unqualified, however. There are large groups of minority citizens—from Turkey, Surinam, South Mulucca, for example—who are not included in the open dialogue between the city and its residents. This communication gap still needs to be bridged.

The municipal government supports Rotterdam's citizen participation effort with funds for personnel to work with citizen organizations. The government identifies neighborhood groups that represent some large part of a community. It supports the "neighborhood group" *(wijkorgaan)* with funds for a project leader whom this group chooses. The project leader helps to facilitate communication, citizen actions, and project follow-through. The municipality also funds the hiring of so-called "external experts" chosen by community members to help on specific projects. In addition, in each revitalization district the municipality supports the rental of a storefront office for the neighborhood group.

These neighborhood participation structures that the municipality established have become part of the ongoing revitalization process. While the people who are supported could be fired and the neighborhood storefronts which are set up could be dismantled, this would not be easy, given the participation processes already established. These positive actions represent a commitment by the municipality to long-term change and demonstrate the success of Rotterdam's citizen participation efforts.

There are two further significant qualities of Rotterdam's program which make it worthy of examination. First, Rotterdam's is a complete program. Citizen participation has begun to affect the city from city council members to local citizens—the whole system. Rotterdam shows how a broad array of pieces can be put together loosely to constitute one complex but influential citizen participation process. Second, Rotterdam's experience shows how large numbers of people can be brought together without creating an unmanageable bureaucracy. It is a decentralized, people-based operation.

The Growth of Citizen Participation in Rotterdam

As in the United States, in the 1960s Rotterdam was solving its urban renewal problems with bulldozers. Complete clearance (*kaalslag* or "cutting it bald") was the worldwide approach of the time—and Rotterdam was with the times. Along with *kaalslag* came the threat that high-income housing would replace middle- and low-income housing and that citizens would organize protest groups in response to this and other redevelopment threats. Top-down redevelopment threatened to create highways and highrises; it did not foster cooperation between government planners and local citizens.

Neighborhood action groups that formed to fight demolition drew some of their expertise from resident students and young professionals taking advantage of low rents. One such neighborhood in Rotterdam was Crooswijk, a neighborhood in which a good deal of demolition took place despite organized citizen opposition. In Crooswijk, a citywide housing corporation which started to plan a housing project for older people asked the local neighborhood council and a group from Delft University, headed by Professor Hugo Priemus, to help organize and develop the project. Such informal efforts eventually led the Rotterdam City Council to enact an ordinance for an Organization of Urban Revitalization (Verordening Organisatie Stadsvernieuwing), making this type of coproduction of urban services official.

The Urban Revitalization Ordinance was not enacted without other major political changes first setting the stage. The first stage-setting change was that the Labor Party gained a majority of Council seats in the municipal election of

1972. This majority was able to make longer-term political commitments than a more fragmented council might have, like Amsterdam's council, held together by a coalition among four political parties. The Labor Party was also committed to citizen involvement in planning, to urban revitalization for local residents, and to housing as a significant social issue.

A second important change was the appointment of Jan van der Ploeg to the position of Alderman for Urban Revitalization. He was committed to a bold strategy for urban revitalization: combining massive urban capital investment with extensive citizen participation. The municipality's capital investment was massive in that it borrowed enough money to buy almost all the existing residential rental property in each of 11 designated urban revitalization districts. As a result, the city of Rotterdam now owns 60% of all residential property in these districts. Owners sold voluntarily because the city was paying 25% over the going market price; perhaps some also felt threatened because they were not maintaining their buildings at a high enough standard. Clearly the unique financial and political power of the municipality makes this particular part of their revitalization strategy difficult to transfer to other contexts.

The city's commitment to citizen participation was as extensive as its financial investment was massive. In each of the 11 urban revitalization neighborhoods, citizen groups formed neighborhood councils and presented themselves to the city council—demonstrating by actions they had taken, like community meetings, that they in fact represented local residents. Once the neighborhood councils were chosen, the city gave them money for space, administrative and community organizing personnel, and professional "external" experts to help in

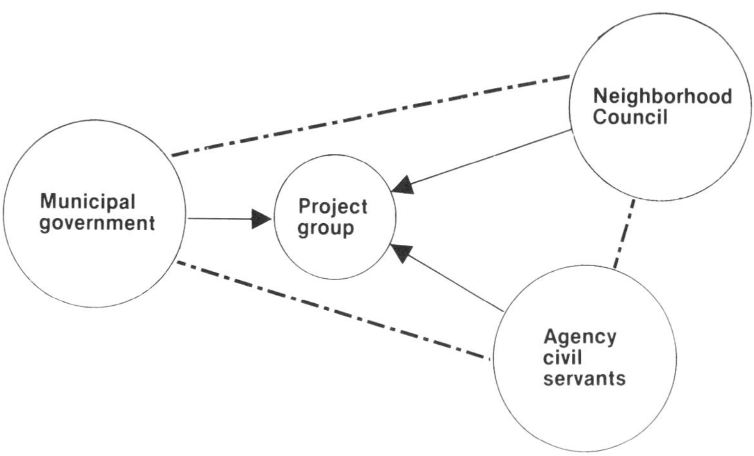

Figure 6. Neighborhood councils joined with municipal officials and civil servants to form urban revitalization project groups in Rotterdam.

planning specific projects. Representatives of these funded neighborhood councils joined with representatives of municipal government and civil servants from affected agencies to form project groups to help guide the implementation of actual development projects: renovations, street rehabilitation, and new construction, as shown in Figure 6.

Project groups directly involved still more residents and other parties, like architects or building corporations, in specific projects. Alderman van der Ploeg and other city council members had decided that participation was worthwhile, although it was by no means a free good.

Neighborhoods

Citizens, project leaders, and external experts working in Rotterdam's diverse neighborhoods are themselves a diverse group. One result is that the lessons of each of the eleven revitalization neighborhoods are different. We have seen that Crooswijk—already partially demolished—offered a training ground for involving external experts and an approved precedent for organizing residents into neighborhood councils. From other neighborhoods, like Feijenoord-Noordereiland, Vreewijk, and Oude-Westen, we learned different lessons.

Feijenoord-Noordereiland. This district, situated on a large island near the center of the city, is a culturally diverse low-income area with some industry along the water's edge. The products of planning carried out here with a great deal of citizen participation shed some light on the criticism by opponents of participation that involving nonprofessionals in planning lowers the quality of its outputs.

In the district there was an abandoned warehouse at the water's edge owned by the Simons factory. The neighborhood project group decided that a family housing complex built on this site would serve several purposes. It could be a rallying point for citizen involvement and eventually a symbol of what might be accomplished; in short, it could make revitalization real for residents. It would serve the needs of a large group of residents now living in low-standard housing, in turn freeing these units for subsequent renovation.

After setting up a storefront office with the assistance of a project leader and after hiring external experts, the project group served as the unofficial spokespersons for the clients of Simonsterrein (Simons area), as the complex become known. The site itself was choice waterfront property providing views for most tenants as well as walkways and places to sit or fish for anyone in the neighborhood. The site plan provided open space for all residents and gardens for ground-floor residents. Buildings were sited to maintain some views to the water for residents living in older houses surrounding the new complex. An example can be seen in Figure 7.

According to project leader Wim van Es:

> With people involved in the planning we got higher quality and more money—because of the action of the people themselves. Some people in the central

Figure 7. Ground floor residents in Simonsterrein housing personalize the territory outside their living-room door.

> government think architectural quality has decreased: drama, systems building, elevator buildings, are missing. To them we seem like we are going backwards to low-rise housing, mixed uses with shops in the houses, and functional diversity.

The question of quality seems to boil down to a question of values: if you hold that trained professional judgment is always better than residents' judgment, then participation may well lead to lower quality. But if you feel that dealing with incorporating citizens' ideas is a good in itself, coproduced urban revitalization products may actually be of higher quality than those produced by professional planners working alone.

Simonsterrein was just the starting point for urban revitalization. Since it started, other projects have begun: traffic has been rerouted to increase street safety, block-by-block rehousing and renovation efforts have begun, local factories have been equipped with new acoustical control devices. The amount of physical planning seems to be on the rise. But this does not mean that all problems

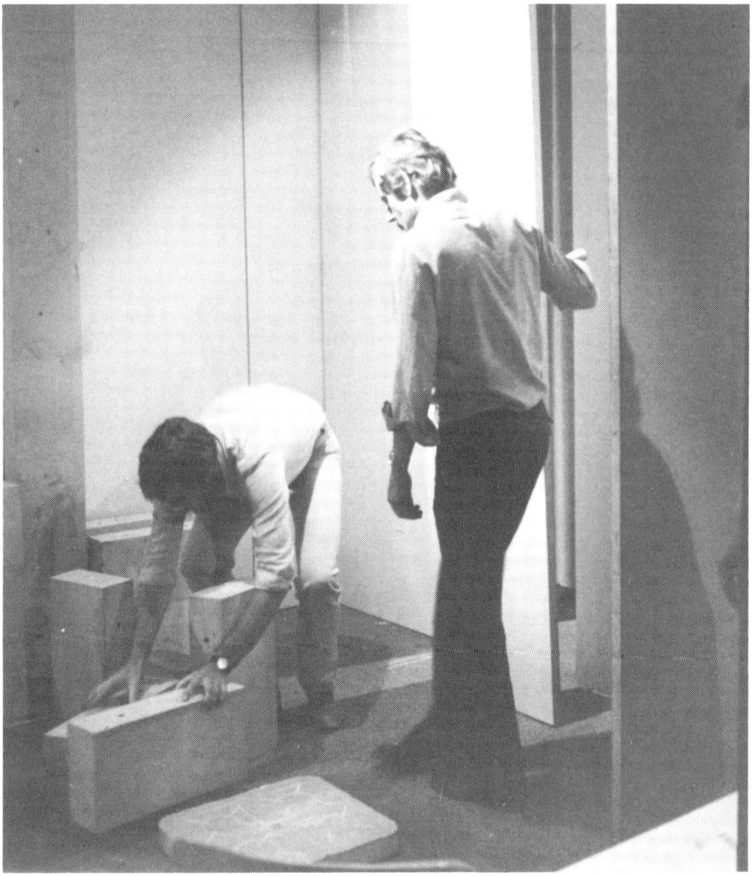

Figure 8. Professionals and community residents move walls in the Delft model to test alternative housing layouts.

are being solved or even planned for. As we shall see more clearly in the Oude-Westen neighborhood, a fundamental dilemma is creeping up: as the majority of citizens gain in power through organization and participation, they have begun to exclude minority members of the community from planning and might well like to see minority groups leave the district altogether.

Vreewijk. Built during 1920–1940, Vreewijk is the only Rotterdam renewal district to have been developed in this century; the others were nineteenth-century products. In Vreewijk major building renovation is taking place and the external expert there is using a unique interactive technique to bring together residents, architects, and university experts to determine residents' desires and needs. This technique makes use of a full-scale architectural-model facility located at Delft University and designed by psychologist Herbert van Hoogdalem. The model has a stable ceiling. Above the ceiling is an electrical grid which can be used to create interior and exterior lighting effects—sunlight streaming into a kitchen window, a light above a stove. Hollow-core wall panels, available in several standardized widths, are designed with pressure plungers at the bottom to hold them in place against the floor and ceiling while also allowing them to be moved easily. To create stairs and other constructions in space, the model is equipped with hollow yet sturdy plywood building blocks. The system is completed with simple pieces of everyday furniture.

The Delft model is not actually complete until it is in use, however. To demonstrate this we will briefly describe how the Vreewijk group used the model. In the Vreewijk neighborhood at community meetings, the external expert discussed with residents what design changes they thought they wanted. Of course, there was not full agreement, because people naturally have different preferences even when they are from the same social background. On the basis of these discussions, the architect, with help from the expert, designed three of the many alternate interior layouts possible within the existing building shell to be renovated. Along with drawings he provided "evaluative tradeoff annotations": lists of both advantages and disadvantages each particular plan offered. When the group arrived at the university to use the model, they first met in a separate room where the expert explained the annotated plans and presented them using an overhead projector. The ensuing discussion enabled residents to point out the advantages and disadvantages they saw, given their own unique experiences. Then the group of experts, architects, residents, and sociologists descended into the model.

There, with a glass of orange juice or a bottle of beer in hand, the residents walked through full-scale models of the plans they had just been discussing. Discussion focused on opinions about the alternatives. No decisions were forced.

After available alternatives were discussed, the professionals began to ask questions about minor changes—making the changes on the spot with the model—as shown in Figure 8. For instance, the architect would ask "What would you think of a half wall between the kitchen and dining area instead of a full opening?" And the half wall was created for comparison. Or he would ask "Would you like to have this partial wall create a corner for a piece of furniture or is this small protrusion for the plumbing acceptable?" And a corner was created.

Finally, the group disbanded to continue its discussion at the next community meeting, after other similar groups had a chance to work with the Delft model. While no agreements were reached, residents seemed to have a good grasp of the tradeoffs they would have to make, individually and collectively.

Vreewijk's use of the Delft model points out a significant paradox. Because individuals are expected to use their imagination and innovativeness in diverse ways, techniques like the Delft model and resources like university specialists are brought into the citizen participation process. For the same reason there appears to be a minimum of sharing between individuals and between participation efforts; the use of the Delft model, for example, is limited to just one Rotterdam neighborhood.

Oude Westen. The large Oude Westen neighborhood is located near the Rotterdam city center. Its dense residential interior is bordered by active traffic and shopping streets. The external expert of Oude Westen is trained as an architect and has lived in the district since 1962. He defines himself as a resident activist. While other neighborhoods make a point of the differences between (political) action groups and the working project group, the external expert in Oude Westen is pleased that they are both the same organization in his district. This reflects what he sees to be the group's continued action orientation. Among the projects the group has developed are building a congregate housing project for older people, coordinating street-by-street house redesign and renovation, and acting as the clearinghouse for rent strikes for rent control against the national government.

The Lion-Corner Congregate House, the neighborhood's first project in Oude Westen, is an innovation in living and design. The project's older residents—some couples, some singles—each have their own small apartments. On the ground floor there are several large living-room lounges and a club bar—a place where people can go in the evening and order a beer. Administratively there are also shared services, including a meals program in which residents can have a hot meal brought to them in their apartments. Communal facilities also include a workshop, greenhouse, and an exercise room. The Oude Westen external expert worked closely with a local philanthropic foundation (Stichting Humanitas) and a firm of architects to see that this congregate house was designed and built so innovatively.

Building renovation in the neighborhood is being carried out just as sensitively. On a single street, the city buys as many houses as it can. With owners who do not want to sell, the city makes a contract: it will renovate the building just as it renovates its own houses, and the owner will maintain a similar low rent. A few owners, of course, do go along with this arrangement, hoping to make more money by selling their building after renovation of the street. The action group coordinates resident activities, primarily moving out and later back in, and redesign. All the residents on a street to be renovated have the option of moving to temporary housing (Wisselwoning) and then back. Of the 32 families on one street, 18 chose to do this. Two families chose to relocate into totally new housing that had recently been built in another part of Oude Westen. Twelve families chose to take relocation allowances of $1,600 (DFL 3,500) and relocated themselves. For those who returned the average rent doubled—from $40 (DFL 95) to $80 (DFL 190) per month for a one-bedroom flat. The renovation plan for the buildings themselves took advantage of the reduced number of residents who were to live there. For every three old units, the renovation plan yielded two larger new ones.

An external expert worked with this group of tenants to develop these plans. Construction is being carried out by a private contractor. In other parts of Oude Westen, residents themselves carry out much of the renovation after the walls have been relocated. Residents may apply most of the finishes—flooring, paint, paneling—and build cabinets and counters.

All together, given the goals of many planners, citizen participation in Oude Westen is a measurable success. The external expert has used a multiple-project multiple-technique approach to involve all residents who want to plan their own urban revitalization programs and control their own urban lives. The percentage of people annually moving into and out of Oude Westen has dropped dramatically from 25% to 5%.

Despite this seeming success, there is still a profound and fundamental problem with the citizen participation process in Oude Westen and in other urban areas: the problem of Dutch residents who have moved to Holland from other countries. These outsiders include Turks, Moroccans, and many seamen from Cape Verde. All the involvement success the external expert could point to excluded these residents. These "non-Dutch" residents represent half the population of 11,000 in Oude Westen. They do not participate actively in urban revitalization. Although the reasons for the separation are clear, they do not justify it. The Dutch residents are interested in better housing and many can afford even the doubling of rents. The non-Dutch residents choose cheaper housing whenever they can. They want low rents and find the unrenovated housing acceptable. One side effect this produces is that the non-Dutch end up concentrated on poorer streets where renovation will be a long time coming. Ghettos are created.

There is little natural opportunity for the Dutch and non-Dutch to interact. The most compelling type of family interaction—at school—does not take place. The Dutch families tend either to be over 65 with no children at home or young with pre-school-age children; the latter are in the neighborhood only to take advantage of low rents and will move to the suburbs for better schools and more space when the children get older. The local schools are then left to non-Dutch children.

Citizen participation has failed to bring these two separate populations together partly because

1. National government policy adopted so far has mistakenly assumed that most of the minorities would return to their home countries after a few years (Schoonenboorn, 1979).
2. Local government and its representatives are not trained or organized to communicate with and respond to the needs of non-Dutch residents. There are few if any non-Dutch external experts, and much outreach is carried out in a formal way—weekly appointments for open discussion—that may be more frightening than inviting.

3. The non-Dutch residents have their own tightly knit social groupings which may not easily accommodate open intercultural dialogue.
4. Many Dutch residents are set in their ways and are not ready or willing to join forces with others whom they see as poorer, darker, less educated, more dangerous, and generally inferior. Planners seem to be dealing with this stand-off so far by focusing on the Dutch population, who participate, respond, and vote. Problems unique to non-Dutch residents—urban crime, drugs, and bad housing—are isolated and sidestepped. But this approach will not do for long. As soon as the problems of one group (such as drugs) can no longer be isolated from the other group, the bridge between the two will have been forged—and those people engaged in citizen participation will have to confront both groups together.[7]

Retooling For Coproduction in the Painters' District of The Hague

During the past 10 years, the urban revitalization process in the Painters' District (Schilderswijk) of the Hague has shifted from strong government control, to conflict, to government–citizen collaboration. In that time the approach to neighborhood renewal was completely "retooled" from centralized government production to decentralized coproduction by municipal authorities, nonprofit housing associations, and a militant residents' organization.[8]

The story of the retooling of renewal in the Schilderswijk district is interesting for a number of reasons. First, the district itself has been the largest residential problem area in the Netherlands, a low-income island of declining housing stock and population with a rough reputation in the midst of the well-to-do and proper national government seat of the Netherlands (Van Dÿck, 1975). Second, partly as a result of the shrewd hanging of black flags in the streets by the district's action group during televised coverage of the visit of the Dutch Minister of Housing, national government housing subsidies for the district were raised to an unheard-of level for the Netherlands. Third, determined opposition by the action group, one of the first in the Netherlands to realize its plan, twice forced the municipal government to stop seeking agreement on specific physical plans and instead, first to seek agreement on a participatory planning process,

[7] This issue is under study at present. A 1979 report on ethnic minorities by the Netherlands Scientific Council for Government Policy recommended a number of steps to increase their participation in decision making (Schoonenboom, 1979).

[8] The account draws primarily on two sources. The first is a case study by H. Van Dijck (1975). The second is an interview with two municipal planners in The Hague, Meester and van Evert (1979).

a radical change from the typical Dutch renewal procedure. Finally, a new level of neighborhood politics has been grafted onto the municipal decision making system, providing power for neighborhood nonprofessionals though not necessarily an ideal form of direct democracy.

Developed in the late nineteenth century as a poorly planned residential area, by the mid-1970s the Schilderswijk housed some 40,000 residents at the very high density of 40 to 42 housing units per acre. Its problems were recognized as early as 1953 by the Hague government, which proposed a plan for complete clearance and redevelopment of the district with relocation of existing residents. Although never carried out, this plan was the basis for the purchase of several hundred acres of potential renewal area by the municipality. Fifteen years later, following the "Provo" protests of the mid-1960s, the government proposed a new plan for the district, but this too was abandoned. Two plans later, in 1971, the municipality agreed to put forward a proposal by a private development company for one neighborhood, the Oranjeplein, within the district and initiated public hearings leading towards its adoption as an allocation plan for redevelopment.

Meanwhile, a Labor party member, Mr. Nuy, was appointed alderman of urban development following the 1970 municipal elections. He promised to help the Schilderswijk and obtained city council approval for his budget, which increased financial support for urban renewal. The municipality, anticipating approval of the new Orangeplein plan, began trying to persuade people to move out of the district by means of an increased relocation subsidy. However, in reaction to a July 1971 newspaper headline stating "Schilderswijk trusts Alderman Nuy," which was seen as a political maneuver to push the plan, a protest meeting was held in the district. The citizens felt that the proposed plan did not provide low enough rents in the reconstruction area and formed an action group called "Reasonable Rents" to speak out for their demands. This action group, initially made up of 30 members representing all the existing groups in the district, developed a counterplan for the Oranjeplein with the help of the National Working Group on Housing Problems, some inside advocates within the government, the builder's office and the architectural firm, and some engineering and town planning students from the Delft University of Technology.

Despite criticism of their counterplan by the municipal experts at public hearings in February 1972, the action group continued to press for lower rents. Alderman Nuy postponed approval of the proposed official allocation plan until talks could be held with district residents. Residents opposed the official plan's high rents, out-of-character architectural style, and potential threats to the existing social structure of the neighborhood. The local government argued that the plan was complete and that further delays would only make the costs higher. To overcome the objections and work out mutually acceptable renewable goals, the local government convened a series of four working-group meetings in 1972

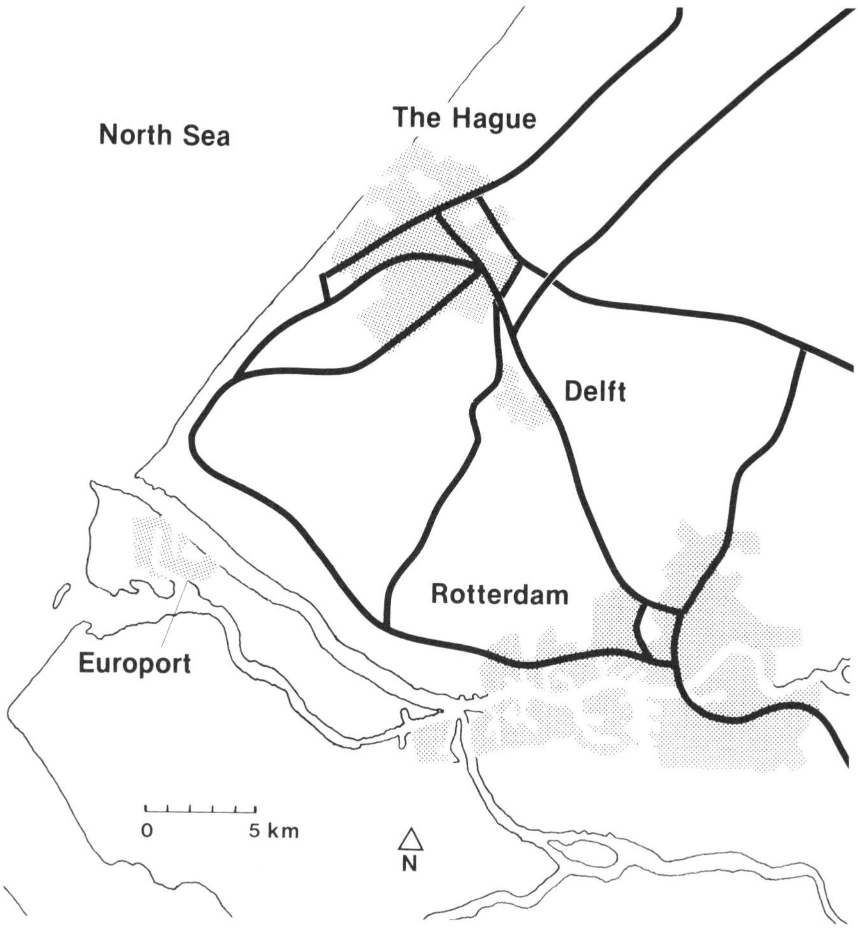

Figure 9. The central coast area of the Netherlands.

between representatives of the residents and the municipal planning agencies. Some 10 members of Reasonable Rents represented the people; Mr. Nuy and officials from the local government departments concerned represented the city.

The primary product of these meetings was a Declaration of Intents, a coproduction contract which set forth both social goals for the plan and an organization structure. Planning goals included rents affordable by the present population, harmony with the present environment, a lively living environment open to the sun and containing improved housing as well as playgrounds and closed to through traffic. As finally agreed on, the organizational structure consisted of a steering committee and a building/design team accountable to it.

Represented on the steering committee were the action group, the municipal government, and two local housing associations and their architects, with the action group holding a majority of the seats. The same groups were on the building/design team, with half of the seats assigned to the housing association and their architects. As Van Dijck (1975) points out, this new structure was a major innovation; previously, "the government and the housing associations drew up a plan (produced a service) and the population depended upon it (used the service)." Under the new arrangement, the population joined the government and the housing associations as coproducers of the plan or service. A new allocation plan for the Oranjeplein neighborhood, based on the Declaration of Intents, was prepared and accepted in 1973. The Oranjeplein housing is now built and occupied, as seen in Figure 10.

While the plan was being debated locally, another event occurred which illustrates the key role of the national government in local renewal. In October 1971, Mr. B. J. Udink, the national Minister of Housing and Physical Planning, visited The Hague to present the final Oranjeplein plan. The action group, feeling that the national subsidies were still too low and thus the Oranjeplein rents still too high, organized the people to put out black flags in all the streets. The resulting television publicity brought national attention. This protest, plus pressure from the municipality and the developer, caused the Minister to agree ultimately to lower rents from an average of 240 gulden ($115) to 210 gulden ($100) per month, a revolutionary event at that time and a dramatic illustration of the power of the action group.

In 1974 the participatory planning begun in the Oranjeplein neighborhood was broadened to the entire Schilderswijk district. The Reasonable Rents action group founded the Inhabitants' Organization of Schilderswijk (Bewoners Organisatie Oranjeplein-Schilderswijk, or BOS). With an initial three-year subsidy from the municipality, BOS hired a resident as a full-time staff person, plus some part-time and volunteer help. At the same time, disregarding the lessons from the Oranjeplein Allocation Plan experience, the city published a proposed Structure Plan for the Schilderswijk district without previous participation by the neighborhoods. Although the city insisted the plan was only a draft proposal for discussion, the people saw a printed document that appeared final to them and they rejected it. BOS and the neighborhood social workers, who had previously refused to work with BOS, joined forces to demand that the plan be put aside and the process be started over with full public participation. After a year of trying to get the structure plan discussed on its merits, in 1975 the municipal government agreed to drop the plan and go back and discuss the initiation of districtwide participation.

Working with BOS and representative discussion groups from 10 small areas within the district, city planners and officials hammered out agreement on a district participatory planning process over the next two years. One of the key decisions was to subdivide the district into 10 small neighborhoods, each with its own discussion group. Because of the great differences in housing type, housing quality, social class, and social problems within the district, this was seen as the only feasible way to get meaningful participation by residents, many of whom do not like being associated with the overall image of the Schilderswijk. While the small area discussion groups had no formal authority, they had the practical power of refusing to participate or of flying black flags and communicating their discontent through newspapers and television.

The resulting district participation proposal was approved by the city council in 1976. It provided for institutionalizing the 10 neighborhood discussion groups, which were open to any residents interested in attending. It set up a two-year five-step plan-making process that included providing information, creating planning organizations, defining problems, identifying alternative solutions, and combining alternatives into one or more plans. This new participatory process was called on in 1978, when the national government set a one-year deadline for producing a new structural plan in order for the district to continue receiving national housing subsidies. Due to the deadline, the process had to be scaled down to one year.

By January 1979, the new structural plan had been prepared and approved by the city council. The plan was published in an innovative format, with "annotations" to highlight key features so that citizens could readily grasp them and with "sequential illustrations" that showed the changes an area would undergo as the renewal progressed through the various rebuilding stages. Four illustrations are shown (Figures 12–15).

The planning process had to face some serious problems, including the short one-year deadline, the effect of inflation on rents (taking them to 290 gulden ($135) in the district and as high as 300–400 gulden ($140–$180) in new projects in other areas), and the uncertainty of the residents, who feared they would not be able to afford to live in the district if it were replanned. The first crisis was one of confidence; the residents asked whether the structure plan was for them or not. The aldermen responded in a letter that they wanted the plan to be for current residents and to provide for low rents. This stand against gentrification allowed the planning process to go forward. The second crisis occurred when the planners tried to discuss general goals for the district and the residents demanded that immediate problems, such as leaking roofs, fire hazards, and traffic congestion, be dealt with first. An emergency team located in the district was created to cope with these immediate problems.

The process was not always smooth; at one time three of the neighborhood groups withdrew from the discussions. However, an active negotiation and exchange process, plus the granting of additional subsidies, eventually enabled agreement on the plan. For each of the five planning steps, a three-stage process took place. First the government planners made alternate proposals, after sounding out the affected groups. These were reviewed and further proposals prepared by "issue work groups" made up of residents, representatives of concerned interests (such as shopkeepers), and civil servants. The five issue work groups were (1) housing (including parks), (2) stores, (3) traffic, (4) schools, and (5) clubhouses (and related facilities). Finally, the 10 neighborhood discussion groups registered their preferences and the planners combined these into one plan.

The plan was approved by the council in January 1979. Throughout the process, opinion leaders from various groups negotiated openly and in private for preferred facilities and locations. If there had been a fundamental disagreement that was not amenable to bargaining (which there was not), a "signal" team would have set up a mediation procedure. That the process was able to work

Figure 10. New low-rise apartment buildings constructed in Schilderswijk.

Figure 11. A storefront office for the inhabitants' organization of Schilderswijk (BOS).

was probably due to active neighborhood politics (Meester & van Evert, 1979) in which "agreement was for us more important than a 'good' plan."[9]

Impacts of participation show up in a comparison of the structure plans produced in 1974 and 1978. As identified by the planners, the most important differences are: (1) gentrification was halted; (2) total demolition was changed to selective demolition; (3) plan implementation was fitted to a concrete time schedule of short, middle, and long range actions; (4) city traffic was re-routed around the district, and (5) neighborhood recreation places related to the worst housing were substituted for a linear districtwide, open space system (Meester & van Evert, 1979). As they note, "the old plan was a blueprint for the future; this is an urban renewal process." They also point out that these gains required a much larger allocation of resources. The 1974 plan was prepared by 2 persons; the 1978 plan involved 150 people.

Another impact of the intense participation has been the formation of a districtwide federation of opinion leaders. This group, Samenwerkende Groepen Schilderswijk, has grown out of the informal neighborhood politics of the structural planning process to become a watchdog on city policy.

Despite its achievement, the citizen participation arrangement of The Hague

[9] For a similar study of the effect of participation on the nature of a U.S. neighborhood renewal plan, see Hyman (1969). What happened in both the Dutch and U.S. cases was that the cartoonlike clarity of a large scale, *de novo* plan was replaced by a more complex and "messy" mingling of small-scale old and new elements in a way that made sense to residents. The everyday logic of the inhabitants was substituted for the abstract logic of a compelling plan.

has some problems. The residents' organization, BOS, does not have regular contacts with the citizens and no longer has unanimous citizen support. Rather than a residents' representative, the BOS staff person has become another community opinion leader, though one with extensive political and media contacts and the power of the black-flag community-mobilization incident to back him up. The elected officials are somewhat isolated, having little contact with the renewal planning process while it is under way and only coming into the picture at the final decision stage. Most serious of all, the process badly underrepresents the growing number of minority groups in the inner neighborhoods.

Despite the provision of interpreters at meetings and translations of notices and plans, few foreigners have taken part. So far, the opportunities within participation to learn about and develop empathy for "different" people have not been used, and strong antiforeign feelings remain among the Dutch residents.

Without minimizing these problems, which are not limited to The Hague, the retooling of the Schilderswijk renewal process stands out as both creative and effective. The lessons of participatory planning had to be learned twice, once for the Oranjeplein allocation plan and again for the Schilderswijk structural plan. However, the learning process left behind a much higher level of awareness

End of 1977: Problems identified

This is the neighborhood. Its condition is bad in many places. The people living here want to work on its improvement. With the municipality they discuss the future of the neighborhood. First they identify their problems on a "balloon map" like the one below.

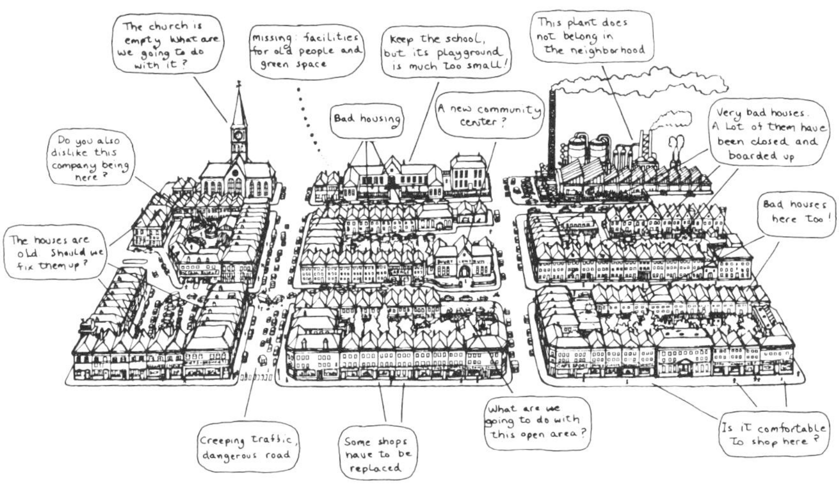

Figure 12. First of seven illustrations from the Annotated Sequential Neighborhood Plan for Schilderswijk (translated from the original Dutch).

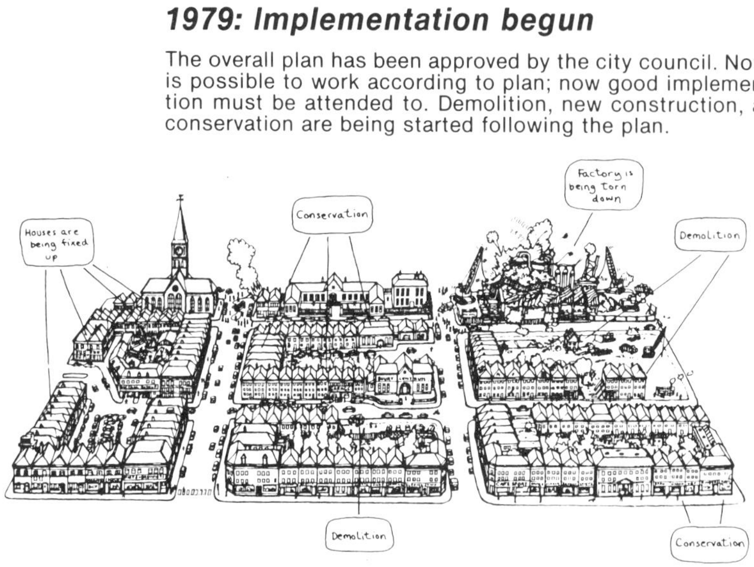

Figure 13. Fourth of seven illustrations from the Annotated Sequential Neighborhood Plan for Schilderswijk (translated from the original Dutch).

about the effects of government intervention in neighborhood rebuilding, a resident constituency represented by a variety of organizations at the district and neighborhood level, and—most important—significant improvements in housing and environmental quality for the existing residents rather than for a replacement gentry.

Overcoming Apathy in Helmond, a Working-Class Factory Town

At the turn of the century, there was a Dutch saying about working-class people: "The pastor of the church kept them stupid and the factory director kept them poor." Changing the attitudes of fatalism and apathy rooted in this history of industrial feudalism has been one of the hardest parts of encouraging citizens to participate in the working-class areas of Helmond, a textile and metals industrial town of 60,000 population in the southern Netherlands.[10]

Making revitalization efforts real in Helmond meant both dealing with doorstep issues and politicizing a previously unorganized population. People had

[10] The information for the Helmond case came primarily from interviews with Raaijmakers, Michielsen, and Tromp (1979) in Helmond and with Eisse Kalk (1979) of Workgroup 2000.

to be shown that they could not only understand but also influence government plans for their neighborhoods. They had to learn how to work in organizations and how to articulate their needs.

Helmond has been designated by the national government as a growth city, one of a dozen or so in the Netherlands projected to grow to about 90,000 inhabitants by 1990. The town is therefore eligible for more money for infrastructure, neighborhood facilities, and organizational needs as well as a speed-up of the plan approval process. However, prior to this, the residents had already begun to strengthen their capacity to manage growth and renewal.

The Helmond case is a good example of the importance of self-organization in Dutch urban revitalization. Workgroup 2000, a national participation organization, provided the outside catalyst that sparked the new citizen awareness in Helmond. Starting in 1973, they worked with local residents to build citizen competence and confidence through the preparation of a master plan. Workgroup 2000 brought in a tool kit of practical participation methods developed during a decade of work in Dutch urban areas. Working from a storefront office, they integrated citizen groups into the traditional master planning process.

A local nonprofit foundation called Information Shop (in Dutch, Informatie

Five to ten years later

The new structure is visibly emerging. An attractive living environment has been attained through the new housing. New facilities are created; at the same time, new areas for demolition are being acquired.

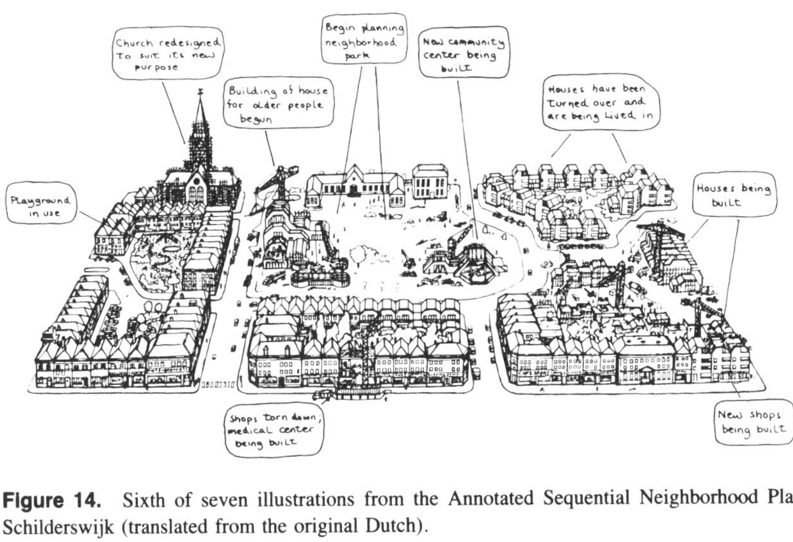

Figure 14. Sixth of seven illustrations from the Annotated Sequential Neighborhood Plan for Schilderswijk (translated from the original Dutch).

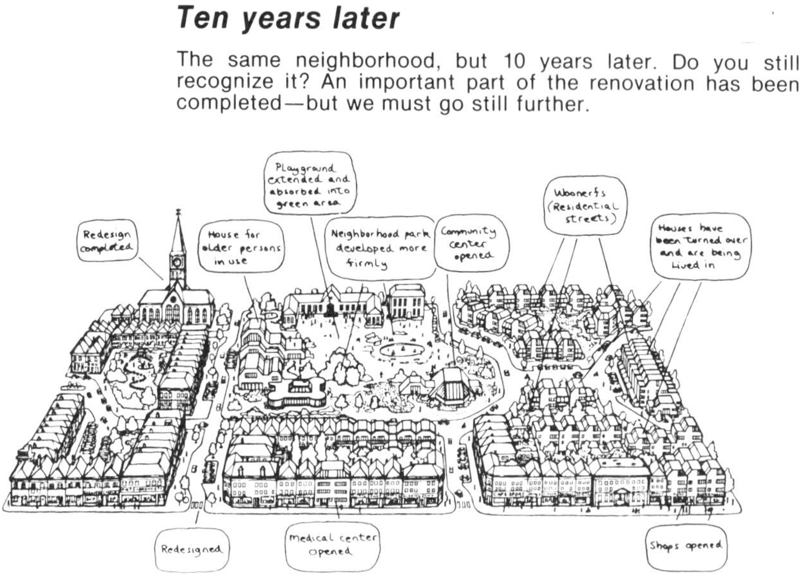

Figure 15. Last of seven illustrations from the Annotated Sequential Neighborhood Plan for Schilderswijk (translated from the original Dutch).

Winkel, or IW) is carrying on the process initiated by Workgroup 2000.[11] Funded half by the national Ministry for Culture, Recreation, and Social Work and half by the local government, IW operates neighborhood information offices in each of the two renewal project areas in Helmond; residents can come to these offices to learn about the details of plans and issues. Staff members are young local residents with degrees in social work and community development who gained experience with Workgroup 2000 methods. They issue monthly information bulletins, hold briefings prior to meetings on street plans, see that newspaper and radio announcements of hearings are made, and provide technical assistance to citizen video programs.

To compensate for its lack of experience in renewal, the Helmond municipal government contracted with DHV, a consulting firm, to provide an external project leader. Serving under a four-year contract with the town, the project leader is a key member of the various renewal organizations.

Government, citizens, and community organizations are brought together in three levels of organization at present, with the possibility of a fourth level in the future (see Figure 17):

[11] The Dutch also have established "science shops" (Wetenschapswinkels) at five universities and "legal shops" in several law schools (Nelkin and Rip, 1979).

Figure 16. A clearly "Dutch" balcony in the new Schilderswijk housing complex.

1. *A steering group,* consisting of the alderman for physical planning as president, the Alderman for social development, the director of the development office, the project leader, the public works director, the social development director, the town clerk, and a secretary.
2. Two *project groups* (one for each of the renewal areas—Binnenstad-Oost and Helmond-West), each consisting of the project leader as president, civil services representatives, citizen and interest group representatives, housing corporation representatives, and an IW staff member.
3. Three *work groups* within each project group, including one for structural plans and allocation plans, one for design of public areas in residential precincts, and one for public and social welfare services. Members are representatives of the concerned civil services, interest and citizen groups, housing corporations, and IW staff members.

4. Possible future *sub-work groups* to handle tasks such as video production.

One Workgroup 2000 participation method that has been used effectively in Helmond is citizen-conducted neighborhood interviews. These citizen self-surveys, recorded on videotape and compiled into television programs for community viewing, are a means of assessing needs prior to planning. A sequence of three research steps is involved. The first identifies problems, the second solutions, and the third priorities, all from the point of view of citizens. After the video programs are viewed, a printed "photo album" can be made as another record of the process.

One popular topic for video programs in Helmond has been the conversion of individual neighborhood streets into *woonerven,* streets where pedestrians, bicyclists, and cars have equal rights. Neighborhood residents can decide which streets they want to be converted to *woonerven*. What was discovered in Helmond was that the video discussions of the *woonerf* became a vital starting point for involving citizens in the planning process. People easily grasped the *woonerf* image from pictures of applications in other cities and were able to debate the merits of having a *woonerf* on their doorstep. The range of their concerns about renewal then moved from their individual house to their street, the first step toward a community consciousness. IW staff saw the *woonerf* debate as more important for community organization than for physical change.

What made the *woonerf* debate lively was that people could foresee some potential negative as well as positive impacts. First, a *woonerf* is expensive to build and maintain. Second, because it limits parking, residents may not be able to keep their cars (which they see almost as members of the family) directly in front of their houses, where they can watch and protect them from vandalism or theft. Third, in some cities the *woonerf* has become a stigma, identified with those older neighborhoods that have social problems. Finally, there is the problem of enforcing the parking rules without calling in the police, a potential social strain for the residents. A *woonerf* is shown in Figure 18.

Although its achievements are not as dramatic or highly publicized as those of other groups, the Helmond renewal process has succeeded on a number of fronts. The working-class residents have shed their reluctance to take part in politics. They overturned a national government decision to demolish and rebuild, instead of renovate, a complex of 175 houses in Binnenstad-Oost by following the Minister of Housing and Physical Planning around during an election campaign and demanding renovation in exchange for their votes. They convinced the housing corporation involved in rehabilitating 147 houses in Helmond-West to settle for minor repairs and no higher rents rather than much more expensive renovation and inevitably high rents. In the Binnenstad-Oost plan, they have gotten the Helmond municipal government to seek national funding for relocating a steel construction industry from a neighborhood, to substitute a *woonerf* for an improved access street, and to extend a neighborhood shopping area. Meanwhile, the citizens themselves have started neighborhood fairs and social activities—a complete change from five years earlier, when there was no neighborhood awareness.

Renewal plans in Helmond now go through an intensive review and discussion process. The IW staff brings together street-by-street meetings of residents and records their ideas and reactions to plan alternatives, annotated to highlight their impacts. Municipal officials comment on the record of citizen

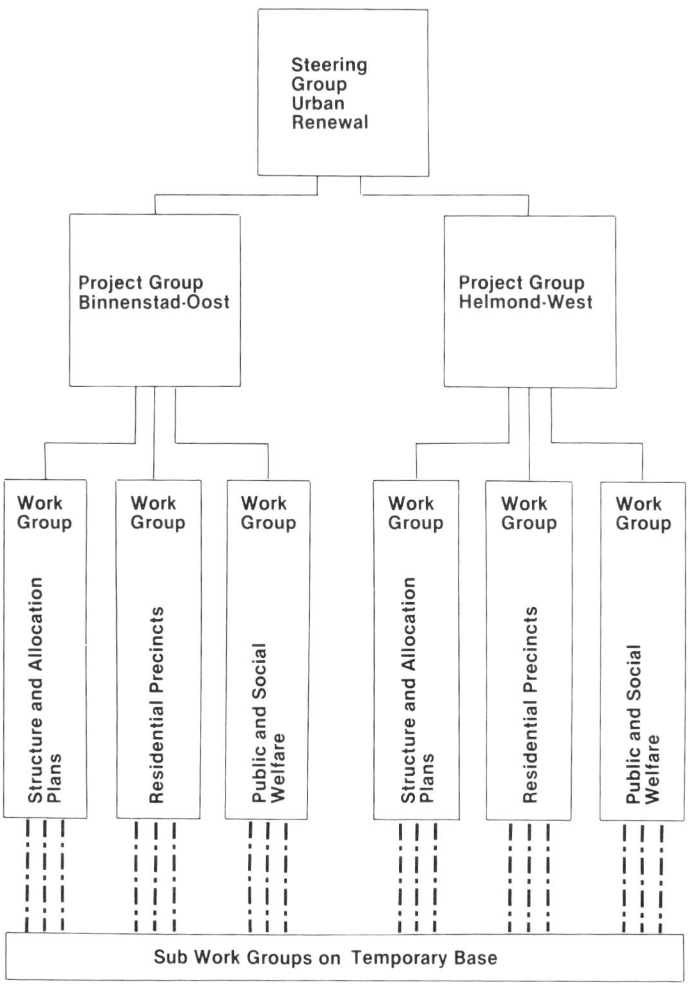

Figure 17. Organizational chart of urban renewal in Helmond.

meetings. Elected officials are furnished with a complete written record of comments on proposals. In a remarkable spirit of give-and-take planning, citizens learn what is going on and government learns citizens' positions on major issues before action is taken.

Not everything has worked out in Helmond. There remain some problems in pulling together the myriad street groups into organized community groups. The burgeoning demand for full information about project costs and benefits at times exceeds the capacity of a small-town government to supply it. And, as in other parts of Holland, minority groups such as gypsies and foreigners have not been well integrated into the participatory planning process. Yet it is clear that a long tradition of industrial feudalism has been turned around with the mobilization of a previously apathetic working-class population into informed and effective community organizations that actively guide the renewal of their neighborhoods.

Lessons For America

What can American public officials, planners, and citizen activists learn from the Dutch efforts to coproduce urban renewal? Despite very significant differences in culture, politics, geography, and government, there appear to be several transferable lessons.

Lesson No. 1

It is possible in some cases to move beyond simplistic ideas of either government-managed citizen participation or action initiated at the grass-roots level to a more advanced partnership between government and action groups, in which the mutual goal is coproduction of neighborhood renewal. For this to happen, a majority of elected officials must be sympathetic to the needs of low- and moderate-income residents and willing to grant them power in project planning. In the Netherlands, this meant that a Labor party majority was necessary. Central and local government must also be willing to invest a lot of resources in renewal products and processes. In the Netherlands, this has meant major expenditures on housing rehabilitation and rent subsidies as well as on support of community organizers and planners working with resident action groups. Finally, government planners must be receptive to resident ideas about desirable neighborhood environments and to bottom-up planning. In the Netherlands, this has meant accepting the demands of low- and moderate-income people that they remain in their present neighborhoods and that government subsidies be used to rehabilitate their houses rather than to underwrite gentrification. Clearly, coproduction is not a free good; it requires a long-term commitment, substantial financial and personnel resources, and a willingness to accept neighborhood priorities.

Figure 18. A *woonerf*, apparently being used and maintained as planned.

Lesson No. 2

If the preconditions for coproduction are met, then the mechanisms developed in the Netherlands to recognize and resolve differences may be transferable, allowing for situational differences. Recognizing and defining the different objectives of the groups affected by renewal plans at the start of a project can provide an early understanding of the potential for conflict resolution. Allowing these groups to organize themselves, with the encouragement and support of government, is a useful way of recognizing the diversity inherent in any community, as is the willingness to let organizations take various forms rather than holding to a single standardized form and structure. The use of written agreements is a useful technique for keeping diverse group objectives visible throughout the planning process while acknowledging mutual goals. Setting up planning processes that systematically pass responsibility for idea generation and review between planners and citizen groups is a useful mechanism for breaking conflicts and problems into manageable pieces and working toward realistic solutions.

Lesson No. 3

The mechanisms for making renewal efforts real to citizens should be most easily transferable. Citizen apathy is often a function of distance from the concerns and impacts of planning. By focusing on doorstep issues—such as rehabilitation and improvement of housing, steets, and public facilities in the immediate neighborhood—citizen interest can be kindled. Such interest can be the basis for activating community organizations and investing citizen energies in planning.

Lesson No. 4

Participation alone, even though adequately supported and aggressively pursued, cannot break through cultural and ethnic barriers to social integration. Participation in the Netherlands has worked most effectively in homogeneous neighborhoods. Minority groups of foreign extraction have not taken part in renewal planning or even attempted to participate as consumers of the rehabilitated housing produced by renewal projects. For solutions to this problem, the Dutch might look at the experience of U.S. renewal, which has had to deal with serious issues of minority participation and integration from the start.

In addition to these major lessons, several further observations about the Dutch experience should be made. First, urban renewal requires a long-term view. Too often, public programs in America are expected to succeed immediately. The history of U.S. urban development legislation is filled with worthwhile programs that were cut off prematurely, before they had really been given a chance to become productive. The Dutch have more patience, especially in the face of short-term setbacks.

Second, coproduction of renewal requires that criteria for measuring success take into account the value of creating and preserving physical and social environments. Objectors to citizen participation often claim that the quality of the physical product is sacrificed for the sake of the social process. Usually these are not the people who will live in the area being planned. They are indifferent to the displacement of long-term residents and favor more expensive architecture. The Dutch planners in Rotterdam, Helmond, and the Hague elected to use criteria generated by the people affected by renewal plans rather than those offered by outside critics. As Haberer (1980) has emphasized, urban renewal in the Netherlands started to be successful the moment it was seen as both a social and a technical process.

Third, the Dutch cities' effectiveness in obtaining larger shares of the national housing and renewal resources suggests possible parallel actions in the United States. Citywide coalitions of local governments and neighborhood groups might unite to lobby state and federal officials for increased subsidies for local renewal. In such cases, it could be profitable for neighborhood advocate groups to overcome their suspicions of government and their fears of cooptation in order to gain the material benefits of coproduction.

Fourth, a major difference between the Dutch and the American contexts is the attitude toward housing and property rights. The Dutch have accepted a severe limitation on the rights of the individual property owner. They have a broader view of the rights of the individual to decent and affordable housing.

Finally, even the casual observer of renewal in those Dutch cities where citizens have been actively involved cannot help being impressed with the valuable service they have provided. They have redirected renewal toward more socially equitable and humane outcomes. They have generated a sense of neighborhood pride and purpose where these qualities had long been absent. And residents have contributed their knowledge and energies toward more sensitive rehabilitation of their houses, streets, and neighborhoods.

APPENDIX

Examples of Agreements for Participation in Dutch Urban Renewal[12]

All agreements are translated from Patries Haberer *et al.*, *De Buurtaanpak* (The Neighborhood Approach). The Hague: State Publisher, 1978.

1. ORDINANCE CONCERNING NEIGHBORHOOD ORGANIZATIONS FOR URBAN RENEWAL

Article I: Scope of the Ordinance

1. This ordinance leaves intact the right to existence of neighborhood organizations not within its framework.
2. It merely regulates the status of neighborhood organizations in neighborhoods declared urban renewal action areas by the city council.

---

[12] The text of the model ordinance is inspired by the neighborhood organization ordinances of the city of Enschede.

Article 2: Types of Neighborhood Organizations

1. The following may be considered neighborhood organizations within the framework of this ordinance:
   a. Neighborhood committee, action-oriented
   b. Neighborhood association, action- and agreement-oriented
   c. Neighborhood council, action-, agreement-, and authority-oriented

2. Neighborhood organizations, i.e., collections of active neighborhood residents and possibly (including) neighborhood shopkeepers and social workers, have the right during the preparation phase to request that the mayor and alderman recognize them as a neighborhood committee or neighborhood association, or that they be recognized by the city council as a neighborhood council.

3. Neighborhood organizations may, during the project, also request to be recognized as a different type (of organization). In any instance, the neighborhood organizations will consult with the neighborhood residents.

Article 3: Neighborhood Committee

1. The mayor and aldermen will recognize the neighborhood organization as a neighborhood committee from the moment the neighborhood is declared an action area and for the length of the project. The recognition will be publicized in the neighborhood.

2. A neighborhood committee will be recognized when the following conditions have been met:
   a. The neighborhood organization requests recognition as a neighborhood committee
   b. The neighborhood organization acts in agreement with the neighborhood residents in requesting recognition as a neighborhood committee
   c. The neighborhood organization actively assures open participation, by at least:
      i. Publicizing meetings except in special cases
      ii. An annual convocation of the neighborhood residents

3. Refusal and cancellation of recognition may be appealed to the council.

4. A recognized neighborhood committee has the right to:
   a. Professional support
   b. Facilities
   c. Information
   d. Regular access to the aldermen of the neighborhood

Article 4: Neighborhood Association

1. Recognition of a neighborhood association by the mayor and alderman occurs the moment the neighborhood is declared an action area and for the length of the project. The recognition will be publicized.

2. A neighborhood association is recognized when the following conditions have been met:
   a. The neighborhood organization has requested recognition as a neighborhood association
   b. The neighborhood organization acts in agreement with the neighborhood population in requesting recognition as a neighborhood association
   c. The neighborhood organization actively assures open participation by at least:
      i. Publicizing meetings except in exceptional cases
      ii. Offering an annual accounting to the neighborhood residents
   d. The board of the neighborhood association will delegate as much as possible to related associations and commissions

3. The refusal and revocation of recognition may be appealed to the council.

4. A recognized neighborhood association has the right to:
   a. Professional support
   b. Facilities
   c. Information
   d. Regular access to the neighborhood aldermen
   e. Enter into contracts with the workers of the neighborhood office in order to prepare:
      i. Research proposals
      ii. Housing plans
      iii. Street plans
      iv. Welfare plans
      v. Annual neighborhood participation report
   f. Enter into agreements, namely:
      i. Mandate agreement (requiring landlords to improve rental housing)
      ii. Cooperation agreement (between home building associations and renters)
   g. Exercise the power to suspend
   h. Evaluate the annual report on neighborhood participation

5. Internal decision making will be regulated in the rules of the neighborhood association. The rules will be presented to the mayor and aldermen for approval.

Article 5: Neighborhood Council

1. The installation of a neighborhood council shall take place in accordance with Article 61, Part 1, Section (a) of the Municipal Law if requested by separate council decision. The installation will be widely publicized in the neighborhood.

2. A neighborhood council is installed the moment the neighborhood is declared an action area if the following conditions are met:
   a. The neighborhood organization requests recognition as a neighborhood council.
   b. The neighborhood council acts in agreement with the neighborhood residents in requesting recognition as a neighborhood council.
   c. The neighborhood organization actively assures open participation, by at least:
      i. Publicizing meetings except in exceptional cases
      ii. Offering an annual accounting to the neighborhood residents.
   d. The members of the neighborhood council with the exception of one member of the city council are chosen by interest groups in the neighborhood, namely:
      i. Tenants of Public housing (per block)
      ii. Tenants of private housing (per owner or per street)
      iii. Owners–residents (per street)
      iv. Shopkeepers, tradesmen, market vendors
      v. Social workers, teachers, health workers
   e. The neighborhood council will regularly consult with the neighborhood residents.

3. Refusal or cancellation of recognition may be appealed to the Provincial Council.

4. The city council may meanwhile dissolve the neighborhood council if it appears that during elections less than 10 percent of the eligible population votes.

5. A recognized neighborhood council has the right to:
   a. Professional support
   b. Facilities
   c. Information
   d. Regular access to the neighborhood aldermen

e. Enter into contracts with workers of the neighborhood office to establish:
   i. Research proposals
   ii. Housing plans
   iii. Street plans
   iv. Welfare plans
   v. Annual resident participation report
   vi. Neighborhood plan
f. Enter into agreements, namely:
   i. Delegation agreement (giving power to plan and issue permits to the neighborhood council)
   ii. Mandate agreement
   iii. Cooperation agreement
g. Exert authority, namely:
   i. Elaboration of the overall allocation plan
   ii. Determination of a street plan
   iii. Determination of a welfare plan
   iv. Power to suspend
   v. Granting of building permits
   vi. Granting of development permits
h. Evaluation of the annual report concerning resident participation
i. The right to speak at meetings of the city council

6. The internal decision making process is decided by an internal regulation. The regulation will be submitted to the mayor and the aldermen for approval.

7. The neighborhood council will report annually to the city council.

8. The city council may nullify decisions of the neighborhood council if they conflict with the law or with the established limiting conditions.

Article 6: Advisory Commission

1. The city council shall appoint an advisory commission for each recognized neighborhood organization consisting of:
   a. Neighborhood aldermen
   b. City council member
   c. Member of the neighborhood organization
   d. Member to be appointed by the neighborhood organization

2. This advisory commission shall advise the city council regarding solutions of differences in interpretation of these ordinances and will propose rules for execution.

3. This advisory commission also advises regarding continuation of "democratization" upon completion of the urban renewal project.

4. The mode of operation of the advisory commission will be regulated by an internal order.

Article 7: Effectiveness

This ordinance will become effective on . . .

2. DELEGATION AGREEMENT BETWEEN A CITY COUNCIL AND A NEIGHBORHOOD COUNCIL

The city council and the neighborhood council agree to the following:

1. The neighborhood council is charged with the preparation and establishment of plans to improve living conditions in the area declared an action area by the city council.

2. Preparation and establishment of the plans is conducted in close cooperation with all residents of the action area, the project coordinator, and the neighborhood office.

3. In the action area the neighborhood council exerts powers granted to the mayor and the aldermen in Article II of the Physical Planning Law.

4. The neighborhood council may grant:
   a. Building permits
   b. Development permits
   c. Exemptions as specified in Article 17 of the Physical Planning Law
   d. Building permits with the limitations specified in Article 49 of the Building Law

5. The neighborhood council can determine the street plan.

6. The city council may nullify decisions which the neighborhood council makes in the exercise of the powers granted to it by, or in accordance with, this agreement if the decisions conflict with the law or the established limiting conditions.

   The mayor and aldermen have similar authority with respect to decisions made by the neighborhood council in its exercise of powers granted to the mayor and aldermen in accordance with this ordinance. Two months after notification of the decision of the neighborhood council, the city council and the mayor and the aldermen can no longer nullify it. The city council and the mayor and aldermen may nullify a decision of the neighborhood council only upon hearing the neighborhood council.

3. COOPERATION AGREEMENT BETWEEN HOME BUILDING ASSOCIATIONS AND RENTERS[13]

In connection with planned home improvements, the home building associations and the residents of the homes, hereinafter referred to as "tenants," agree to the following:

Under the conditions described below, the lessor transfers the following rights to the tenant:

1. The selection of an architect and placing an order with a building expert for making drawings, descriptions, and calculations as well as the contracting, directing and supervision.

2. The choice of the builder and placing an order for improvement and overdue maintenance work, the execution of the rent level. This and other rights under the following conditions:
   a. The tenant shall be advised by the neighborhood office prior to entering any agreement with a third party which would entail financial or other obligations on the part of the lessor.
   b. The neighborhood office will inform the lessor and the tenant of this advice. In the case of a favorable judgment, the tenant has the authority to enter into the respective agreement insofar as he has procured the approvals as required by law. In the case of an unfavorable judgment, the lessor or his delegate has the power to suspend. The tenant then has no right to act in the name of the lessor for a period of one month. In each case the power to suspend may be used only three times.
   c. If there is continuing disagreement, the city council commission will act as an arbitration board. It should announce a decision within one month.
   d. In this context the lessor shall agree to the following maximum costs:
      i. Design costs
      ii. Building costs (contracting costs)

---

[13] Inspired by the agreement between the residential quarters Crooswijk and Partimonium in Rotterdam.

The tenant will allow inspection of documents through the neighborhood office to prove that no conditions are violated.

4. LENDING CONTRACT FOR WORKERS OF THE NEIGHBORHOOD OFFICE TO ASSIST THE NEIGHBORHOOD ORGANIZATION

Within the framework of the urban renewal action neighborhood. . . . , the city council has decided to establish a neighborhood office for the benefit of residents and contractors, where for the period of one year the following workers will be available:

1. A construction designer (one-half time)
2. A construction supervisor/calculator (full time)
3. An administrative aide/secretary (one-half time) and, in addition, as many experts as deemed necessary by the council

On the basis of this decision the following parties conclude the following lending contract:

The parties which conclude this contract are the city, the home builders association, the architectural firm . . . etc., hereinafter referred to as the employer; the office worker made available to the neighborhood office, hereinafter referred to as the employee; the neighborhood association or neighborhood council, hereinafter referred to as the neighborhood organization.

The parties agree that for the length of one year:

1. The employer shall make the employee available to the neighborhood organization.
2. The employee will be assigned to the neighborhood office.
3. The employee shall assist the residents and the contractors in general with the design and control of the execution of improvments of homes and the residential environment, and he shall be especially responsible for the following tasks:
   a. The worker shall report to the neighborhood organization.
   b. The employer will replace an employee if the neighborhood organization rejects the employee as a worker in the neighborhood office.
   c. This agreement is renewed each time for the period of one year unless otherwise decided by the city council.

5. ORGANIZING CONTRACT[14]

The neighborhood organization, welfare institution, and organizing worker conclude the following contract:

1. The welfare institution and the organizing worker underwrite the goals of the action program of the neighborhood organization.
2. The neighborhood organization underwrites the goals of the welfare institution and the organizer of the project.
3. The neighborhood organization can for the period of three years be assured of support by the organizer. This support shall include:
   a. Advice with respect to open participation
   b. Advice with respect to types of organization
   c. Skills training

[14] Inspired by the cooperation agreement between the Cooperation Union, Organization Work, Groningen, and the Consulting Group, Korrewegdijk.

4. The organizer will report regularly to the neighborhood organization regarding his activities in the neighborhood.
5. The organizer does not speak for the neighborhood organization.
6. The organizer makes a proposal for a procedure for the residents' participation. Upon request of the neighborhood organization, the organizer will draft an annual report concerning the residents participation. The neighborhood organization may accept or reject this report. These rights of the neighborhood organization do not apply to the neighborhood committee.
7. The neighborhood organization will annually review the action program. The organizer will annually review the welfare plan, based on an evaluation report. The neighborhood organization, the welfare institution, and the organizer may annually renew the organizer's contract.
8. In the case of conflicts between the neighborhood organization and the organizer, the neighborhood organization may request that the welfare institution replace the organizer.
9. The organizer and the welfare institution are charged with the careful completion of the organizational work, as stipulated in the welfare plan and in consultation with the neighborhood organization.

6. TASK DESCRIPTION FOR A PROJECT COORDINATOR

1. The first task of the project coordinator is to anticipate the consequences of urban management regarding renewal, as well as to anticipate the consequences of neighborhood renewal regarding urban management, and to respond accordingly. The project coordinator will enforce the limiting conditions set by the city.
2. The project coordinator shall assist with the formulation of the overall neighborhood plan in the preparatory phase. During the planning phase, he shall work on the elaboration of the neighborhood plan and he shall also prepare other plans, such as the business plan, street plan, and housing plan.
3. The project coordinator shall assist the neighborhood organization in clarifying or indicating options and in outlining alternatives in the realization of the action program and the various plans.
4. The project coordinator coordinates the incremental approach to private lessors (cooperation, summons, buying, expropriation).
5. The project coordinator shall report to the neighborhood alderman if the neighborhood organization is a committee or association. The project coordinator shall report to the neighborhood organization if this organization is a neighborhood council.
   Each type of neighborhood organization may request that the neighborhood alderman replace the project coordinator with another.
6. Depending on the agenda, the project coordinator shall have access to the municipal urban renewal commission. Before this commission, the project coordinator shall defend the result reached in consultation with the neighborhood residents.
7. The project coordinator shall maintain close contact with the neighborhood alderman, the deconcentrated services, the welfare work, the home building association, the neighborhood office, and the neighborhood organization. He will participate in administrative neighborhood deliberations. He will inform the heads of the services regarding the progress of the project, so that they will be able to consider this in their planning.

8. The project coordinator will have great independence within the established limiting conditions and may place them before the neighborhood alderman. Persistent differences will be presented to the city council.

9. The project coordinator works at the (city) secretariat with a team of project coordinators. The team activities are under the responsibility of the coordinating alderman for urban renewal.

ACKNOWLEDGMENTS

Information for this study was gathered by means of interviews and site visits in the Netherlands during June and July 1979 as part of the project entitled "Citizen Participation in Neighborhood Revitalization in Europe" sponsored by the German Marshall Fund. The authors are extremely grateful for the assistance of the following Dutch planners and public officials:

Frans Vonk, Housing Division, Ministry of Housing and Physical Planning, The Hague (with TNO at the time of the study).
Eisse Kalk, Workgroup 2000, Amersfoort
Wim van Es, Project Leader, Rotterdam Urban Renewal
Patries Haberer, SAR Foundation for Architectural Research, Eindhoven (with Ministry of Housing and Physical Planning at the time of the study)
Professor Gerard de Zeeuw, University of Amsterdam
Yap Hong Seng, Director of RIGO (Research Institute for the Built Environment), Amsterdam
Herbert van Hoogdalem, University of Technology at Delft
Hans Meester and Gerard van Evert, Municipal Town Planning Department at The Hague
Hylke Tromp, Project Leader Helmond Urban Renewal, and Wim Raaijmakers, Leon Michelson, and Bertus Vd Leeuw, IW staff, Helmond
Frits van der Burg and H. Van Vliet, Landelijk Ombudsteam Stedsvernievwing, Alkmaar
Professor Hugo Priemus, University of Technology at Delft
Professor Andreas Faludi, University of Amsterdam
Alderman J. van der Ploeg, Rotterdam
Dries Ouwerkerk, Advisor to Alderman van der Ploeg, Rotterdam
Harry Eijkelenboom, External Expert, Oude Western Action Group, Rotterdam
John Carp, SAR Foundation for Architectural Research, Eindhoven
Junte Lohman, Ministry of Public Health and Environmental Protection, The Hague (at the time of the study)
Nico van den Dool, External Expert, Vreewyk, Rotterdam

Without the information and guidance of these knowledgeable people, we would not have learned as much as we did in such a relatively short time.

We also appreciate the assistance of Marc Draisen, MIT graduate student, who gathered follow-up information for us during his study trip to the Netherlands, October 1979 to January 1980.

Bibliography

Arnstein, Sherry. "A Ladder of Citizen Participation." *AIP Journal,* July 1969.
Beck, Barbara. "Holland: Too Good to be True?" *The Economist,* 29 May 1976.
Berg-Elderling, Lotty van den. "Moroccan Families in the Netherlands." *The Netherlands Journal of Sociology,* 15, 1979. pp. 73–82.

Burie, Hans, Turpijn, Wouter, and de Wit, Cees. *Buurtvernieuwint: Analyse en aanpak van een maatschappelijk probleem*. ('s Gravehnage: Staatsuitgeverij, 1978).

Frieden, Bernard J., and Kaplan, Marshall. *The Politics of Neglect: Urban Aid from Model Cities to Revenue Sharing* (Cambridge, Mass.: MIT Press, 1975).

Haberer, Patries, and Vonk, Frans. *Urban Revitalization: On the Changing Urbanization and Government Intervention in North-West Europe, with Emphasis on the Nethlands*. Background Report for the 8th Johns Hopkins University International Fellows Conference, Noordwijk, The Netherlands, 25–29 June 1978.

Haberer, Patries, de Kleijn, G., Nicolas, F., and de Wit, C. *De Buurtaanpak: Verbetering van oude buurten door en voor bewoners (The Neighborhood Approach)* ('s-Gravenhage: Staatsuitgeverij, 1978).

Haberer, Patries. Letter to John Zeisel, 25 March 1980.

Hale, Robert L., Jr. "In the Netherlands Rehabilitation Is a National Effort with New Ideas for the United States." *Journal of Housing*, November 1978.

Hale, Robert L., Jr. "Rehab in the Netherlands: Tackling Both Housing and Neighborhoods." *Habitat*, no. 1 (1979), pp. 26–32.

Heeger, H. P. "The Dutch Solution to the Problem of a Residential Environment." *Planning and Development in the Netherlands*, no. 1 (1979), pp. 3–16.

Hoogdalem, Herbert van. Het Laboratorium voor Experimentele Ruimstudie (Laboratory for Experimental Spatial Study). Afdeling Bouwkunde Technische Hogeschool Delft, mimeo, 1975.

Hyman, Herbert. "Planning with Citizens: Two Styles." *AIP Journal*, March 1969, pp. 105–112.

Kalk, Eisse. Interview by David Godschalk and John Zeisel in Amsterdam, 20 June and visit to Work Group 2000 offices in Amersfoort, 21 June 1979.

Leeuw, Bertus Vd. Interview by David Godschalk and John Zeisel, Helmond, 26 June 1979.

Meester, Hans, and van Evert, Gerard. Interview conducted by David Godschalk and John Zeisel at The Hague municipal planning department on 29 June 1979.

Ministry of Housing and Physical Planning. *Housing in the Netherlands*. (The Hague: Information Department, undated).

Nelkin, Dorothy, and Rip, Arie. "Distributing Expertise: A Dutch Experiment in Public Interest Science." *Bulletin of the Atomic Scientists*, May 1979, pp. 20–23.

Ostrum, Elinor *et al*. "The Public Service Production Process: A Framework For Analyzing Police Services." *Policy Studies Journal*, special issue: Symposium on Police and Law Enforcement Policy, 1978.

Percy, Stephen L. "Conceptualizing and Measuring Citizen Co-Production of Community Safety." *Policy Studies Journal*. special issue: Symposium on Police and Law Enforcement Policy, 1978.

Priemus, Hugo. *Stadsvernieuwing: Problemen en Perspektieven: Analyse van de Rotterdamse aanpak* (Alphen aan de Rijn: Samson Uitgerverij, 1978).

Raaijmakers, Wim, and Michielsen, Leon (IW office in Binnenstad-Oost) and Tromp, Hylke (Project Leader). Interview by David Godschalk and John Zeisel, Helmond, 26 June 1979.

Sharp, Elaine B. "Citizen Organizations and Participation in Law Enforcement Advocacy and Coproduction: The Role of Incentives." Ph.D dissertation, Department of Political Science, University of North Carolina at Chapel Hill, 1978.

Turpijn, Wouter, and Veenema, Henk. *Bewonersparticipatie (Citizen Participation)* (Deventer: Van Loghum Slaterus B. V., 1979).

Van Dijck, H. "Reconstructing the 'Schilderswijk': A Study of an Attempt to Introduce Innovations into the Making of an Urban Reconstruction Plan." (Paris: Organization for Economic Cooperation and Development, mimeo, 1975).

Whitaker, Gordon P. "Citizen Participation in the Delivery of Human Services." Conference on Participation and Politics, Tutzing Akademie fuer Politische Bildung, 5–9 June 1978.

Whitaker, Gordon P. "Coproduction: Citizen Participation in Service Delivery." *Public Administration Review*, Vol. 40, No. 3, May/June, 1980.

Citizen Participation in the Netherlands: Some Comments

FRANS VONK
Planner with the Ministry of Housing and Physical Planning, The Netherlands.

David Godschalk and John Zeisel have presented in their paper a correct analytic description, optimistically interpreted, of one type of participation in the Netherlands. In this commentary, I would like to make clear the context for this participation within Dutch society and to examine some achievements and problems with the neighborhood approach to citizen participation. I take as a starting point that citizen participation has to be considered as an integral part of the total decision-making structure.

The Dutch Welfare Society

European policy makers operate within political cultures that are more conducive to public planning than those in the United States. In the opinion of the Dutch, the welfare state has to provide a number of minimum services and benefits to its citizens. Over the course of time, the number of activities in which the welfare state has become involved has grown enormously. Rising expectations made it necessary to focus on more and more aspects of society. This expansion is also due to new interpretations given by people who run it. This took place in a period of rapid economic growth which made it possible to pursue a range of wishes and expectations. Both factors have contributed to a tremendous growth of bureaucracy, which is charged with carrying out the task of the welfare state as elsewhere in the world; a growing professionalization and specialization of those involved in the bureaucratic system has also emerged.

As one consequence of the foregoing, we have come to a situation which makes it difficult for most citizens to comprehend the system.[1] Moreover representative government can no longer adequately control the system. It has become more disconnected from grass-roots needs and ideas. A growing dissatisfaction with the products and procedures of the welfare state has emerged. Because of a growing discrepancy between expectations and achievements of welfare state actions, we now face a declining belief in the welfare state.

While the legal basis for physical planning in the Netherlands was established in 1965, its roots go much farther back. The physical planning system is a three-tier system. It is strongest at the national and local levels and weakest at the regional level, under provincial government jurisdiction. Financial power in the public sector is vested in central government, but physical development is largely regulated by local government plans and controls. Through a very complex system of revenue sharing, local authorities are financially dependent on the central government. No doubt this factor has contributed to the emergence of local resistance and opposition and has helped to emphasize the need for more grass-roots involvement to meet local needs.

The period of planning for (economic) growth is over, however, and a new set of conditions requires a new interpretation of planning and implementation. Dutch society is increasingly characterized by a declining consensus and a rejection of planning as it used to be. As in other Western countries, we notice the emergence of what is called "interest-group liberalism." A growing number of interest groups want to make their needs known and are willing to achieve their goals by using means not generally accepted until recently. The explicit presentation of these needs is taking place in a period of no growth. This implies a growing competition and increasing tension between groups.

Neighborhood Approach

Urban policies formulated in the sixties by national and local governments concentrated on city centers and their economic functions. The renewal of the residential function—still an important one in many city centers and adjacent areas—received less attention. Governments favored total, large-scale redevelopment and the improved accessibility of the core area by car to strengthen the economic functioning of the city. The approach was highly technocratic and physically oriented.

Between approximately 1968 and 1973, urban policies underwent a radical change. The population reacted against large-scale clearance and the construction of modern dwellings (usually too expensive) by local governments. Urban renewal nomads, people moving from one renewal area to another, became vocal.

[1] It has also produced a new function: the "subsidiologist" or "grantsman" who studies government to find out what the organization he or she is working for can get from government.

Opposition grew as emancipatory forces in Dutch society made people and groups aware of their position in society.[2] Accelerating physical deterioration of the inner cities; the ongoing, selective in-migration of minority groups; dissatisfaction with living conditions in recently built residential areas on the city outskirts; and, finally, new ideas concerning the conservation of resources and the role of planning and the planner—all these factors too account for the new approach.

The neighborhood approach[3] described in the Godschalk–Zeisel paper is but one reaction to this new situation. The key words for the new urban renewal policies are *rehabilitation, renovation,* and the *neighborhood approach*. The policies have a strong social emphasis which *inter alia* implies planning and implementation with and by the people as well as more attention to housing and the human and financial consequences of housing improvement.

Acceptance of the neighborhood population as a genuine partner in the decision-making process in many cases implied the immediate stopping of clearance programs. The income structure in many zones designated for urban renewal is skewed: in general, these areas are inhabited by lower-income groups. A good deal of the neighborhood discussion on urban renewal therefore concerns the new rent level. The sociocultural amenities that provide favorable conditions for the livability of the neighborhood are a second major topic. Particularly in the larger cities of western Holland, many potential urban renewal areas have become very attractive for low-income minority groups, mainly from Mediterranean countries. The urban renewal problem is an ethnic problem.

Effects, Problems, and Questions

Comparative research should not be limited to the analysis of achievement; the effects and problems that go with it must also be considered. I think these aspects are less developed in the Godschalk–Zeisel paper. Therefore, this final section of my comments will examine the neighborhood approach from several different angles.

[2] See, for example: M. Castells, *The Urban Question* (London: Edward Arnold, 1977), and C. Pickvance, ed., *Urban Sociology* (London, Tavistock Publications, 1976).

[3] The term *neighborhood approach* tends to be an umbrella concept referring to a wide variety of attempts made to involve the local population in the decision making on their own neighborhood. However, the publication with the same title sets out a specific approach in which the neighborhood as an organized group defines through a self-survey the problems, goals, and instruments for changing the situation. On the basis of a tentative plan approved by the local authority, the national government and the local authority create a financial agreement for carrying out the plan. The national government provides a lump sum (a block grant) to the neighborhood organization through the local authority. This approach gives both the local authority and the neighborhood organization some latitude in determining priorities, changing parts of the tentative plan, and so on. Moreover, it implies less *ex ante* central government control. Currently, some experiments are being set up to test the approach formulated in the publication *The Neighborhood Approach*.

What are the achievements of the neighborhood approach? First, it has demonstrated people's interest in their neighborhoods, their attachment to an area, their needs, and their priorities. This, in turn, revealed conflicts, both within the neighborhood itself and with the local and national government. In some cases it has speeded up the urban renewal activities by changing priorities and/or making public authorities aware of existing problems. The neighborhood approach helped stop large-scale clearance, led to a greater emphasis in preservation and rehabilitation, and revealed the power structure and the often implicitly biased actions of traditional decision makers. Finally, through successful action, people have developed faith in their ability to influence the stream of events and to develop their own communities.

Yet, the experiences have also shown some problems and have given rise to a number of questions. First, the neighborhood approach described in the Godschalk–Zeisel paper turns out to have a strong emphasis on housing, particularly housing for the neighborhood population. In many neighborhoods, residents sought the removal of economic activities which cause noise, pollution, or attract traffic. During a period of economic stability or economic decline, this goal is hardly acceptable from an employment point of view. Moreover, a relocation of economic activities—even if economically feasible—may result in more and longer commuting, and consequently generate traffic problems and greater energy consumption.

Second, the existing distribution of population is partially the result of filtering processes that produced more or less socially homogeneous neighborhoods. Ghettoization of minority groups and low-income people is one result. By allowing neighborhoods to focus on their needs, we may prolong the existence of this undesirable phenomenon. Of course, an urban policy could set the conditions for the redevelopment of different neighborhoods, but this would cause additional problems. A flexible *and* integral urban policy at the level of both the municipality at large and the various neighborhoods is difficult to produce. More interest groups are involved. Moreover, time perspectives differ: while the city tends to emphasize a long-term perspective, local residents emphasize the immediate effects of new rent levels.

Third, the norms of current residents and their explicit desire to keep rents at a "reasonable" level are producing rehabilitations which are frequently of questionable value. The small apartments may, in the near future, be insufficient for more than one-or two-person households. The improvements are likely to be considered inadequate by the next generation. Housing quality may continue to decline.

Finally, the interests and wishes of the neighborhood population must be reconciled with those of homeowners. Landlords of many if not most houses in urban renewal areas own only a small number of dwellings. The low rent level of their property makes them unable or unwilling to invest the necessary amount of money for proper upkeep. In the city of Rotterdam—as mentioned by Godschalk and Zeisel—the local government has tried to buy a substantial number of these privately owned dwellings. For this to be successful, however, the local authority must be willing both to undertake such action and to invest the money needed for implementation. Large amounts must be invested for rehabilitation, clearance, and the construction of new housing.

10

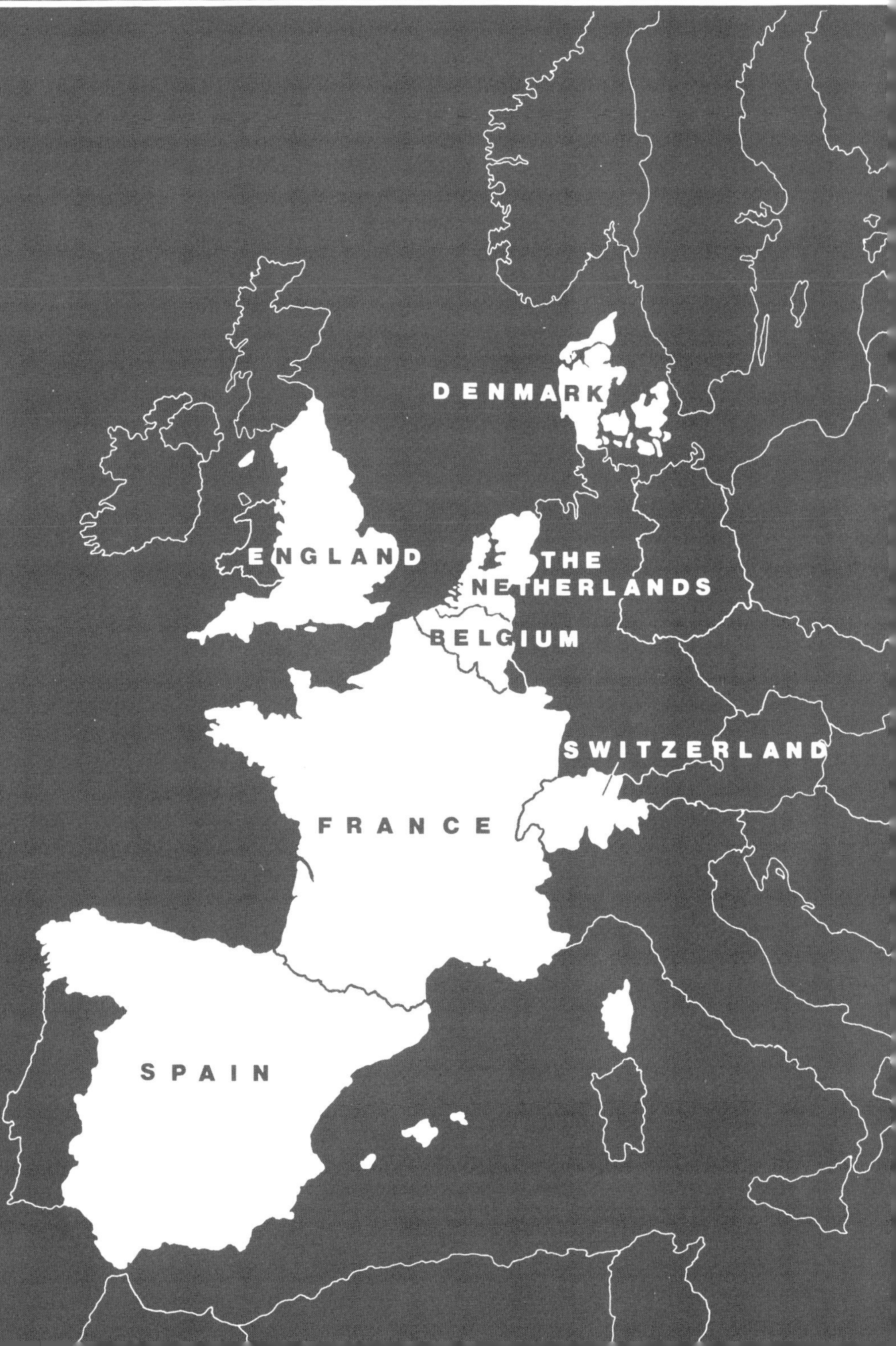
DENMARK
ENGLAND
THE NETHERLANDS
BELGIUM
SWITZERLAND
FRANCE
SPAIN

10

EPILOGUE

When we began this inquiry in 1979, we assumed that public officials and citizen activists in the United States would find recent participatory efforts in western European cities instructive. We also assumed that the lessons learned in Europe could be applied in the United States, even though cultural differences and political traditions make direct transfer of tactics and strategies difficult. Finally, we assumed that a team of researchers with different views and disciplinary backgrounds could study a range of communities and cultural contexts and still find an effective means of pooling their findings. Now, in 1981, as our work draws to a close, it appears that all three assumptions were correct.

In April 1980, we presented our findings at a symposium sponsored by the German Marshall Fund of the United States. This symposium was attended by 65 citizen-activists, public officials, and scholars from all parts of the United States. Each member of our research team presented his or her findings. Comments were immediately provided by the European counterparts with whom we had worked the preceding year. The symposim discussions were extremely animated. For two days we attempted to thrash out a consensual view. The results of these deliberations were incorporated into the final versions of the papers that appear in this volume. We have tried to make these written reports just as evocative and instructive of the European experience as the attendees found the discussions to be.

We worked hard to overcome the obstacles to cross-cultural comparative research. Given the claim that researchers are often blind to what they do not want to see or unable to understand what they do see in another culture, we tried to build a team with extensive knowledge about and experience in the countries we studied. Reliance on European counterparts every step of the way was a crucial part of our strategy. Their presence at the concluding symposium, and their overall support of our interpretation of events and outcomes, boosted our sense of accomplishment.

Given the claim that cross-cultural comparative research cannot be cumulative because of the difficulties involved in generalizing across differing countries, we tried to work as a team in developing a summary of our findings. That is, each member of our group took responsibility for ensuring that overall generalizations did not stretch his or her case study findings beyond a reasonable point. In the end, what we were able to agree upon was a set of indicative findings rather than generalized results or proven hypotheses. The generalizations presented in this volume are not conclusive. Our case studies do not provide enough comparable evidence for that. As a team, though, we were able to agree on certain provocative ideas.

Our primary focus, or unit of analysis, was resident or consumer participation in local activities. In each instance, we tried to determine whether participation enhanced the efficiency of government and whether it increased resident satisfaction with the operation of local government. Not surprisingly, objective measures of increased efficiency were impossible to derive. Thus, we were forced to look at the outcomes in each case (through the eyes of all the participants) and to develop a rough estimate of the extent to which the outcomes could or would have been achieved (to the same or a greater extent) if there had been less participation, more participation, or a different kind of participation. We also tried to determine the extent to which residents and consumers involved in each case felt that the process of participation gave them more of a sense that government officials were being responsive to their needs.

It should be obvious from our case studies that no simple conclusion can be drawn. More participation (of every kind) does not always enhance the efficiency of government. Nor does the offer of participation or the process of involvement always increase resident or consumer satisfaction. Politics and government processes are far too complex for that.

In some instances citizen action and in other instances government-invited participation did produce surprisingly efficient outcomes and substantial feelings of citizen efficacy. Sometimes citizen satisfaction increased while government efficiency was lost because participation dragged on and on. Sometimes both citizen satisfaction and government efficiency increased because participation produced agreements (and eliminated opposition). Finally, some particular efforts to enhance participation may not have increased efficiency in the short run, but they improved relations between government and neighborhoods (or resident groups), so that greater efficiency is realizable in the long run.

Our rather straightforward initial question about the relationship between participation and outcomes (i.e., efficiency and resident satisfaction) did not yield a straightforward answer. What is clear, though, is that the repertoire of participation strategies is much richer than past American experience would suggest and that the possibilities of both enhanced government efficiency and increased resident satisfaction through citizen action and government-sponsored participation are real.

The individual case studies in this volume are important because they may encourage a public official or a neighborhood activist in the United States or Europe to consider a fresh approach to participation. We need to continue experimenting with new ways of empowering citizens and new approaches to strengthening the operation of local government.

INDEX